Preferred Plants and Nectar Collecting Times for Honey Production

Preferred Plants *and* Nectar Collecting Times *for* Honey Production

A South Carolina Beekeeper's Guide

DAVID MACFAWN

JENNIFER M.K. O'KEEFE

SOPHIE WARNY

CAROL WYMER

JUNE PONDER

Clemson University Press

Clemson, South Carolina

ISBN 978-1-63804-167-2
eISBN 978-1-63804-206-8

LCCN 2025941475

Published by Clemson University Press
in association with Liverpool University Press

Cover design by Lindsay Scott
Interior design and typesetting by Percolator Graphic Design

This book is dedicated to two amazing palynologists, without whose mentorship and training, the detailed taxonomic analyses needed to carry out the statewide study would have been impossible. Dr. Jean-Pierre Suc, now emeritus at Sorbonne Université (France) is a gifted taxonomist who has dedicated his life to understanding the marine and floral development of the Mediterranean region and mentoring young scientists. The late Dr. Vaughn M. Bryant Jr., was the founding father of melissopalynology in the United States and was well-known in the beekeeping, forensics, archaeological, and quaternary palynology communities. He trained or mentored many of the palynologists active in the United States today. With this work we honor his memory and their legacy of sharing palynology with future generations.

Contents

Acknowledgements

This project was made possible by the South Carolina Department of Agriculture (SCDA) via the U.S. Department of Agriculture's (USDA) Agriculture Marketing Service through Specialty Crop Block Program grant 21SCBPSC1010. Betsy Dorton, Grants Administrator, and Brittany Jeffcoat, Grants Coordinator, at the SCDA were particularly helpful throughout the project.

The support of the South Carolina Beekeepers' Association Executive Board was instrumental for the success of this project: Danny Cannon (President 2020–2022), Susan Jones (President 2022–2024), Wendy Gray (Treasurer 2018–2022) and Henry Campbell (Treasurer 2022–2023). Special thanks to Dr. Angelina Perrotti for the identification of multiple unknown pollen grains.

Thanks to Ben Powell, Clemson University Extension Associate with Apiculture & Pollinator Program, for helping select the 19 honey-collecting sites throughout South Carolina and contributing to the purple honey manuscript.

Thanks to Dr. Mary Kate Fidler, Clemson University Department of Environmental Engineering and Earth Sciences, for generating the map used in this book.

Thanks to Brett Kahley and FedEx for special sample shipping arrangements.

The authors contributed a variety of unique skills to the project; first and last authors represent the funded grant while palynologists are listed alphabetically.

About the Authors

David MacFawn is a North Carolina Master Craftsman Beekeeper (1997), Eastern Apicultural Society Master Beekeeper (2019), co-founded the South Carolina Master Beekeeping Program, was awarded 1996 & 2020 South Carolina Beekeeper of the Year, incorporated the South Carolina Beekeepers Association as a 501-C-3 Non-Profit Corporation, and has published over 60 articles in the *American Bee Journal, Bee Culture,* and *Beekeeping: The First Three Years*. David is a 2021 CIPA EVVY™ Awards Book Second Place Competition Winner and has published four books for beekeepers. He provided expert advice to Bee Downtown identifying bee colony characteristics they could apply to their programs which improve organizational dynamics by drawing parallels with bee colony dynamics. David has a Bachelor of Science degree in Electrical Engineering and a Master of Business Administration with concentrations in Finance and Operations Research. He was in the computer business for over 30 years.

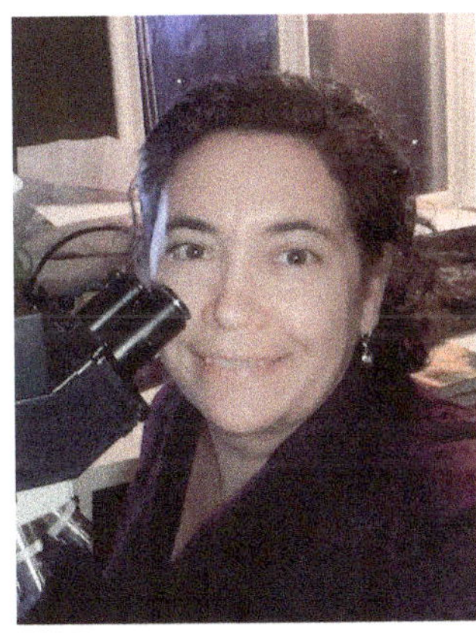

Jen O'Keefe is a Professor of Geology and Science Education at Morehead State University in eastern Kentucky. She received a Bachelor of Arts degree in Geology from Cornell College in Iowa, a Master of Science in Geology at Texas A&M, and a PhD in Geology from the University of Kentucky. She studied palynology under the late Dr. Vaughn M. Bryant, Jr. at Texas A&M, where she was introduced to melissopalynology as part of her training, and the late Dr. William C. Elsik, who introduced her to fungal palynology. Jen leads the OPaL Lab at Morehead State University, which focuses on training high school and undergraduate students to answer a wide variety of questions about past ecosystems and climates and environmental issues using palynology, organic petrography, and geochemistry, as well as pollen typing of local and regional honey samples. She publishes collaborative research in several journals, including *Science of the Total Environment, International Journal of Coal Geology, Climate of the Past,* and *Palynology*. Dr. O'Keefe is a Fellow of the Geological Society of America and served simultaneously as chairman of the Energy Geology Division of the Geological Society of America and as President of AASP-The Palynological Society (AASP-TPS) in 2015. She is currently a member of the Board of Trustees of the AASP Foundation, which is tasked with supporting palynological research and dissemination. For these and other efforts she received AASP-TPS's Distinguished Service Award in 2023. In that same year, Dr. O'Keefe received an Apple Award from the Morehead State University Student Council for mentoring students both in and outside of the classroom and was named the Morehead State University Distinguished Researcher.

Sophie Warny is the AASP Endowed Chair Professor in Palynology, Associate Chair of the department of Geology and Geophysics, and a Curator at the Museum of Natural Science (MNS), all within Louisiana State University in Baton Rouge, Louisiana. She grew up in Grasse, France and Namur, Belgium. She received a Bachelor of Science degree in Geology from the Université Catholique de Louvain (UCL) in Belgium, a Diplôme d'études approfondies in Oceanography from the Université Libre de Liège (in Belgium), and a PhD from UCL in Marine Geology/Palynology in 1999, working under the direction of Dr. Jean-Pierre Suc (Université de Montpellier in France). She is the director of the AASP-The Palynological Society Center for Excellence in Palynology (CENEX). CENEX focuses on various aspects of palynological research, from the use of pollen, spores, and algae in biostratigraphic studies (biosteering applications) in collaboration with the industry, to the use of pollen in forensic applications and melissopalynology when she partners with the USDA or beekeepers. Her research also focuses on using palynological records for paleoceanography and paleoclimate reconstruction, including investigations of the palynological record to decipher past sudden climatic events and climate variability to help constrain their triggering mechanisms. She received an NSF CAREER award in 2011 and has published in journals such as *Science, Nature, Nature Geoscience, PNAS, Geology, Gondwana Research, Climate of the Past,* and of course the society's journal, *Palynology*. In 2016, she served as the Vice President of the Gulf Coast Section of the SEPM society and was one of six AAPG Distinguished Lecturers for 2019/2020. She is currently the President of AASP-The Palynological Society. The most touching honor of her career was to have been nominated by her former graduate students for the society's Medal for Excellence in Education in 2021. She has supervised 30 theses and dissertations since starting at LSU in 2008.

Carol Wymer is partially retired. Her time is divided between teaching, melissopalynology, and volunteer endeavors. She received Bachelor of Science and Master of Science degrees in Biology at the University of Dayton and a PhD in Plant Physiology at The Pennsylvania State University. She completed post-doctoral research at The Pennsylvania State University and in a joint appointment to the University of East Anglia and the John Innes Centre in England. The common theme throughout these varied research experiences—spent studying hormones, the cytoskeleton, the cell wall, and intracellular calcium—was physiological aspects of plant cell growth. In 1998, Dr. Wymer established her own lab at Morehead State University (MSU), where she continued to study different aspects of plant growth including propagation and rooting of the invasive plant species Japanese knotweed. Over time, her professional focus moved to Biology Education, and she served as a leader on multiple grants; this allowed her to play a role in the improvement of regional K-12 science instruction. After 16 years at MSU, she retired as a full professor and her family moved to northwest Ohio. Since that time, she has taught at several local universities and is an adjunct faculty at MSU and Bowling Green State University. Dr. Wymer's collaborations with Dr. Jen O'Keefe began in education and expanded into palynology. This occurred first through the development of an enzymatic method for processing honey and other materials to extract pollen, and then while training to identify pollen grains (with Dr. O'Keefe as a mentor). In addition to teaching and a little research, Dr. Wymer takes an active role in local plant conservation. She volunteers in the conservation departments at the Toledo Zoo & Aquarium and Wood County Park District growing native plants that support butterfly conservation, local habitat restoration, and the expansion of native species in residential neighborhoods.

June Rampy Ponder is an active backyard beekeeper who has kept bees and sold honey since 1996, when she joined the Oconee County SC Beekeeper's Association, where she has served a variety of officer roles. She is a South Carolina Journeyman Beekeeper and served one term on the South Carolina State Beekeeper's Board of Directors. She was awarded the 2003 South Carolina State Beekeeper of the Year and 2014 Oconee County SC Beekeeper of the Year. June has a Bachelor of Science in Mathematics from East Tennessee State University and a Master of Business Administration with a concentration in Entrepreneurship from Clemson University. She initially taught high school math at Daniel Boone High School in Gray, TN and at Elizabethton, TN High School. She spent 30 years working in software development at Clemson University, participating in the development of the South Carolina Medicaid System (1977–1986), streamlining processes for campus-wide file backups/restores and userid issues, contributing to the year 2000 conversion (evaluation, installation, and maintenance) of university systems for Finance and Human Resources, and managing the bulk mail software and user training for mass mailings. She finished her career with Clemson University, making enhancements to the South Carolina Medicaid System (2003–2007) for user friendliness. June then spent 12 years as an Aflac Insurance Agent before retiring.

Honey Collectors

This project would not have been possible without the dedication of the 19 beekeepers who collected honey samples throughout the 2022 nectar flow. These individuals were widely distributed across South Carolina and because of their participation, we were able to analyze honey from all 5 of South Carolina's ecoregions and 17 of South Carolina's 46 counties.

Locations of Participating Apiaries.

Map ID	Apiary Location	Apiarist
1	Hilton Head Island, Beaufort County	David Arnal
2	Ehrhardt, Bamberg County	Susan Chewning
3	Mt Pleasant, Charleston County	Tim Liptak
4	Ladson, Dorchester County	Woody Weatherford
5	Loris, Horry County	Karen Hilbourn
6	Aiken, Aiken County	Robert Abshire
7	Aiken, Aiken County	Hank Smalling
8	Smoaks, Colleton County	Tom Parker
9	Camden, Kershaw County	Rosalind Severt
10	Florence, Florence County	Ben Powell

Map ID	Apiary Location	Apiarist
11	Lakewood, Sumter County	Paul Wise
12	Greenwood, Greenwood County	Kathy Murray
13	Columbia, Richland County	Larry Coble
14	Saluda, Saluda County	Lizanne Melton
15	Rock Hill, York County	Bennie Copeland
16	Rock Hill, York County	John Williams
17	Whetstone, Oconee County	Rannie Bond
18	Easley, Pickens County	Brett Kahley
19	Powdersville, Anderson County	Barbara Tate

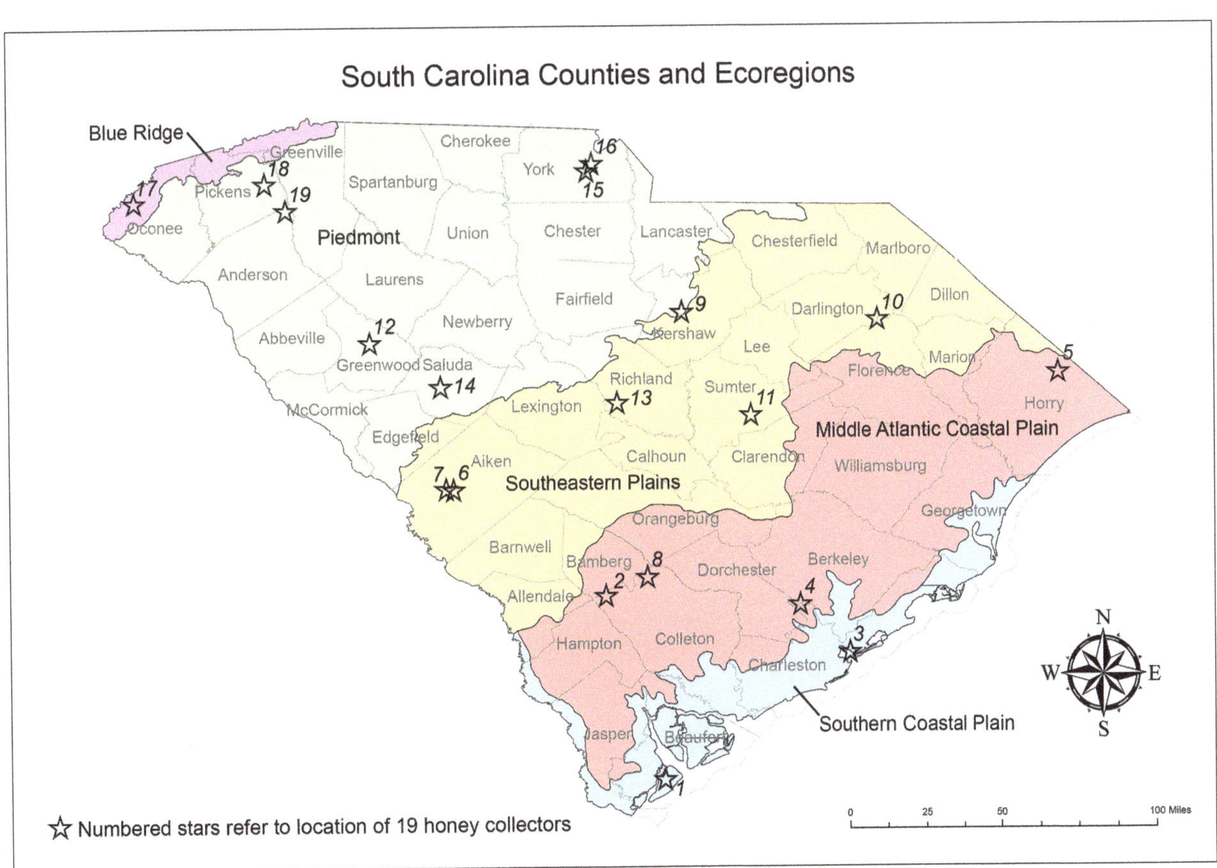

Level III ecoregions of South Carolina showing apiaries involved in the study.

Introduction

This book is intended primarily for beekeepers, non-beekeeping homeowners who want to grow bee-friendly plants, and stakeholders participating in Bee Campus and Bee City endeavors. Honeybees feed from a variety of sources, ranging from nectar to honeydew and pollen to spores. While many beekeepers in South Carolina are keen observers of what their bees are browsing, it is difficult to follow every trip to and from the hive—and it is often impossible to tell if a bee is visiting a plant as a pollen source, a nectar source, or both. Therefore, we often rely on little more than anecdotal evidence to know which plants our bees are using for which purpose.

Adult bees use nectar as their primary source of carbohydrates; pollen intermixed with this nectar provides vital protein, fats, and other nutrients. While pollen collected in a bee's pollen baskets is used primarily to feed nurse bees and produce royal jelly for larvae, nectar, in the form of honey, provides for adult bees' nutritional needs year-round. It is therefore vitally important to know which plants bees are using for nectar forage. While there are published lists of plants frequented by bees for most states, these were generated more than 50 years ago. These lists do not necessarily capture current bloom times or plant distributions, which seem—whether due to urbanization or changing climate—to have changed significantly in our lifetimes. For instance, a preliminary study of a single hive from a single bee yard in the Congaree National Biosphere Reserve in 2021 strongly suggested that the published lists of bee nectar forage plants and bloom times of those plants are not accurate.

Understanding nectar foraging habits and probable bloom times of nectar sources is critical for honey production and colony management. Beekeepers need to know when to install honey supers (boxes) for floral and multifloral honey. They also need to know the nectar flow's start time, stop time, and length to allow for hive and super management. Finally, beekeepers need to know when to move hives to a specific region to capture specific nectar flow if they intend for a specific type of honey to be produced. By understanding the nectar flows, premium honey (light, dark, floral, multifloral, and flavor) can be produced.

The nectar source of honey is identified primarily through the process of melissopalynology, or examination and identification of the pollen contained in honey. This can be done one of two ways: using traditional transmitted light microscopy and/or via DNA metabarcoding of genetic material remaining within the pollen. While DNA metabarcoding shows significant promise and will likely overtake melissopalynology in the future as the preferred method for determining botanical contributions to nectar, it has not yet reached that stage due to its cost and the sheer number of taxa not yet in sequence databases. Traditional melissopalynology, too, will likely change in the future, as machine-learning techniques improve and the number of pollen grains recognizable using a computerized microscope increases. As of this writing, however, transmitted light microscopy with a highly trained palynologist as analyst remains the industry standard. This time-consuming and painstaking process requires highly skilled practitioners who are well versed in the preparation and identification of pollen. While there are quite a few palynologists in the United States, few are trained to work with honey samples, which can be highly variable and contain pollen types rarely encountered elsewhere.

Building upon a preliminary study undertaken in fall 2021, the South Carolina Beekeeper's Association, led by David MacFawn, designed and implemented a research project to determine what bees across South Carolina are using for nectar sources and how—or if—the plants used vary across the state. Prior to the current study, most of the existing information on preferred nectar sources was anecdotal and based on direct observation of which plants bees were visiting. The current book focuses on developing a baseline of the regional nectar ecosystems in the state to determine the impact of urban sprawl and other factors affecting these ecosystems (such as climate change). A lot of discussion will ensue about the data in this book. This is a good thing.

1.1 History of Honey Production and Analysis in South Carolina

No records exist for the arrival of honeybees in South Carolina or the initial development of beekeeping, but it stands to reason that they were brought to South Carolina at some point between the initial recorded shipments to the Virginia Colonies in 1622 and the first records of the industry in Georgia in 1743 (Oertel, 1980). Early beekeepers struggled with American Foulbrood disease until the 1850s, when development of the Langstroth hive allowed improved colony inspection and care (Nelson, 1971). By 1859, hundreds of thousands of pounds of honey were produced in South Carolina (Aldrich, 2015). The first beekeeping organization in South Carolina, the Oconee County Beekeeper's Association of South Carolina, was established in January 1921 to support development and education within the beekeeping industry. Other clubs soon followed, and the South Carolina Beekeeping Association was established on March 22, 1975. The primary goal of both the early clubs and today's clubs is beekeeper education, training, and support. Proper management is vital to the success and health of honeybee colonies in both large and small apiaries and, by extension, the success of pollination-dependent agriculture.

Because of this, beekeepers as a group are keenly interested in honeybee health and dietary support, as well as the identification of honeys of economic value. Honeybee foraging preferences and plant availability in a roughly 3-km radius of hives (Eckert, 1933) have a direct impact on bee health and on the economic value of produced honeys. Both foraging preferences and plant availability can be monitored in part by determination of the botanical composition of the contents of honey and, where appropriate, pollen cells in drawn comb or in bee pellets collected upon hive entry. This composition is typically accomplished through the process of melissopalynology, or the identification and quantification of pollen in honey (de Almeida-Muradian et al., 2020).

Melissopalynology is used globally to determine the floral origin of honey, identify honeybee foraging behaviors, and monitor actual bloom times. The United States has a long but sparse history of using melissopalynology, lagging behind the rest of the developed world; most US palynologists use their skills for oil exploration or Quaternary paleoecology rather than forensics (e.g., Warny et al., 2020) or melissopalynology. Young's 1908 analyses of a hundred honeys that had been previously displayed at the world exposition in 1903 is considered the first melissopalynological study in the United States. It was followed by Pellett's 1930 "American Honey Plants" and Pammel and King's 1930 "Iowa Honey Plants," which noted the characteristics of honeys produced from specific plants and, in the case of Pammel and King, descriptions of pollen morphologies. While a handful of other researchers contributed minor reports from the 1930s through 1960s, Todd & Vancell's 1942 paper on pollen grains in honey stands out as the only major contribution until the 1970s. At that time, Meredith Lieux—then a faculty member at Louisiana State University—began her studies of Louisiana and other regional honeys and published extensively on the topic (1970, 1972, 1975, 1977, 1978, 1980, 1981). Lieux's work informed Vaughn Bryant's, which began with a 1975 contract with the USDA to identify potentially adulterated honey. These types of fraudulent honeys have long been of concern both for the negative economic impact they have on regional beekeepers and for the human health risk presented by some substances added to the honey (Young, 1908; Jones and Bryant, 1992; Schneider, 2011; McNeil, 2012). Despite the concerns surrounding adulterated honey, it was only in March of 2018 that the FDA issued guidance for truth-in-labeling of honey samples. This guidance necessitates that any honey labeled with its botanical origin have information available to support this claim (FDA, 2018). This guidance, six years old at the time of writing, has resulted in a significant increase in need for melissopalynological and other honey analyses—and with it, an increase in the number of labs providing melissopalynological services (Angelina Perrotti, personal communication, 2023). It is not only for truth-in-labeling purposes that melissopalynology is so important; melissopalynology is also a means of monitoring honeybee nutrition (Lau et al., 2022) and the impacts of habitat changes and/or fragmentation (Bottero et al., 2023; Lau et al., 2023). Habitat change can take a number of forms, from anthropogenic habitat changes due to increasing urbanization to changes in regional plant communities due to climate change.

Many beekeepers in the United States follow a regional schedule of activities in part dictated by published bloom, and thus nectar flow, calendars. Most beekeeping associations publish a regional calendar, and South Carolina is no exception (see: https://scstatebeekeepers.com/beekeepers-calendar/). This calendar is based on a very old blooming schedule produced for the 1962 edition of "Beekeeping in South Carolina," with only minor revisions made in the 1990 edition (Purser & Sparks, 1962; Hood et al., 1990). It was clear to regional beekeepers by the late 2010's that this blooming schedule no longer aligned with observed bloom times, and in 2021, a significant calendar update was released (Smith, 2021).

Prior to the 2021 calendar release, we completed an extremely successful pilot study (2020) comparing bloom times on published charts with pollen content in freshly collected honey extracted from squares of comb originating from a single apiary in the Congaree Biosphere Reserve. Not only did this preliminary study identify significant variations in bloom time (and thus nectar production), but it also documented the pollen of numerous nectar-producing plants present in the honey that had not been previously reported in South Carolina (MacFawn & O'Keefe, *in review*). This localized study set the stage for the new, statewide study discussed herein.

The present study started in early 2021, when the South Carolina Beekeepers Association (SCBA) began laying the foundation for a South Carolina statewide honey pollen analysis. Melissopalynology *per se* was not new to the statewide beekeeping community; several individual producers had been having their honey periodically analyzed for years, primarily by Dr. Bryant at the Texas A&M Pollen Lab. But no state has completed a statewide analysis of pollen in honey since Lieux's work in Louisiana nearly 50 years ago. Unlike Lieux's work, the present study was crafted to examine honey from each ecoregion of the state and is thus the first of its kind for the United States in terms of capturing the weekly dynamics during seasonal nectar flows for an entire year.

The study aimed to help South Carolina Beekeepers and Clemson University Extension agents identify bee forage throughout the state on a region-by-region, month-by-month basis. We know from the preliminary study in the Congaree Biosphere Reserve that the published lists of bee forage plants are out of date, and conversations among master beekeepers at statewide meetings suggest that the problem is not limited to the Congaree. In addition to variations in bee forage plants, we have noted significant changes in bloom timings; it is becoming more and more difficult to determine when to add supers (honey boxes) onto a hive for specialty honey collection. Specialty honey varietals command a higher market price, and thus more accurate timing of varietal honey collection brings higher income potential to our regional beekeepers. Additionally, knowing plant sources and when pollen-rich honey is produced aids in identifying honeys that may be prone to crystallization. We also need to know much more about early season honey, especially the impact of the maple (*Acer*) bloom on nectar production.

The data analysis in this book focuses on variations within and among South Carolina's five Level III ecoregions: the Blue Ridge, the Piedmont, the Southeastern Plains, the Middle Atlantic Coastal Plain, and the Southern Coastal Plain (Griffith et al., 2002). Level III ecoregions were chosen deliberately, as South Carolina contains such exceptional ecological and biological diversity at the finer scale (Level IV) that meaningful comparisons between study sites would have been virtually impossible. Each Level III ecoregion contains similar plant communities that have developed upon similar bedrock, with similar water flow styles both above and below ground (hydrogeology), and with similar landforms (physiography). These ecoregions are used for regional ecosystem and environmental monitoring and assessment across state and federal agencies, as well as for informing regional environmental policies and decision-making.

1.2 Organization and Planning for Proposal Submission for Inclusion in the South Carolina USDA Block Grant

It is never easy to be the first. No study of this type has been explicitly designed in the United States before. However, with increased concern about hive health and long-term colony viability, especially in light of rapidly-changing environmental conditions (Lau et al., 2023), we are unlikely to be the last. Indeed, we hope that this will be the first of several similar studies in South Carolina as we continue to monitor the pollen content in honey and hive mass changes through the annual nectar flow. Here we present an overview of our grant-development processes, in the hopes that our experiences will aid other states in generating a similar wealth of data.

We started out thinking BIG. Of course every beekeeper would want to be involved! Of course there would be plenty of money for everyone's time! Of course we could travel to every small bee club many times, not only to pitch our project, but also to share our results. Of course we could pay team members for the significant time involved in bringing this project to fruition. Indeed, our original budget, which included the pollen analyses, labor, and travel, came to $400,000.

We pitched the project as initially planned to colleagues at the South Carolina Department of Agriculture, and reality quickly came crashing down. USDA Block Grants simply are not large enough to support that much of the budget being allocated to a research project. With a dramatically revised budget target of $50,000 in hand, we had a lot of hard decisions to make. It was only after we consulted many Master Beekeepers statewide that we decided to proceed, as the only way we could conceive of staying within our budget was to sacrifice most travel and beekeeper support in favor of the vital melissopalynological analyses and final publication of results. Melissopalynology is expensive, and like everything else, inflation has taken a toll: detailed analyses like those we needed would run at least $115 per sample. Publishing is, likewise, not cheap. We all took a deep breath and decided to make this a true labor of love, because we had so many questions that needed to be answered. Even then, we had to go back to the negotiating table before agreeing to a final revised budget of $70,000.

David MacFawn and the leadership of the South Carolina Beekeeping Association co-wrote the proposal, and David was the chief project manager. June Ponder, a long-time beekeeper in upstate South Carolina, agreed to help co-manage the grant. Our very first challenge beyond recruiting beekeepers was establishing a means of unified processing and analyses. We chose Global GeoLab Ltd. in Medicine Hat, Alberta, Canada—already used for other national-level projects and trained in the melissopalynological processing methods of the late Dr. Bryant—to process the honey samples and generate the pollen slides for melissopalynological analysis. This meant finding a cost-effective method of shipping honey to Canada and the slides and residues back to the scientists chosen as project analysts. This could not have been accomplished without Brett Kahley, a district manager for FedEx, who was instrumental in coordinating all project shipping. At the time this grant was written (mid-2021) and the research accomplished (2022), no single lab had enough capacity or expertise in regional pollen to take on the melissopalynology.

Thus, we chose a team of three analysts: Dr. Sophie Warny (CENEX, Louisiana State University), Dr. Jen O'Keefe (OPaL Lab, Morehead State University), and Dr. Carol Wymer (C&S Science Consulting). Brett and June devised a system to collect honeys from across the state for transport to the honey processor and then distribute the processed pollen samples to the three analysists in the most cost-effective manner possible. June and David developed strict timelines to keep the entire team on track.

Our next challenge was in recruiting beekeepers. It is not easy to commit to opening your hive every single week to make an inspection, much less pulling a frame and cutting out a square of freshly-filled comb every week. We had hoped for greater participation across the state. Our suggestion for future studies would be to plan for the minimum number of beekeepers needed for successful completion of the study and consider any participation above that as good news. That said, fewer participating beekeepers meant that we were able to reallocate funds to support use of BroodMinder hive scale systems as an independent monitor of nectar flow, so in the long run, having fewer reporting honey collectors was a good thing for us.

We did not, however, plan well for the unexpected. There was a sample that burst in shipping. There were inevitable counting slowdowns as samples were not shipped to Canada in a timely manner and thus arrived at the analysts' labs outside of their defined working windows. COVID repeatedly struck the analyst team, resulting in delayed analyses. One university was hacked, resulting in delays in manuscript preparation. And of course, the cats adored sitting behind (or on) microscopes, keyboards, and analysts(!). Lesson learned: write in an extra six months of time just in case—and maintain a sense of humor throughout!

We originally organized the project around a South Carolina ten-region economic management map. This map exemplifies the thinking of many beekeepers as to what constitutes local honey, but it has no bearing on the plant communities that constitute the bees' foraging areas. We were very lucky that our plan could be converted to the Level III ecoregions. We recommend the use of ecoregions, rather than development or economic management districts, so that data can be integrated seamlessly with other ecological studies. Remember that in the end, that's what we are looking at: the local and regional ecology that impacts and is impacted by our bees.

Speaking of maps, it is always good to have an expert in the use of Geographic Information Systems (GIS) on board to make your maps. We were very lucky to have found Dr. Mary Kate Fidler at Clemson University, who generated the map showing the locations of the honey sampling sites used in this book.

1.3 Project Framework

This project had three main objectives:

- To determine **what** plants are nectar sources for honeybees by region across the state of South Carolina via analysis of pollen from collected honey. These data serve as a baseline to determine how the bee-preferred vegetation has changed at the sampling locations.

- To determine **when** these nectar plants bloom across the state by regular, routine collection of honey samples for one calendar year.
- To determine whether the maple (*Acer*) bloom (at the end of January/early February) is **both** a pollen source and a honey/nectar source.

To meet these objectives, we set out to collect weekly fresh honey samples from 19 individual hives throughout the year 2022, capturing nectar flow variations through changes in hive weight and changes in plant contribution to the nectar through identification of the pollen in the honey. This yielded an immense amount of information about variations in the timing of nectar flows both within and between the five Level III ecoregions and which plants were contributing to those flows.

This book is designed to be consulted as a reference for beekeepers and other interested stakeholders. As such, it is not meant to be read chapter by chapter, but rather on an as-needed or as-interested basis. Vast amounts of data were collected and the analyses presented here are simplified for the audience. Detailed analysis of the hive weight data alone would be another book, and we hope that future scientists are able to utilize these data to more fully explore hive health in the region.

Beyond this introduction, the book is organized into the following sections:

- Methodology: provides a detailed description of how the data were generated;
- Identifying Pollen in Honey: explains the theory and practice of extracting and identifying the pollen in honey samples;
- Results (broken into sub-sections by ecoregion): provides a general overview of data analysis, then highlights key results in each ecoregion;
    - Ecosystem overview—a description of the geology, soils, and plants typically found in the ecoregion;
    - Nectar flow periods—based on hive weight changes;
    - Plants used as nectar sources—an overview of which plant pollen types occurred in the honey across the year and when these pollen types appeared;
    - Monofloral vs. multifloral honeys—an examination of when and where monofloral (varietal) honey occurred and a discussion of multifloral honeys;
- Comparison of Ecoregions: comparison of the plant taxa between ecoregions to reveal plants that bees prefer that are unique to an ecoregion vs. plants bees are using across multiple ecoregions;
- Concluding Remarks: summary of the results relative to the aims of the project and suggestions for future monitoring activities;
- References

CHAPTER

2

Methodology

2.1 Study Sites

Our study sites included 19 apiaries in South Carolina (Table 2.1; Figure 2.1). One hive from each apiary was included in the study. Most apiaries contained additional hives, but they were not included in our study. Of these 19 sites, 15 sites represented rural apiaries, and 4 were urban. Research began in January—February 2022 and continued until November—December of the same year. Timing of honey collection was based on nectar flow, as determined by the first presence of new, uncapped honey at the site (see below). Each participating beekeeper agreed to monitor their hive using BroodMinder technologies, obtain comb samples during nectar flow, and make observations about plants blooming in the vicinity of their hive as described below. Each beekeeper used 10-frame Langstroth hives.

A study hive in Apiary 14. The hive is shown on the BroodMinder® hive scale in January of 2022.

2.2 Hive Weight Monitoring

Hive weight monitoring has been used to measure hive health and productivity since the 1950s (McLellan, 1977; Lecocq et al., 2015), and in this study, it provides an invaluable measure of nectar flow; significant increases in hive mass without the addition of supers are indicative of high honey production (and thus high nectar flow).

The hive in each apiary used for this project was fitted with a BroodMinder hive scale in January or February 2022 following directions posted at www.mybroodminder.com (last accessed 12/19/2023). Automatic data collection via www.mybroodminder.com began upon installation and continued through 1 December 2022. Hourly plots of hive mass were available to the beekeeper and grant leadership team throughout the project period. Data for all 19 beekeepers were downloaded as Excel spreadsheets in early 2023.

Hourly data, while fascinating, is unwieldy and, in many cases, adds noise to interpretations. For this reason, the final data set for each beekeeper was simplified to a daily midnight reading only, following the method of Sponsler et al. (2020) (see Figure 2.2).

Table 2.1. Locations of Participating Apiaries.

Map ID	Apiary Location	Apiarist
1	Hilton Head Island, Beaufort County	David Arnal
2	Ehrhardt, Bamberg County	Susan Chewning
3	Mt Pleasant, Charleston County	Tim Liptak
4	Ladson, Dorchester County	Woody Weatherford
5	Loris, Horry County	Karen Hilbourn
6	Aiken, Aiken County	Robert Abshire
7	Aiken, Aiken County	Hank Smalling
8	Smoaks, Colleton County	Tom Parker
9	Camden, Kershaw County	Rosalind Severt
10	Florence, Florence County	Ben Powell
11	Lakewood, Sumter County	Paul Wise
12	Greenwood, Greenwood County	Kathy Murray
13	Columbia, Richland County	Larry Coble
14	Saluda, Saluda County	Lizanne Melton
15	Rock Hill, York County	Bennie Copeland
16	Rock Hill, York County	John Williams
17	Whetstone, Oconee County	Rannie Bond
18	Easley, Pickens County	Brett Kahley
19	Powdersville, Anderson County	Barbara Tate

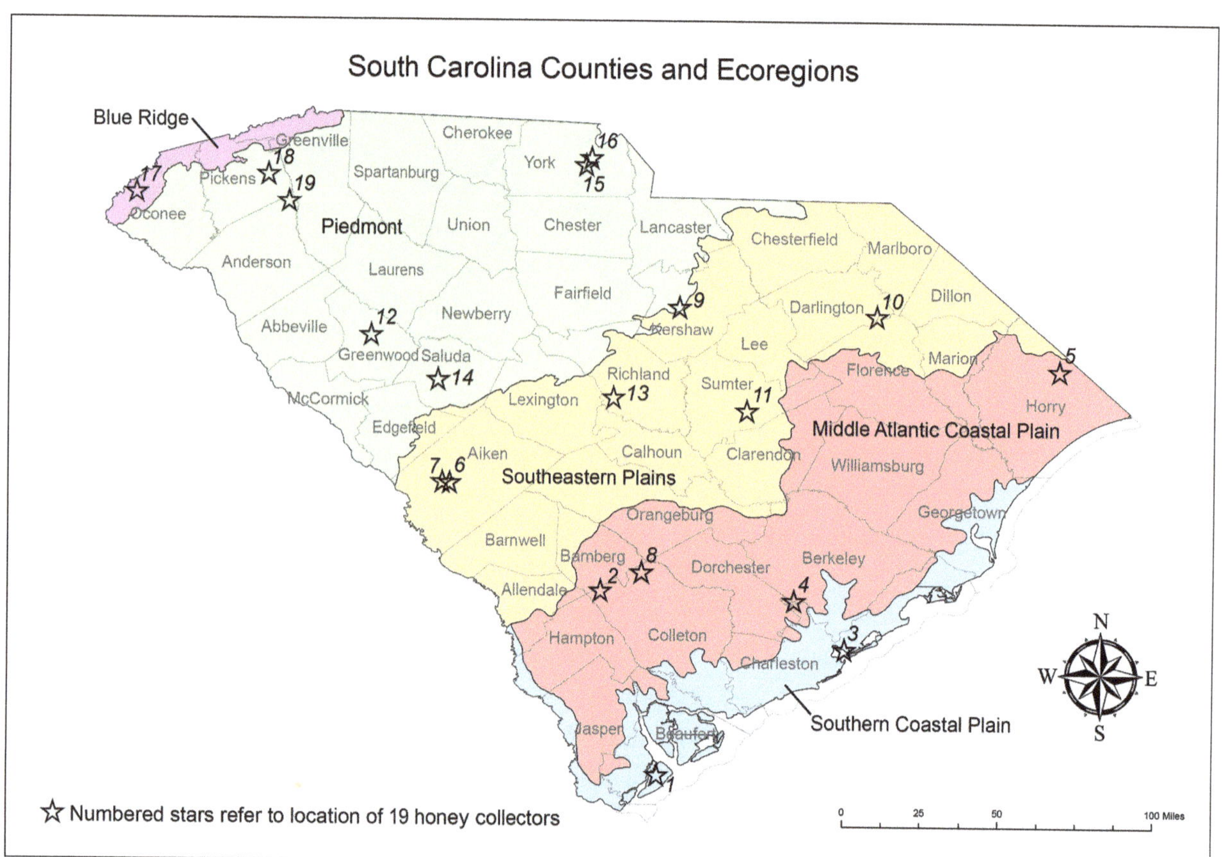

Figure 2.1. Level III ecoregions of South Carolina showing apiaries involved in the study.

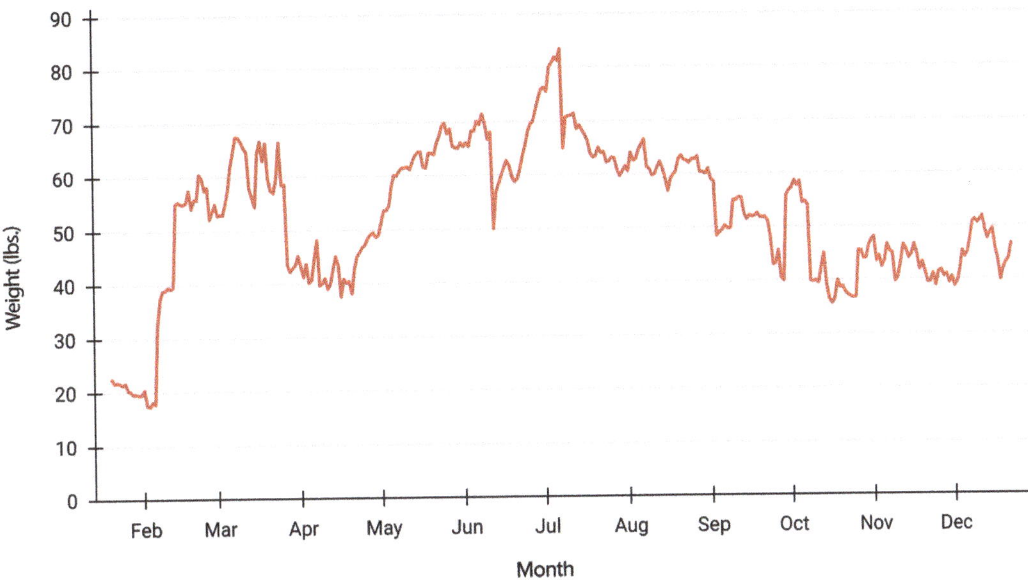

Figure 2.2. An example of a simplified hive mass graph constructed from data collected from Apiary 18.
Note: Labels indicate the first day of each listed month.

2.3 Honey Collection

Beekeepers were directed to use supers with drawn comb, not supers with foundation or foundation frames interspersed with drawn comb. This was necessary because it generally takes 8.4 pounds of honey to produce a pound of beeswax if the wax cells need to be drawn out from the foundation. Hence, honey yield would be impacted by foundation or foundation frames. Those using plastic foundation were directed to heavily coat the plastic with beeswax to get the bees to draw out comb from the plastic foundation rather than use incoming resources to produce beeswax to draw out the comb.

Honey was collected weekly during periods of nectar flow. Nectar flow was identified using a four-pronged approach: hive scale showing weight gain; observation of fresh nectar in the comb; observation of white, new wax; and observation of bees flying with a sense of purpose. Care was taken to avoid sampling during periods of sugar syrup feeding. We utilized the following observational test to help the participating beekeepers differentiate between a sugar-sourced honey and a nectar-sourced honey. We instructed them to do the following:

> *Turn the comb upside down and if "nectar" shakes out, it is fresh and has not been cured by the bees. If you are feeding sugar syrup and the bees are still taking the syrup, this liquid is syrup, and you should not take a sample. If you are feeding sugar syrup and the bees are not taking the syrup, this is fresh nectar, and you should take a sample.*

Once comb was determined to contain nectar-sourced honey, beekeepers were directed to collect samples around the brood nest during the maple (*Acer*) bloom and in the upper part of the super stack during the spring and subsequent nectar flow periods.

Honeybees store surplus honey above the brood nest and typically store surplus nectar in the center of the super and work their way outward as the nectar flow proceeds. When the super is full (except around the outermost frames), beekeepers typically add another super. How quickly the frames are filled is a function of the number of bees in the hive and how strong the flow is. It was critical that the participating beekeepers sampled honey from the outermost frames of the top honey super to capture the most recently produced honey.

Beekeepers were directed to sample a 2-in. square of fresh comb if possible, using a clean, sharp knife or hive tool to slice the comb from the frame (Figure 2.3). If the honey cells were already capped, they were directed to keep them intact if possible. They were also directed to avoid sampling pollen cells, though if the pollen cells were obtained along with the comb, they could be worked around if the cells were capped.

Comb samples were immediately placed in sterile urinalysis containers and labeled with the latitude and longitude of the apiary, sample number, hive number, date, beekeeper's name, and project identifier (Figure 2.4). Lids were then firmly attached and preferably taped down to prevent any opening in transit. Samples were shipped in batches to Global GeoLab Ltd.

Figure 2.3. A comb sample containing fresh, uncapped honey and no pollen cells obtained from Apiary 3 on 1/30/2022.

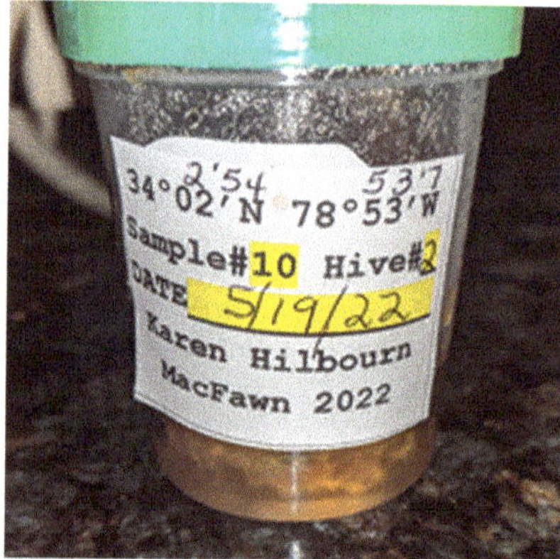

Figure 2.4. Comb and honey in a sample cup showing label information.

2.4 Beekeeper Bloom Observations

In addition to obtaining weekly comb samples during nectar flow, beekeepers were directed to observe plants in bloom at the time of sampling and record their observations in standardized log sheets ("Checklist of Honey Sample Dates and Blooming Plants") (Figure 2.5). Copies of these log sheets were mailed along with the comb samples. The primary purpose of these bloom observations was to aid the melissopalynologists in identification of pollen grains in the sample, particularly when non-native plants contained in pollinator gardens might be present in the vicinity of the hives, or to help differentiate plants to a lower taxonomic level when pollen morphology is not sufficiently unique to identify various species from a genus.

Checklist of Honey Sample Dates and Blooming Plants

You may log Blooming Plants even when Samples are not taken as indicators of not producing honey. Try to take samples about the same day of week if possible. For Example, avoid doing a Saturday and then a Monday right afterwards. But better to do it that way than to skip a week in honey flow. Weather may alter schedule.

2022

WEEK	DATE Sample Taken	Hive Weight	Plants Blooming
Jan Wk1			
Jan Wk2			
Jan Wk3			
Jan Wk4			
Feb Wk1			
Feb Wk2			
Feb Wk3	2-15-22	105.5	Henbit, dandelion, Rosemary, Tea olive, Camellia
Feb Wk4	2-22-22	106.24	Maples, Red Deadnettle, Grey Field-speedwell
Mar Wk1	3-03-22	105.2	Burfordii Holly, Winter Honeysuckle, Mayberry, Plums, Cherries
Mar Wk2	3-08-22	104.9	Blueberry, white clover, elm, yellow jasmine, Calla, Saucer Magnolia, Pear
Mar Wk3			
Mar Wk4	3-24-22	98.70	Apple, Pear, Mayberry, Blackhawk, Azaleas, white clover
Mar Wk5			Bartlet Pear, Southern Dewberry, Bristle Thistle, Black Cherr
Apr Wk1			Longbract Wild Indigo, Scots Broom, Rosemary
Apr Wk2	4-12-22	88.24	Blackberry, white clover, Viburnum opulus, Apple, Pear
Apr Wk3	4-20-22	98.16	Crimson clover, white clover, blackberry
Apr Wk4	4-27-22	139.38	privet hedge, crimson clover, blackberry, white clover, Pyracan
May Wk1	5-04-22	193.62	Small's Ragwort, Japanese Honeysuckle, Spiderwort, China Ros
May Wk2	5-10-22	201.32	white + Crimson clover, magnolias, dandelions, Wild Garli
May Wk3	5-16-22	172.2	spider wort, Sneezeweed, Plains Fleabane, wild onion
May Wk4	5-24-22	181.06	Oakleaf hydrangea, poppy, Horsenettle, Sparkleberry, Pasture Ro, Dandelions, Clover, Rocket Larkspur, Catalpa, squash, persimmon, Queen Ann's Lace
Jun Wk1	5-31-22	181.76	Queen Ann's Lace, St. John's Wort, white clover, dandelions, Magnolias
Jun Wk2	6-9-22	129.6	Mimosa, Crape Myrtle, Gardenia, Oliander, Corn, Day Lillies
(* Hive #2 Jun Wk3)	6-16-22	108.2	Tree Privet Ligustrum lucidium, Common Hibiscus, American Pokeweed, Prickly Pear Cactus
No sample Jun Wk4	6-24-22		Chineese Talow, Mimosa
Jun Wk5			
Jul Wk1	7-2-22	128.82	Black-eyed Susan, Wild Carrot, Crepe Myrtle
Jul Wk2			~~St. John's Wort, Crepe Myrtle~~
Jul Wk3	7-13-22	117.34	White Topped Aster, Chinese Bush Clover, Oakleaf Fleabane,
Jul Wk4			Leafy Elephants foot, Linnaea, Magnolia grandiflora,
Aug Wk1	8-1-22	134.16	Marsh woundwort, purple top Vervain, Garlic chive, Butterfly bush
Aug Wk2			Flax leaved Horseweed
Aug Wk3	8-14-22	81.82	Woodland sunflower, Carolina Desert Chicory, Brazilian Vervain
Aug Wk4			white Rain Lilly, Sweet Everlasting, Hyssop leaf Thoroughwort
Aug Wk5	8-31-22	84.54	Buttonweed, Shining Sumac, Rosemary, Slender Bush Clover
Sep Wk1			Sensitive Pea, Bitterweed, Partridge Pea
Sep Wk2	9-14-22	88.38	Buttonweed, Quail grass, Cypress vine, Autumn Clematis
Sep Wk3			Osmanthus fragrans, Bearded Beggarticks (Bidens), White morning glo
Sep Wk4			Sweet Goldenrod, Maryland Golden aster, blazing Star (Liatris)
Oct Wk1	10-6-22	107.8	Blue mist flower (Conoclinium) Lantana, Allium
Oct Wk2			Camphor weed
Oct Wk3			sunflowers
Oct Wk4	10-28-22	94.84	Sasangua Camellia, Sweet Goldenrod, Wild Asters
Nov Wk1			
Nov Wk2			
Nov Wk3			
Nov Wk4			
Nov Wk5			
Dec Wk1			
Dec Wk2			
Dec Wk3			
Dec Wk4			

Figure 2.5. Example checklist of honey sample dates and blooming plants, provided by Apiary 14's beekeeper.

2.5 Palynology Processing

Extraction of pollen from honey typically uses one of two methods: that of Louveaux et al. (1970), which relies on dilution, centrifugation, and pelletization or filtration to separate honey sediment from fluids (von der Ohe et al., 2004; de Almeida-Muradian et al., 2020), or that of ethanol dilution followed by acetolysis (Jones & Bryant, 2004)—although enzymatic processing (O'Keefe & Wymer, 2015) can also be used. For this study, the ethanol dilution followed by acetolysis technique was chosen. This technique is favored in the United States because it is relatively rapid (enzymatic processing takes hours rather than minutes) and permits significantly more accurate identification of individual pollen grains, as no nuclear or external material remains attached to the grain. Additionally, this method causes the grains to swell by up to 30%, making the surface ornamentation of each pollen grain easier to see and interpret.

For consistency's sake and speed of throughput, we chose an external lab to complete processing for this project. Global GeoLab Ltd. in Medicine Hat, Alberta, Canada is the premier geological palynology processing laboratory in North America. When Dr. Bryant relinquished part of his contractual melissopalynology work to Dr. Sophie Warny at the Center for Excellence in Palynology (CENEX) at Louisiana State University, they worked together to train colleagues at Global GeoLab Ltd. in the art of processing bee pollen pellets and honey. The method used by Global GeoLab Ltd. is virtually identical to that of Dr. Bryant (Jones & Bryant, 2004; Bryant personal communication to O'Keefe, 1999).

A small sample of honey was extracted from the comb (or from the cup, in cases where the comb was uncapped or had leaked) and diluted with an equal amount of distilled or deionized water. Then, a ten-fold amount of denatured ethanol was added to the sample and mixed well. As an example, for 10g of honey, 10g of water was added and mixed well, followed by 100g of ethanol. To this mixture, a *Lycopodium clavatum* spore tablet (Lund University, Batch #280521291, including 13,761 spores per tablet) dissolved in 5 mL of 10% hydrochloric acid was added. The sample was then centrifuged to sediment the pollen and *Lycopodium* spores into the bottom of the tube. This permitted the honey/water/ethanol mixture to be poured off and disposed of. The pollen residue pellet was resuspended in ethanol, re-centrifuged, and liquid poured off twice more, to ensure that all sugars were eliminated; essentially, the pollen was washed and concentrated to a fraction as pure as possible. The sample was then dehydrated with glacial acetic acid and placed in a heating block or bath set to 90°C. Acetolysis solution (a 1:9 mixture of sulfuric acid and acetic anhydride) was then carefully added to the mixture, stirred, and allowed to react for 9 min before the reaction was stopped with glacial acetic acid, centrifuged, and poured off repeatedly—first with glacial acetic acid, then with distilled or deionized water—until the residue was pH neutral. This process removes the cellular contents of the pollen, making the wall of the pollen grain more readily visible. This is a very important step to allow taxonomic identification of the pollen types, which is done on the basis of the pollen wall morphology (see Chapter 3 for details). At this point, a few drops of the cleaned pollen were allowed to dry on a microscope cover slip before being glued to a microscope slide with Castin'Craft® clear polyester resin. Each slide was given a printed label that matched the information on the original comb sample container. Boxes of slides and vials of residues were shipped from Global GeoLab Ltd. to the three analysts along with the copied pages of reported plants in bloom and a card listing the mass of honey processed and quantity of *Lycopodium* spores added (as this information is required for analysis) (see below).

2.6 Pollen Analysis

Of the 302 samples analyzed for this project, 171 were sent to Sophie Warny at CENEX, Louisiana State University; 93 samples were sent to Jen O'Keefe at the OPaL Lab, Morehead State University; and 38 samples were sent to Carol Wymer of C & S Science Consulting. Identification of grains was coordinated between the three laboratories for consistency and completed using comparisons with material from reference collections housed in CENEX, the OPaL Lab, and the VM Bryant, Jr. digital pollen reference collections, among others. Frequent consultations were held to ensure accuracy of identifications. When possible, pollen grains were identified to the taxonomic level of plant species; more frequently, the pollen grain was only identified to the plant genus or family level. This was because many pollen types do not have strong identifying characters below these broader taxonomic levels (see Chapter 3).

Slides were examined at variable magnifications as needed to observe the pollen grains' key identifying features (see Chapter 3) on the analysts' microscopes, ranging from 400–1250x (600x and above using oil immersion optics) using transmitted white light. Photomicrographs of all identified and unknown pollen types were made using digital cameras and software. All data were recorded and calculations completed in Excel® spreadsheets.

Samples were counted until at least 200 pollen grains were encountered or a self-calculating pollen rarefaction curve reached saturation (Louveaux et al, 1978; Birks & Birks, 1980; Rull, 1987; Jones & Bryant, 1998, 2007, 2014; Jones et al., 1998; Lau et al., 2018). *Lycopodium* grains encountered during the analyses were co-counted, permitting calculation of concentration values using the following equation:

$$C = \frac{(Pc \; x \; Lt \; x \; T)}{(Lc \; x \; W)}$$

where C is the concentration value (number of palynomorphs per gram of honey), Pc is the quantity of non-tracer palynomorphs counted, Lt is the number of *Lycopodium* spores per tablet, T is the total number of *Lycopodium* tablets added per sample, Lc is the number of *Lycopodium* spores counted, and W is the mass of honey processed in grams (Benninghoff, 1962; Maher, 1981). Concentration values were then multiplied by 10 to match the reporting standard of concentration per 10 grams of honey.

Concentration values were used to place honey samples into categories following the recommendation of Maurizio (1975): Category I contains less than 20,000 grains/10g; Category II contains 20,000–100,000 grains/10g; Category III contains 100,000–500,000 grains/10g; Category IV contains 500,000–1,000,000 grains/10g; Category V contains greater than 1,000,000 grains/10g. Category I honeys extracted directly from the comb tend to represent those from nectar sources that produce little pollen or from sugar-fed or honeydew-feeding bees. Category II honeys are produced globally from most floral sources. Category III honeys extracted directly from comb where pollen cells were omitted indicate nectar sources that are high pollen producers. Category IV and V honeys are produced from nectar sources that are extremely rich in pollen (Paredes & Bryant, 2020).

Relative abundances of the different recovered pollen types were calculated for all pollen types, whether from nectar-producing plants or not. In standard practices, the two types of pollen are separated (de Almeida-Muradian et al., 2020); however, this is not widely done in the United States. These percentages permitted the pollen to be assigned to pollen classes following the recommendations of Louveaux et al. (1978): D, I, M, or L classes. D, the predominant pollen types, make up more than 45% of the sample; I, the secondary pollen types, make up 16–45% of the sample; M are important pollen types which make up 3–15% of the sample; L, the minor pollen types, make up less than 3% of the pollen in the sample. Monofloral (varietal) honeys contain one D class pollen and may contain an I class or multiple M or L class pollens. Bifloral honeys tend to contain two I class pollen. Multifloral (heterofloral/plurifloral) honey tends to contain no D class pollen. Use of the 1978 classes is chosen over the 1970 classes (de Almeida-Muradian et al., 2020) because they are the class designations widely used by the late Dr. Vaughn M. Bryant, Jr., and thus more familiar to beekeepers in the United States than the original designations of predominant (PP), accessory (AP), important minor pollen (IMP), and minor pollen (MP).

CHAPTER

3

Identifying Pollen in Honey

3.1 Pollen in Honey

One of the things beekeepers frequently ask melissopalynologists is if they can visit the lab for a day or two and learn how to identify pollen in their honey (Bryant, 2017a). Many come, but in all our years of practice, very few have been willing to invest the time it takes to really ***learn*** how to identify pollen. Indeed, for those of us who trained with the late Dr. Vaughn M. Bryant, Jr., it was only near the end of our second semester of training (i.e., 10 months), when we could identify over 100 pollen grains by sight, that we were handed an "easy honey" to analyze. Vaughn considered the analysis of honey samples to be the "greatest research challenge any pollen analyst could ever face" (Bryant, 2017a). In the United States alone there are an estimated 20,000 native nectar-producing plants that are attractive to bees and an ever-increasing number of non-native transplants. Acquiring the skill to recognize all 20,000 is not something most beekeepers would consider; indeed, it is daunting even to most palynologists. It also takes years of practice to know whether a pollen has a morphology that is unique enough to identify it to the species, genus, or only family level with certainty.

However, learning the basic features used to identify pollen is possible, and working toward being able to key out the 30–40 locally occurring pollen types that regularly occur in your honey is empowering. This level of engagement with the science is something we strive to introduce at regional bee schools whenever possible. You never know which budding beekeeper will choose to undertake advanced training in botany or geology (some pollen experts study fossils of pollen and spores) and join the ranks of professional palynologists! While the field has come a long way since Dr. Bryant's 2017(a) lament that there were not enough melissopalynologists, more are needed to fill the growing need for analyses to support the FDA's truth-in-labeling guidelines.

Melissopalynology is the study of pollen in honey. Pollen identification is grounded in botany with a focus on plant reproductive biology—pollen are plant male gametes (sex cells), after all, and plant families produce a variety of different shapes and sizes of pollen (Figure 3.1). When it comes to pollen in honey, though, you also need a basic understanding of honeybee ecology, especially related to foraging behaviors and digestion (Figure 3.2). Most pollen grains that a honeybee ingests are filtered out of the honey stomach (crop),

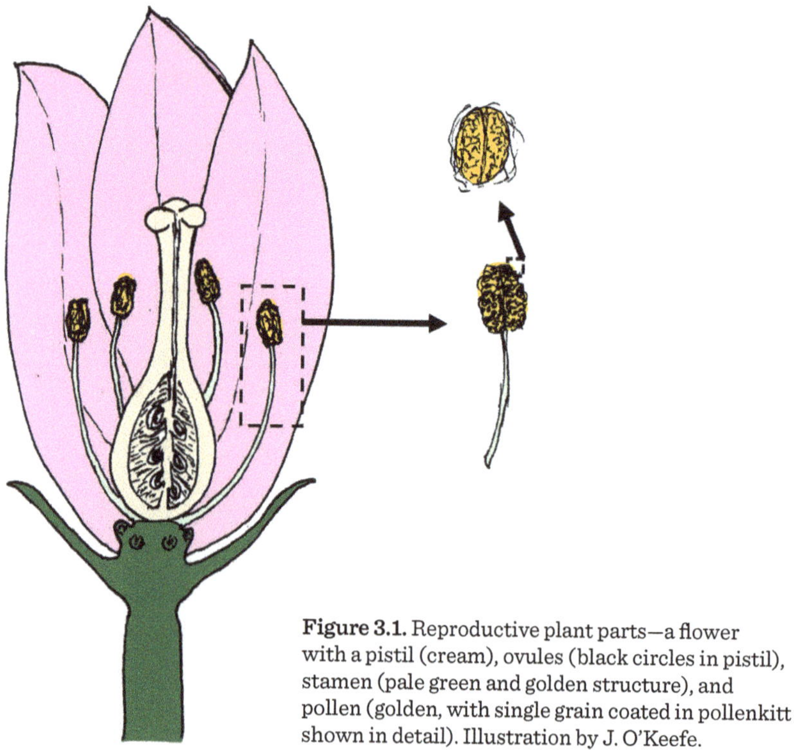

Figure 3.1. Reproductive plant parts—a flower with a pistil (cream), ovules (black circles in pistil), stamen (pale green and golden structure), and pollen (golden, with single grain coated in pollenkitt shown in detail). Illustration by J. O'Keefe.

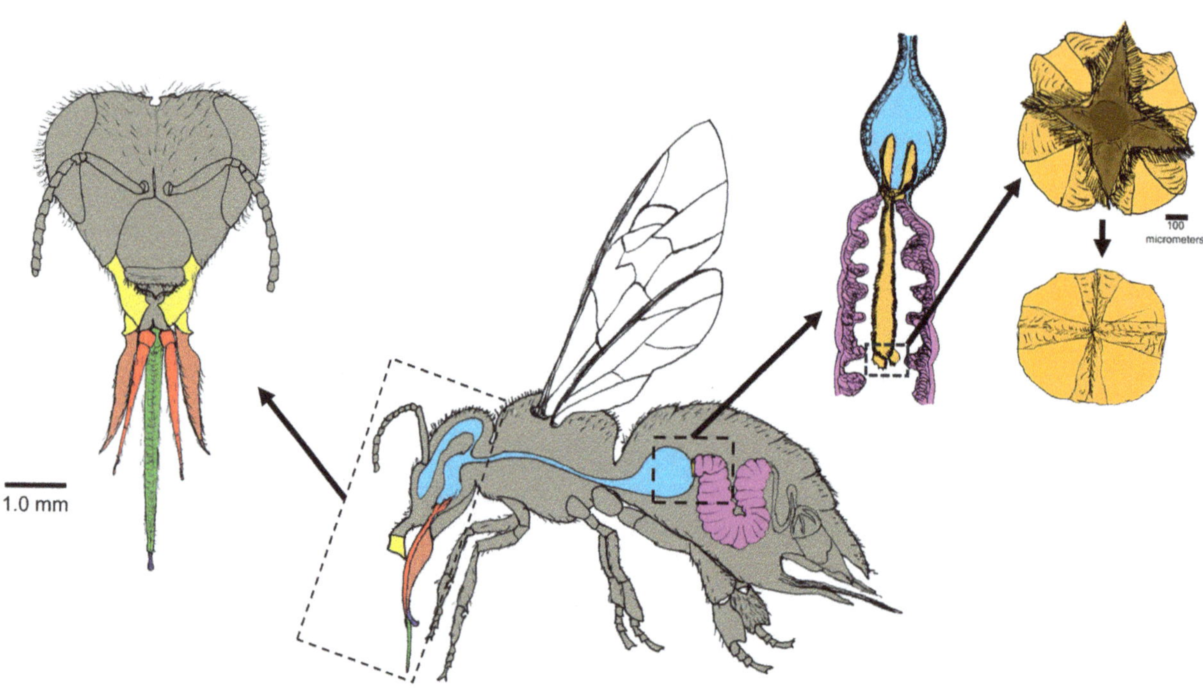

Figure 3.2. Parts of the honeybee digestive system: mandible (yellow), maxilla (orange), labial palps (red), proboscis (green), glossa (dark blue); esophagus and crop [honey stomach/foregut] (pale blue); proventriculus (golden yellow; upper view shows the proventricular valve open, lower shows it closed); ventriculus (purple). Illustration by J. O'Keefe.

packaged into lumps in the proventriculus (Bailey, 1952), and emptied into the ventriculus, where they are digested and exit the bee's alimentary canal (Peng & Marston, 1986). How long this takes varies by individual bee (Bailey, 1952), but it is generally accepted that more pollen grains are lost the further from the hive a bee is foraging (because of this filtering and digestion process). It is the combination of how much pollen flowers produce, how much of this pollen is typically incorporated into the nectar of that plant, and how much of this pollen remains in a honeybee's honey stomach upon its return to the hive that dictates how much pollen is preserved in honey.

How and when the preserved pollen is extracted from honey dictates what a melissopalynologist sees when analyzing a honey sample. If a honey is pre-filtered through anything smaller than a window screen (Bryant, 2015 & 2017b)—or even if it is filtered through a window screen that is not frequently cleaned to remove wax and bee parts—pollen is lost, and the analysis becomes skewed. This is especially true because it is most often the larger pollen grains, such as tulip poplar (*Liriodendron*) and sourwood (*Oxydendron*), that are preferentially trapped by debris during filtration. Indeed, we collected comb for this project specifically to avoid this problem!

How the honey is processed upon reaching the lab also impacts what is seen in the analysis. Once a sound honey sample reaches the lab, the pollen must be liberated from the honey. This is done through dilution; first with water, then with alcohol (either ethanol, which we used for this study, or isopropanol); these steps must be done in order, or an insoluble residue is produced. Once the pollen is separated from the honey, the individual grains need to be cleaned of any remaining pollen kitt and/or tryphine. Cytoplasmic and nuclear material must also be removed from the pollen grains. Some, but not all, of this cleaning is done in the bees' digestive tracts. Cleaning allows the pollen grain's ornamentation and aperturation to be observed (see Chapter 2 for details on chemical processing). This cleaning (see Figure 3.3) is accomplished either using acetolysis (as we did for this study), enzymatic treatment, or—more rarely—a tiny bit of potassium hydroxide. Acetolysis must be done in a fume hood/fume cupboard and produces hazardous waste, necessitating that it be completed in a laboratory, not in a home. Grains that were processed using acetolysis tend to darken enough that they do not need to be stained, but those processed through other methods do need to be stained, as most pollen grains are such a pale yellow that they are nearly transparent. We typically stain pollen pink using an acidified solution of Safranin-O or Basic Fuchsin. In Europe, Basic Fuchsin is widely used to stain un-treated pollen (von der Ohe et al., 2004). Once the darkened or stained pollen is isolated, it is ready to be mounted on a microscope slide underneath a cover slip and sealed in. Clear media like glycerin USP with phenol or silicon oil are widely used to mount the residue, and slides are sealed with clear nail polish, although choosing the correct polish can be tricky (Caffrey & Horn, 2012). Alternatives to glycerin include permanent mounting media like the Castin'Craft® clear polyester resin used in this study. Processing honey and mounting it on slides is both an art and a science, requiring specialized equipment and practice. We recommend contacting a commercial lab such as Global GeoLab Ltd. (as used in this study; see Chapter 2) or MelliFloral (https://www.mellifloral.com/) for commercial preparations.

An inexpensive transmitted light biological microscope with 10x, 40x, and 65x or 100x (oil) objectives can be used to analyze your pollen. It is much better to have a digital camera attached to the microscope (the set up used in this study), but it is possible to simply take pictures through the ocular lens (eyepiece) using your cell phone. One thing the microscope must have, however, is a way to measure the pollen grains you see. This could be an ocular reticule (a measuring bar in one of the oculars) or a stage reticule that allows you to calibrate software to make measurements. This is a key feature, as one of the key pollen identification components is size. After the sample preparation, we're finally ready to take a look at the pollen grains (Figure 3.4).

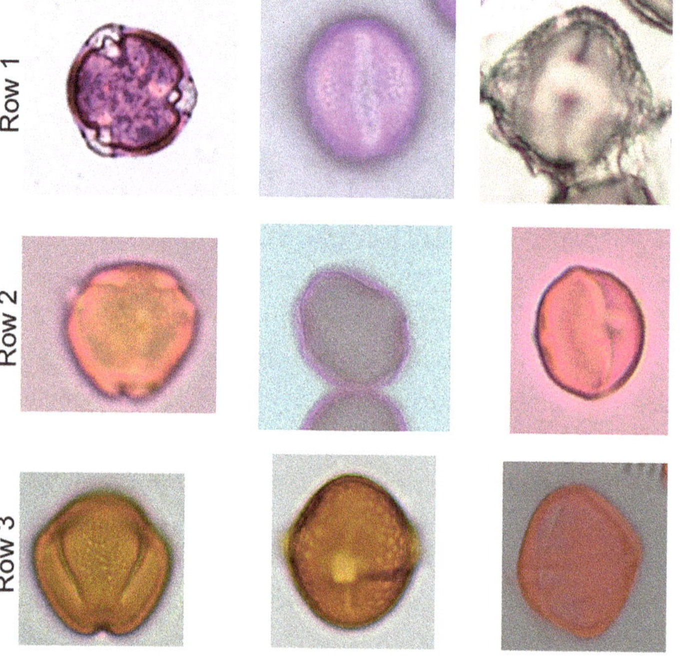

Figure 3.3. A comparison of the effects of cleaning treatment on willow, clover, and California pepper tree pollen.

Row 1: no cleaning treatment; Row 2: Enzymatic Treatment; Row 3: Acetolysis. Note how much easier it is to see the ornamentation following cleaning treatment.

Images from the J. O'Keefe collection.

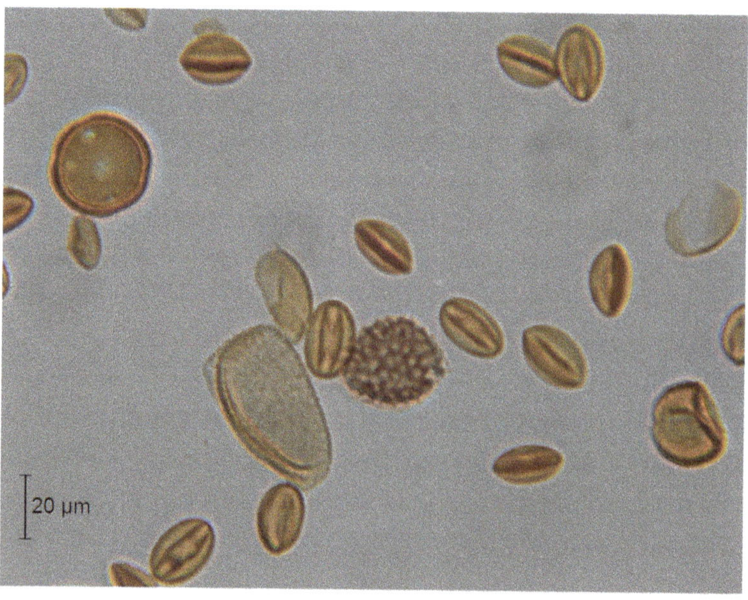

Figure 3.4. A typical field of view from a pollen slide produced from a mid-March 2022 honey.

3.2 How to Identify Pollen

Pollen identification and classification is based on five properties: grain size, grain shape, aperture type, number of apertures, and ornamentation type. Detailed descriptions of these features can be found in Kapp et al. (2000), along with line-drawings and descriptions of key pollen types. It is a great starter book and can be purchased directly from the publisher for $30: https://palynologyshop.org/product/pollen-and-spores-second-edition-kapp-et-al-2001-and-2014/. Halbritter et al. (2018) is also a good option. Here we provide an overview that is largely specific to pollen types found in honey.

Pollen grains found in honey range in size from very tiny (approximately 5 micrometers) to very large (approaching 200 micrometers), although pollen grains this large are rarely ingested by bees because they become caught in the bees' outer mouthparts (Peng & Marston, 1986). Pollen grains are grouped into five size range categories based on the length of their longest axis: very small (<10 micrometers), small (10–25 micrometers), medium-sized (26–50 micrometers), large (51–100 micrometers), and very large (>100 micrometers) (Halbritter et al., 2018). Most pollen grains found in honey are medium-sized, or about half the diameter of a human hair.

Pollen grain shape is generally spherical to ellipsoidal for single-celled pollen, though pyramidal, cubic, pentagonal, and kidney-shaped forms do occur. Some pollen is multicellular, either due to the presence of sacs (sacci) attached to the main body (corpus) of the pollen grain or because they occur in tetrads (groups of four cells) or polyads (groups of many cells) (Figure 3.5).

Aperturation refers to the openings that pierce a pollen grain's wall (Figure 3.6). These openings allow the grain to expand and contract according to osmotic pressure, provide sites for infolding of the grains, and provide emergence sites for the pollen tube following contact with the pistil during pollination (Wang & Dobrista, 2018). Apertures occur in two basic forms: a slit—which may be called a "sulcus" or a "colpus" depending upon how the slit is orientated—and a hole, called a "pore." Sometimes a slit and a pore occur together as a "colporous." If only one aperture is present, the prefix "mono-" is used to describe it; if two are present, "di-"; if three, "tri-"; if four, "tetra-"; if five, "penta-"; if six, "hexa-"; if seven or more, "stephano-". A single colpus that wraps around a grain may be referred to as a "syncolpus," while a pollen grain with no apertures is referred to as "inaperturate."

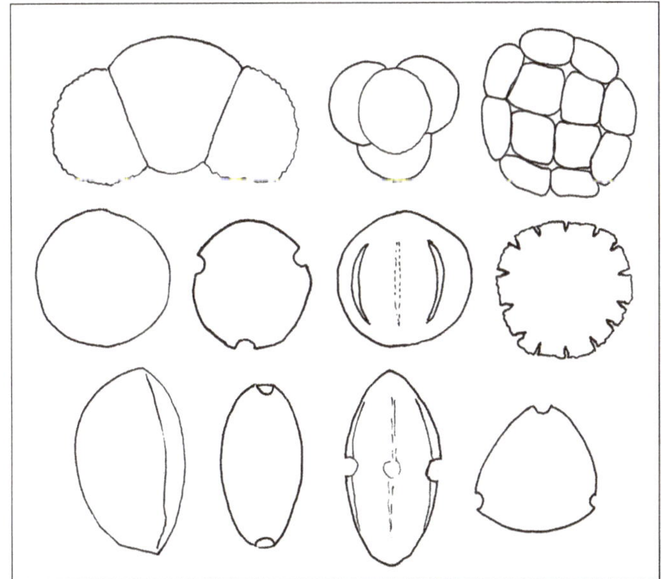

Figure 3.5. Shapes of pollen types commonly found in honey samples. Illustration by J. O'Keefe.

Gymnosperms like pine, fir, and spruce, which do not produce nectar but do produce large amounts of pollen that blow into hives and onto other flowers, generally have either no apertures or a single pore located between two sacci. Angiosperms (flowering plants) have a much wider variety of aperturation. The pollen of monocots like grasses, lily, and magnolia tends to range from inaperturate, to monocolpate, to monoporate. In dicots like maple, mustards, and clover, pollen tends to be inaperturate, tri- to periporate, di- to stephanocolpate, to syncolpate, or tri- to stephanocolporate (Figure 3.6).

Ornamentation refers to the structural features located on the surface of pollen grains. The surface may be nearly smooth (psilate) or exhibit a variety of bumps, lumps, spines, and/or ridges (Figure 3.7).

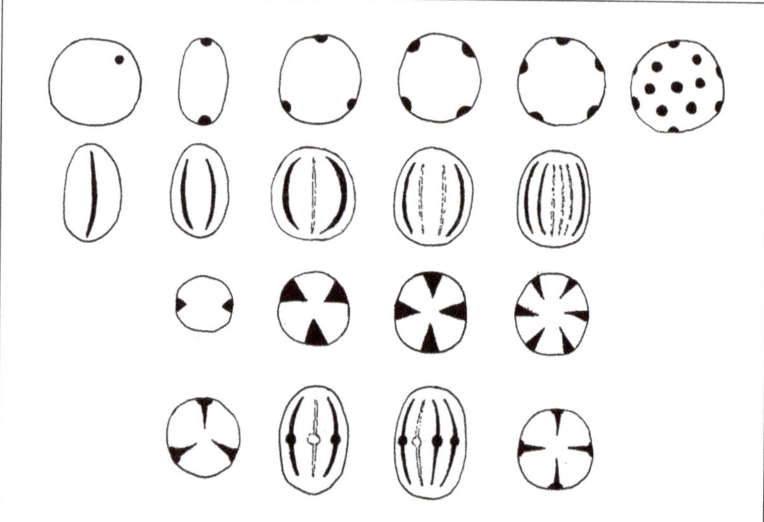

Figure 3.6. Examples of opening types (apertures) in pollen frequently found in honey samples.

Top row, left to right: Monoporate, Diporate, Triporate, Tetraporate, Pentaporate, Periporate.

Second row from top: Equatorial Views of Monocolpate, Dicolpate, Tricolpate, Tetracolpate, Hexacolpate.

Third Row: Polar Views of Dicolpate, Tricolpate, Tetraclopate, and Hexacolpate.

Bottom Row: Polar View of Tricolporate, Equatorial View of Tricolporate, Equatorial View of Tetracolporate, Polar View of Tetracolporate.

Illustration by J. O'Keefe.

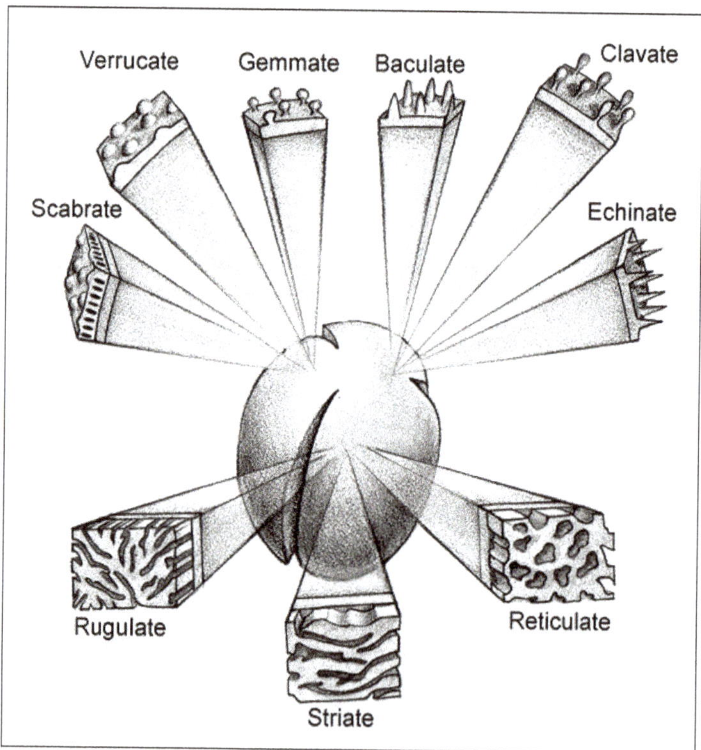

Figure 3.7. Pollen surface structural features (ornamentation) commonly seen on pollen found in honey. This figure was originally published in Kapp et al., 2000 and is used with permission of AASP—The Palynological Society.

The process of identification typically begins by sorting pollen grains by aperturation, then by ornamentation, then by shape, then by size. To illustrate the process of identification, let's use a large pollen grain of the same taxon observed in one of the honey samples collected in mid-March (Figure 3.8). It can be hard to see, but this grain has a single, long sulcus down the left side, making it **monosulcate**. The pollen grain's surface is covered in a rough, irregular bumpy pattern. This, especially with those enlarged bumps, is described as **verrucate**. It's roughly **ellipsoidal** in shape with more-or-less pointy ends, and it is **large**. With this information in mind, we can go to a key such as those found in Ronald O. Kapp's *Pollen and Spores, Second Edition* or at www.paldat.org and use this information to reach a tentative identification. From here, we compare our grain with examples from a type collection (a set of slides made from known pollen types) (Figure 3.9). Would you agree that these are both the same thing? While not identical (no two pollen grains are identical)—the purple one stained with basic fuchsin from the Morehead State University collection is a little smaller—both grains are *Liriodendron* (tulip poplar) pollen.

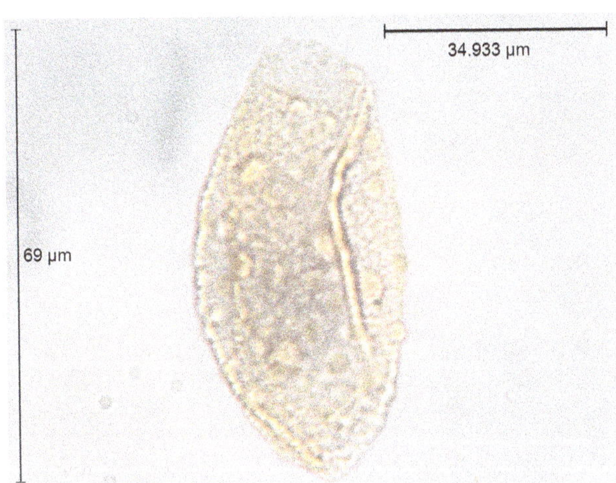

Figure 3.8. A large, bumpy pollen grain with a single slit-like aperture along the left side and a fold along the right.

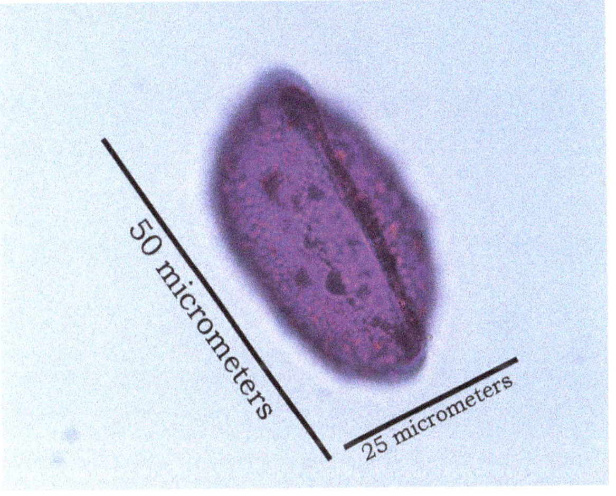

Figure 3.9. *Liriodendron tulipfera*, collected by J. O'Keefe in 2015. Note the single sulcus extending from end to end, verrucate texture, ellipsoidal shape, and medium-large size.

Let's look at another example (Figure 3.10). This pollen grain was observed in a honey sample collected at the end of March from the Southeastern Plains (Aiken County). This grain has two long furrows on the upper surface. The one on the right is broken, with a pale area to the right of it, indicating a pore. The one on the left has a paler area surrounding the midline. There is a third similar slit and-pore coupling on the back. This is an example of a **tricolporate** pollen grain. Note the network-like pattern of paler (on the right side of the grain) and darker (on the left side of the grain) ornamentation; this is what **reticulate** ornamentation often looks like. The grain is **ellipsoidal** to **ovoid** in shape and is **medium-sized**. Again, we take this information to a pollen key, narrow down likely suspects, and compare this grain to one from our type collection (Figure 3.11). This grain from the type collection is at a slightly different orientation, but you can clearly see the third colporous through the grain, as well as the two on top, and the reticulate ornamentation. It is almost exactly the same size as the one from our honey. Would you agree these are the same? They are *Trifolium* (clover), which was counted as *Melilotus/Trifolium* in our study because these two species can be difficult to differentiate with full accuracy.

Another common pollen found in many honeys is this one (Figure 3.12). This one is spiny (**echinate**) and angular, with big ridges (**reticulae**). It's almost impossible to see the aperturation, although it almost looks like there are big gaps or windows through the pollen. We call this combination of characteristics **lophate**. Lophate pollen grains with spines (echinae) on the ridges are characteristic of several members of the Asteraceae (aster/daisy family), including *Taraxacum* (dandelion). Compare this specimen with an example dandelion pollen from our collections (Figure 3.13). Grains of this pollen type are typically counted as Asteraceae (dandelion/endive-type) because so many members of the family make similar pollen. Identifying them to a unique species level is only possible if a hive is located within a monospecific field of Asteraceae.

Figure 3.10. A pollen grain with two long slits interrupted by paler or broken areas indicating pores on the upper surface; there is a third similar aperture on the reverse. It has a bumpy, almost network-like surface, is elliptical in shape, and is medium-sized.

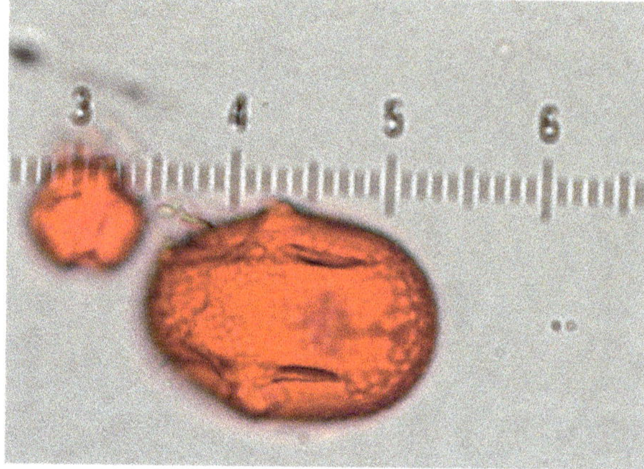

Figure 3.11. *Trifolium* sp. from the Vaughn M. Bryant, Jr. digital pollen collection. Note that it is tricolporate with reticulate ornamentation, ovoid in shape, and approximately 48 micrometers long.

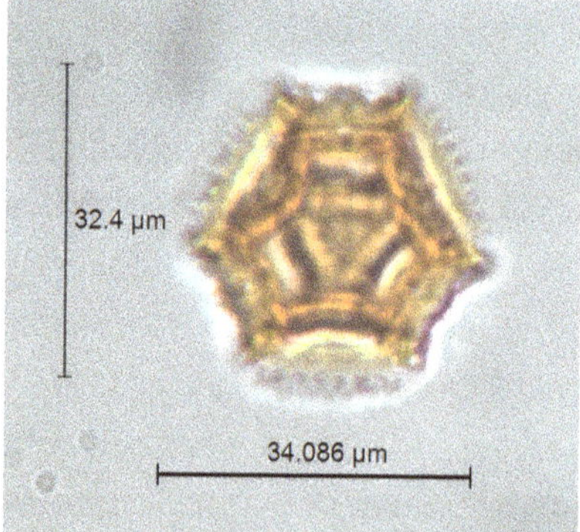

Figure 3.12. A pollen grain characterized by a series of broad reticulate topped by echinae. Gaps between the reticulae give the appearance of openings or "windows" through the grain.

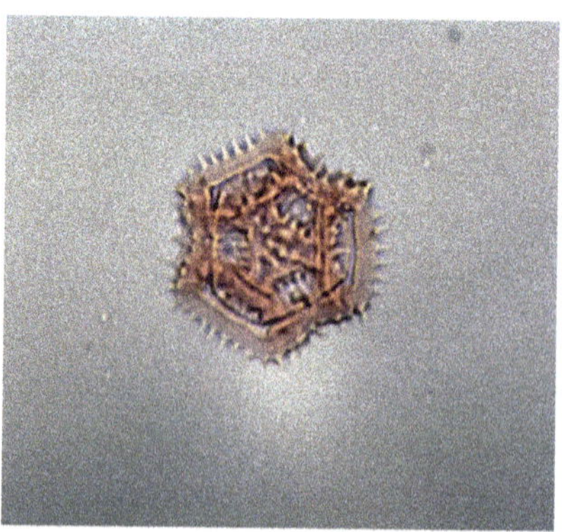

Figure 3.13. *Taraxacum* sp. from the J. O'Keefe Collection. Note the echinae on the reticulae and lophate texture.

Identification of these pollen is fairly straightforward—they are all laying on the coverslip such that we can see defining characteristics. Not all pollen in honey is this pristine, however. Pollen grains can be broken as they pass through honeybee mouthparts and fractured during the internal churning and filtration processes that occur in a bee's crop as pollen is strained out of the honey and concentrated as lumps in pockets in the proventriculus. It takes a long time to develop the skill to recognize pollen that is torn, folded, or mashed into unusual shapes (Figure 3.14), and there are some pollen grains that simply cannot be identified due to damage. Those that are whole are counted as "indeterminant" pollen, while those that are broken are counted as "partial palynomorphs." Sometimes, regardless of the skill of a melissopalynologist, there are grains that cannot be identified with certainty. These are listed as "unknowns."

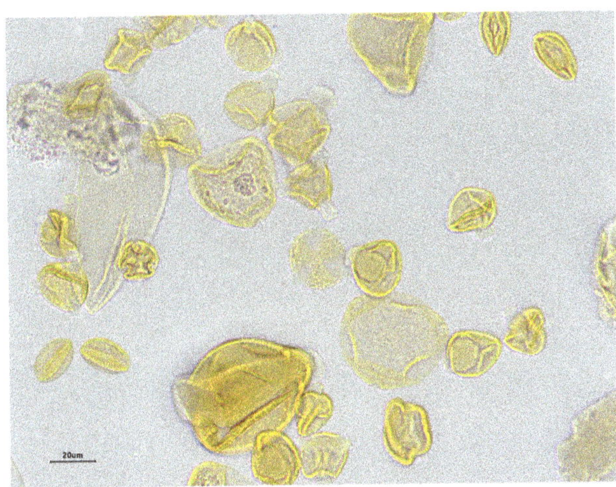

Figure 3.14. Pollen grains that are so distorted that they are indeterminant. Image from honey collected mid-April 2022.

3.3 Pollen Types Found in South Carolina Honeys

During the course of this project, 141 unique pollen types were recognized. Of these, 22 were identified to family and 6 could not be identified; they are listed as unknowns because they were not found in our collections. The remainder were identified primarily to genus, as the pollen between species is challenging to differentiate in many cases. The majority of these taxa are from nectar-producing plants. Several, including *Alnus* (alder), *Betula* (birch), *Fraxinus* (ash), *Pinus* (pine), *Quercus* (oak), and Poaceae (grass family), are from plants that lack nectar-producing organs (nectaries). It is likely that this pollen had blown onto other plants or into the hive rather than being a source of nectar used in honey production. Here we present photomicrographs of the pollen encountered during this study. Their distribution among the ecoregions is discussed in subsequent chapters.

A series of detailed plates were created to illustrate most key types of pollen recovered during the course of this study. All plates are organized by pollen aperture types, i.e., from inaperturate, monoporate, triporate, periporate, stephanoporate, fenestrate/lophate, echinulate, monocolpate & sulcate, tricolpate, stephanocolpate, syncolpate, stephanocolporate, tricolporate, tetrad & polyad, spiraperturate, and vesiculate (see pages 26–66 for illustrations).

INAPERTURATE

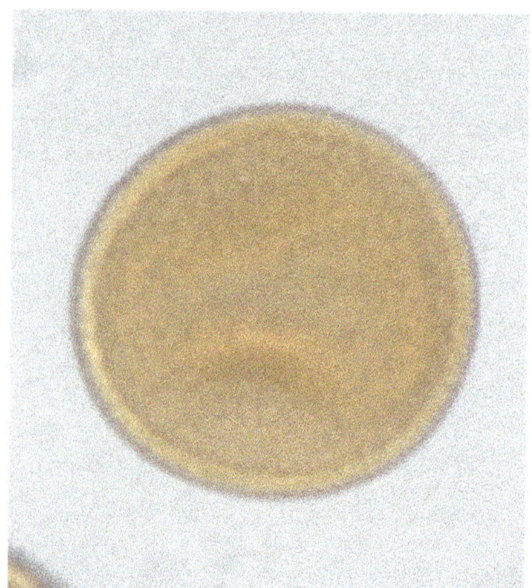

Populus

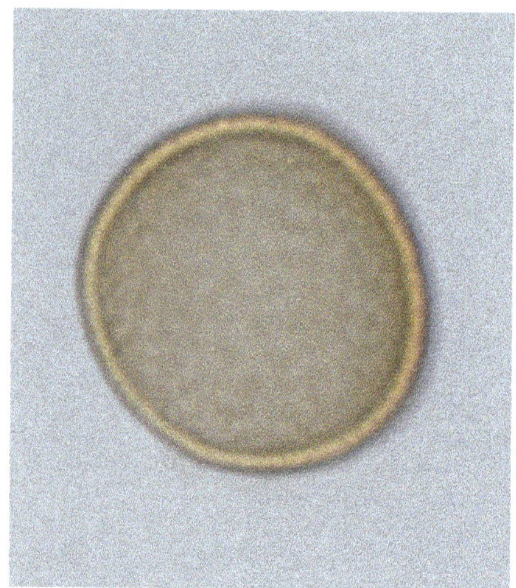

Populus

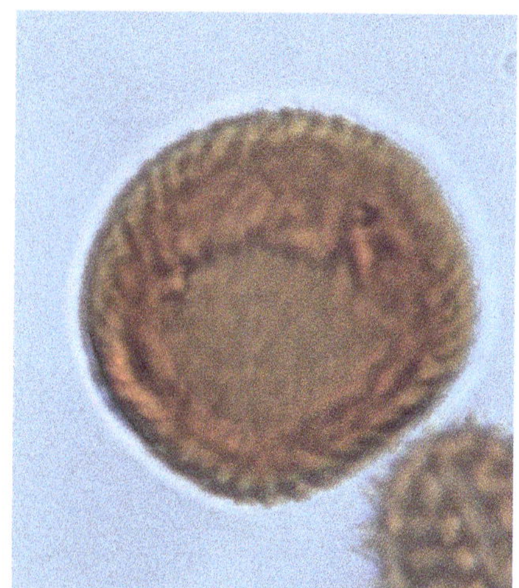

Spathiphyllum

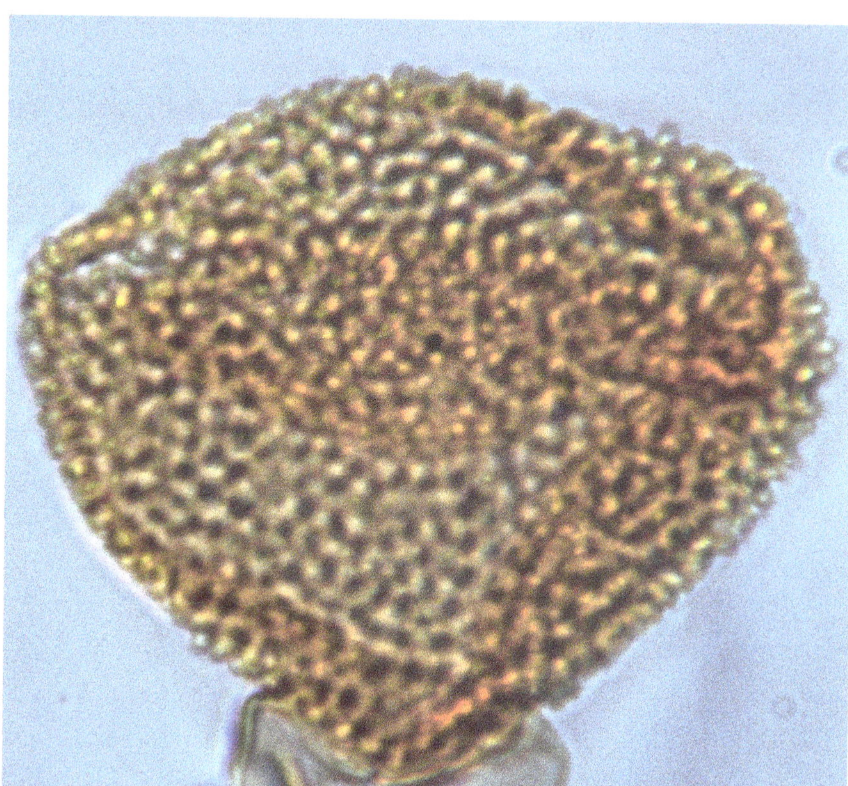

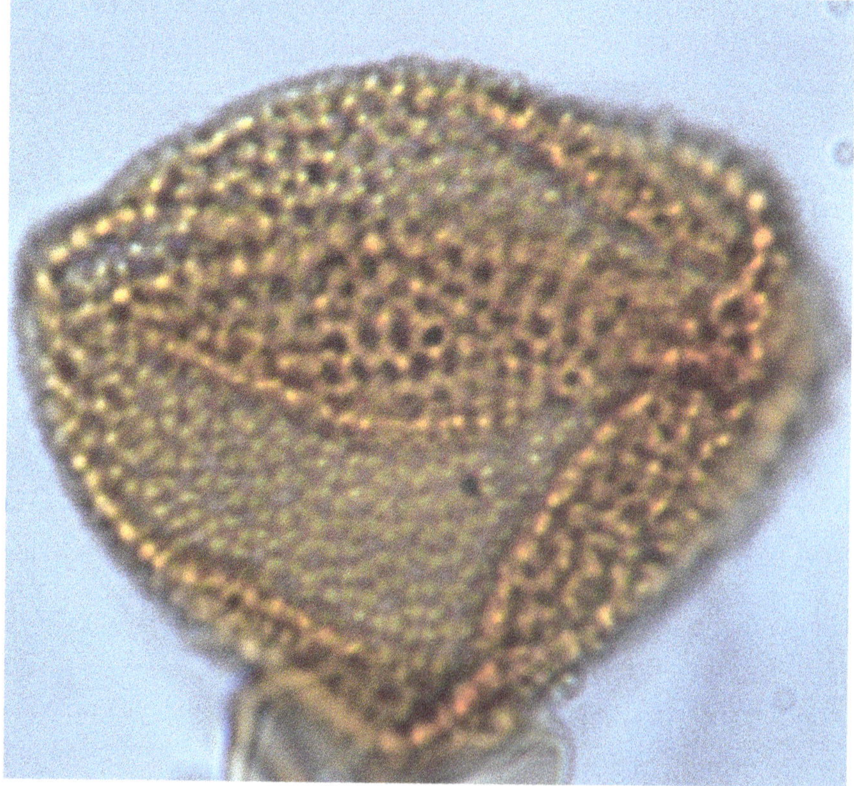

Croton

MONOPORATE

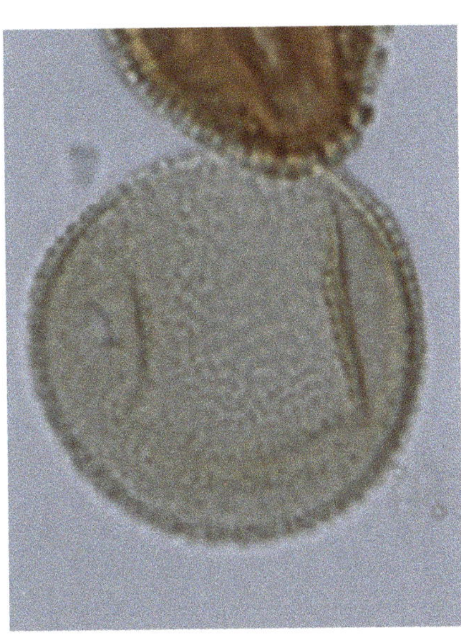

Typha

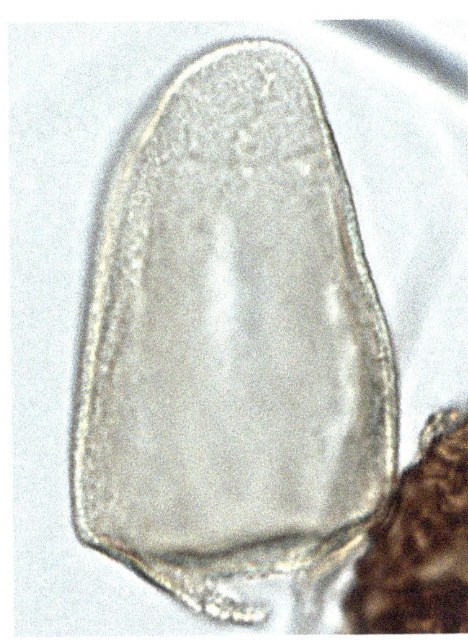

Carex

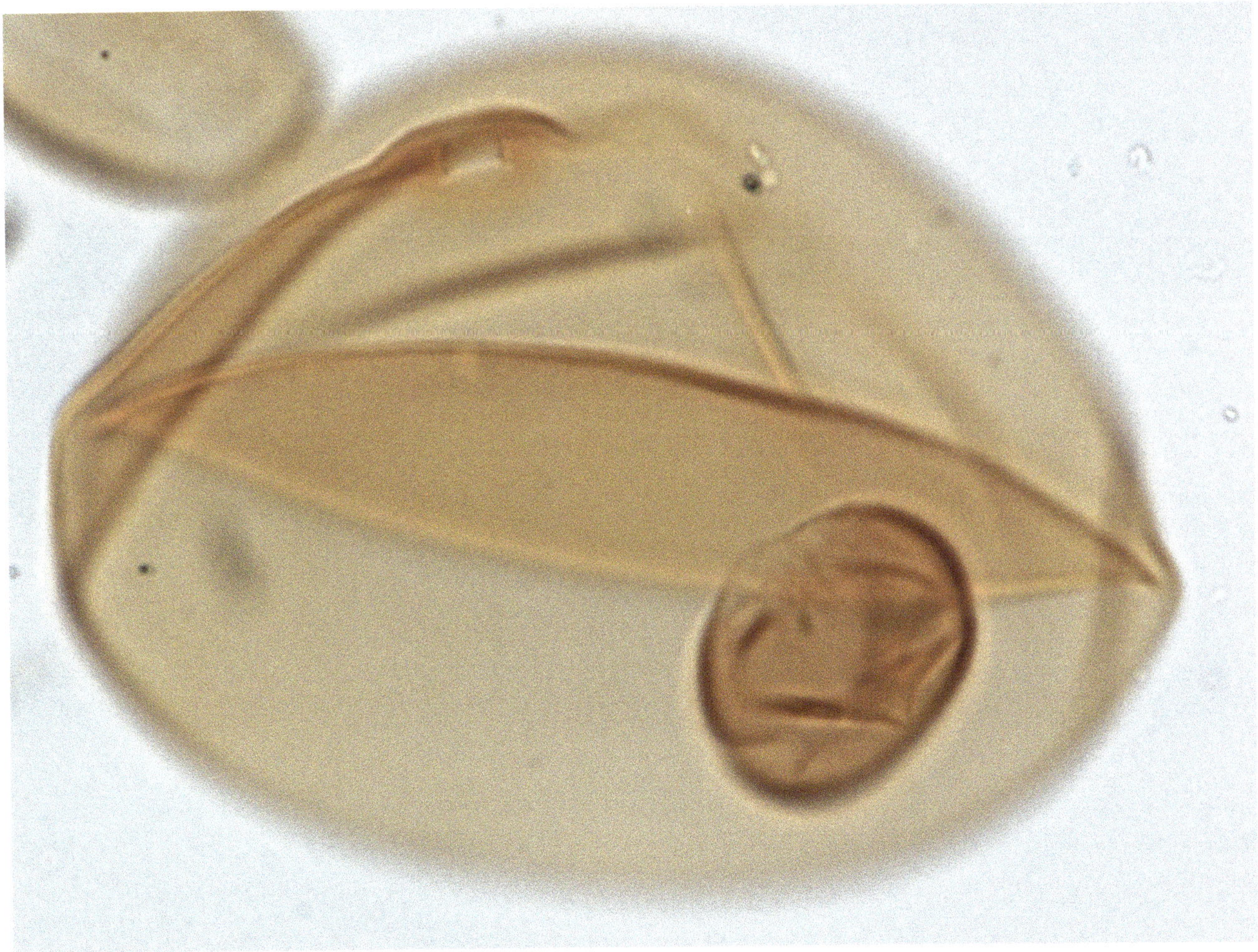

Zea mays (85 micrometers)

Zea mays (100 micrometers)

Poaceae

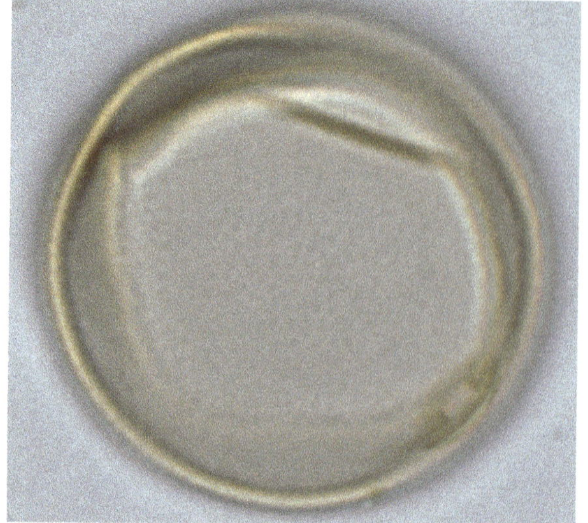

Poaceae

TRIPORATE

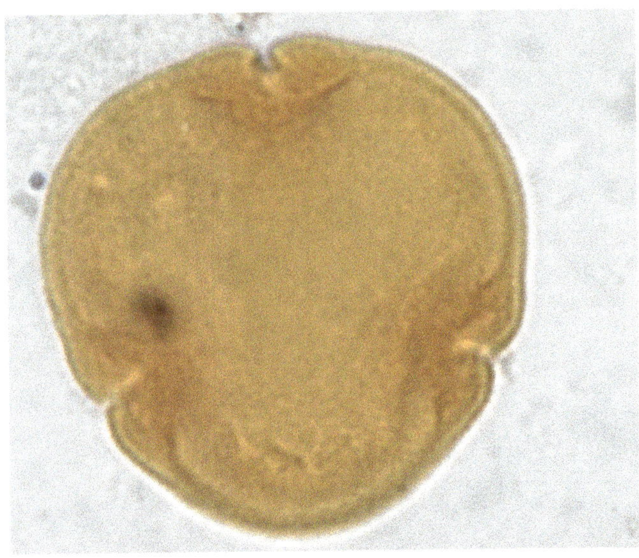

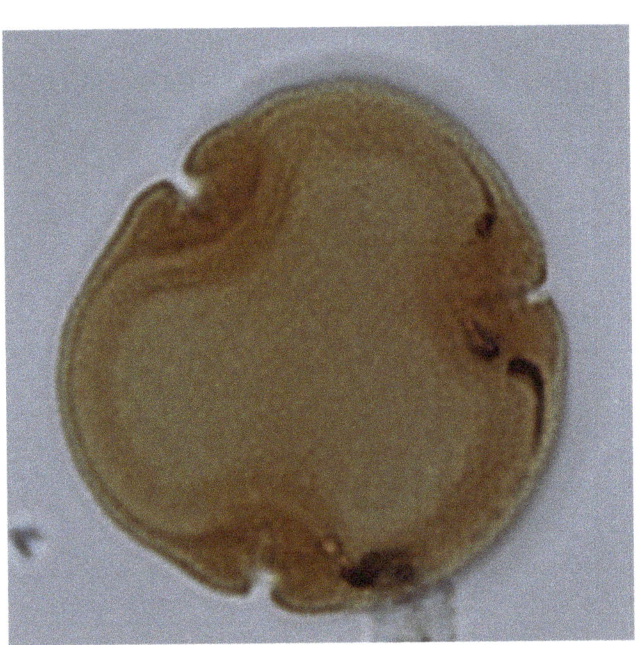

Tilia: two specimens with a range of sizes

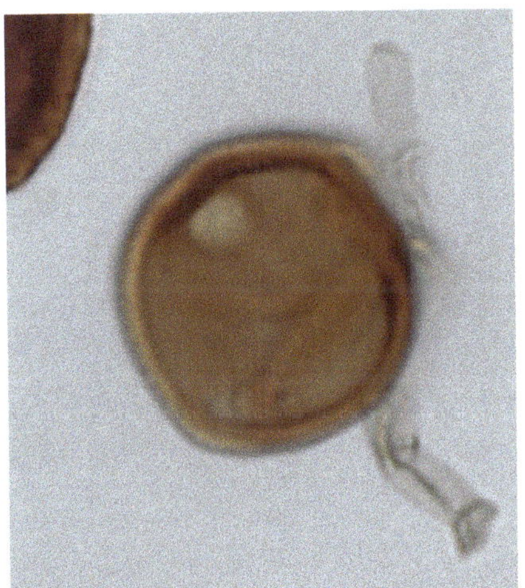

Glycine max

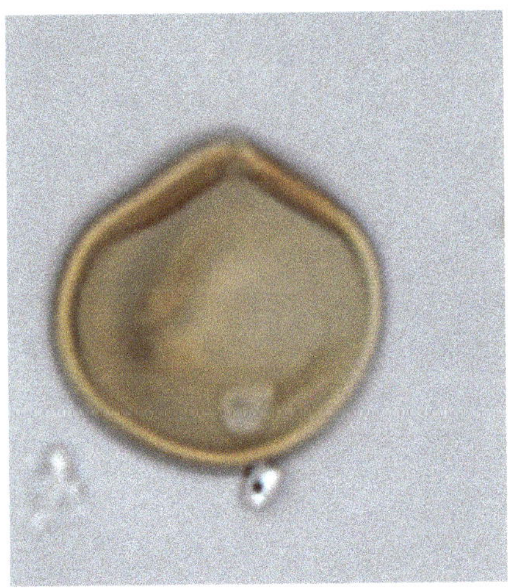

Glycine max

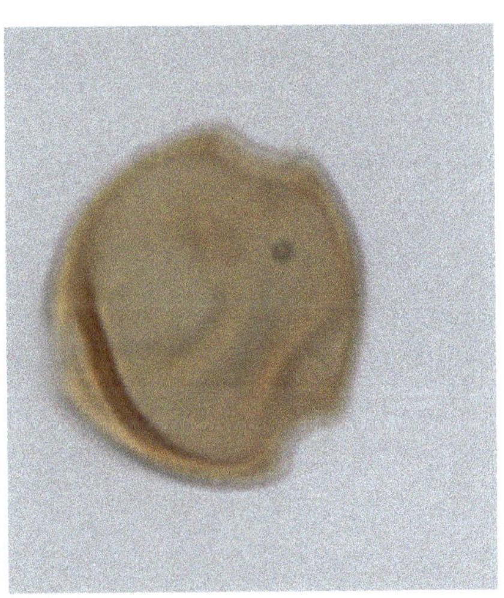

Glycine max

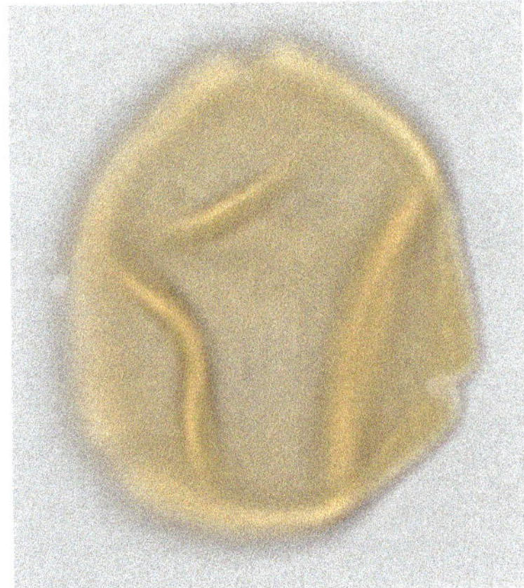

Corylus

Corylus

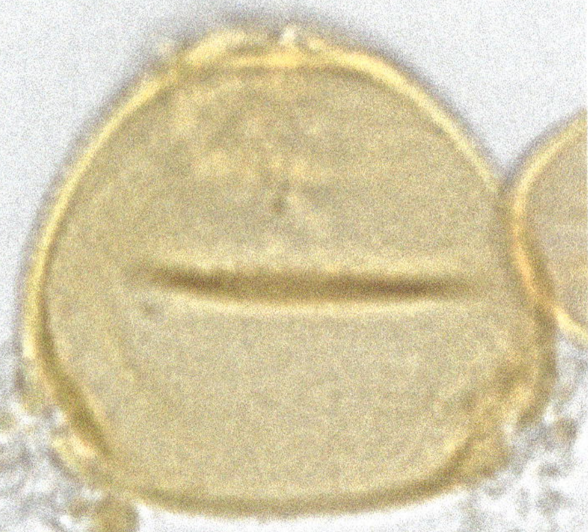

Symphoricarpos

Betula

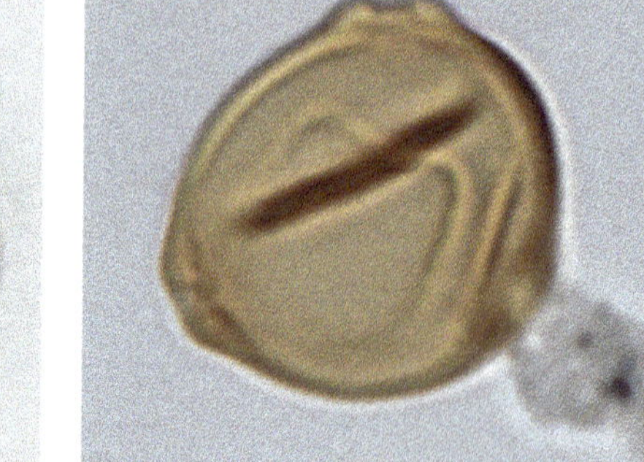
Betula

Betula

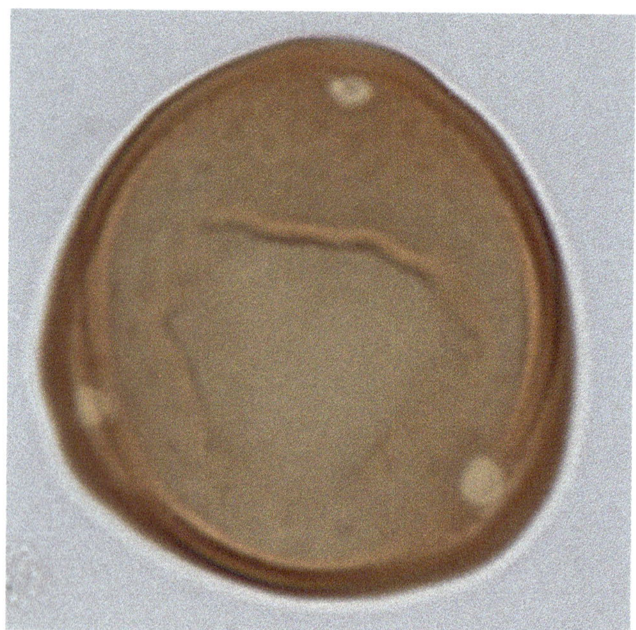
Carya

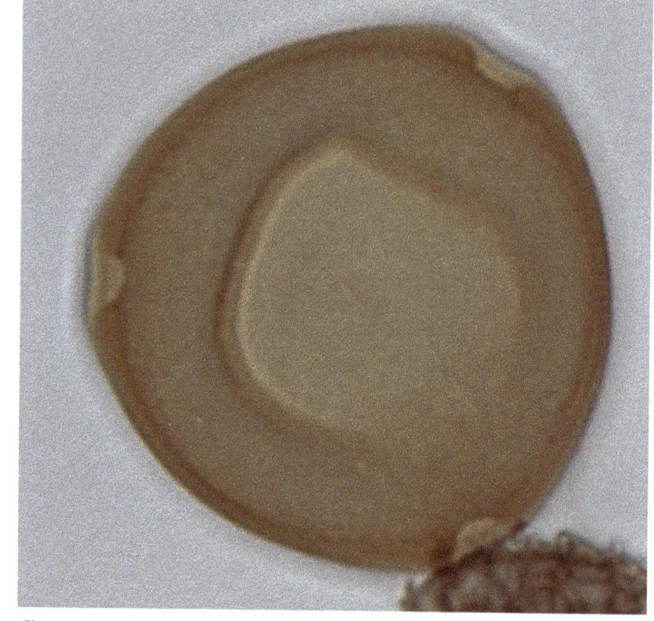
Carya

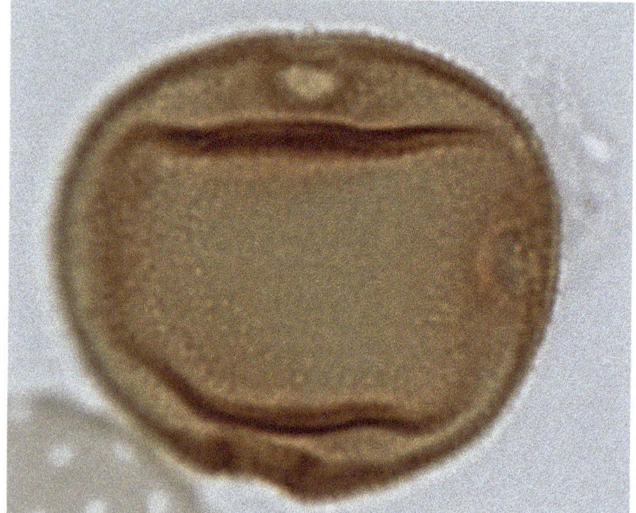
Cucumis

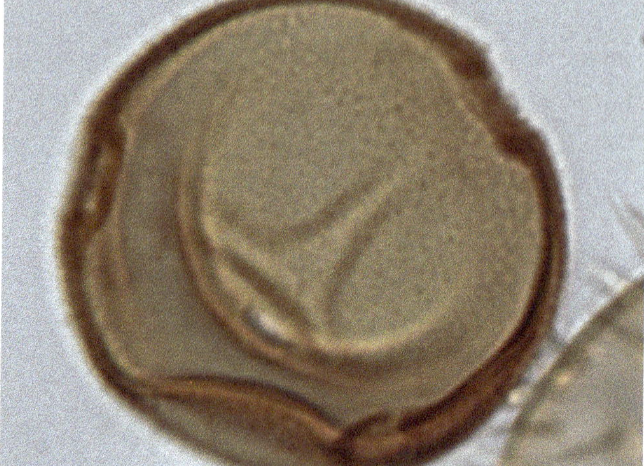
Cucumis

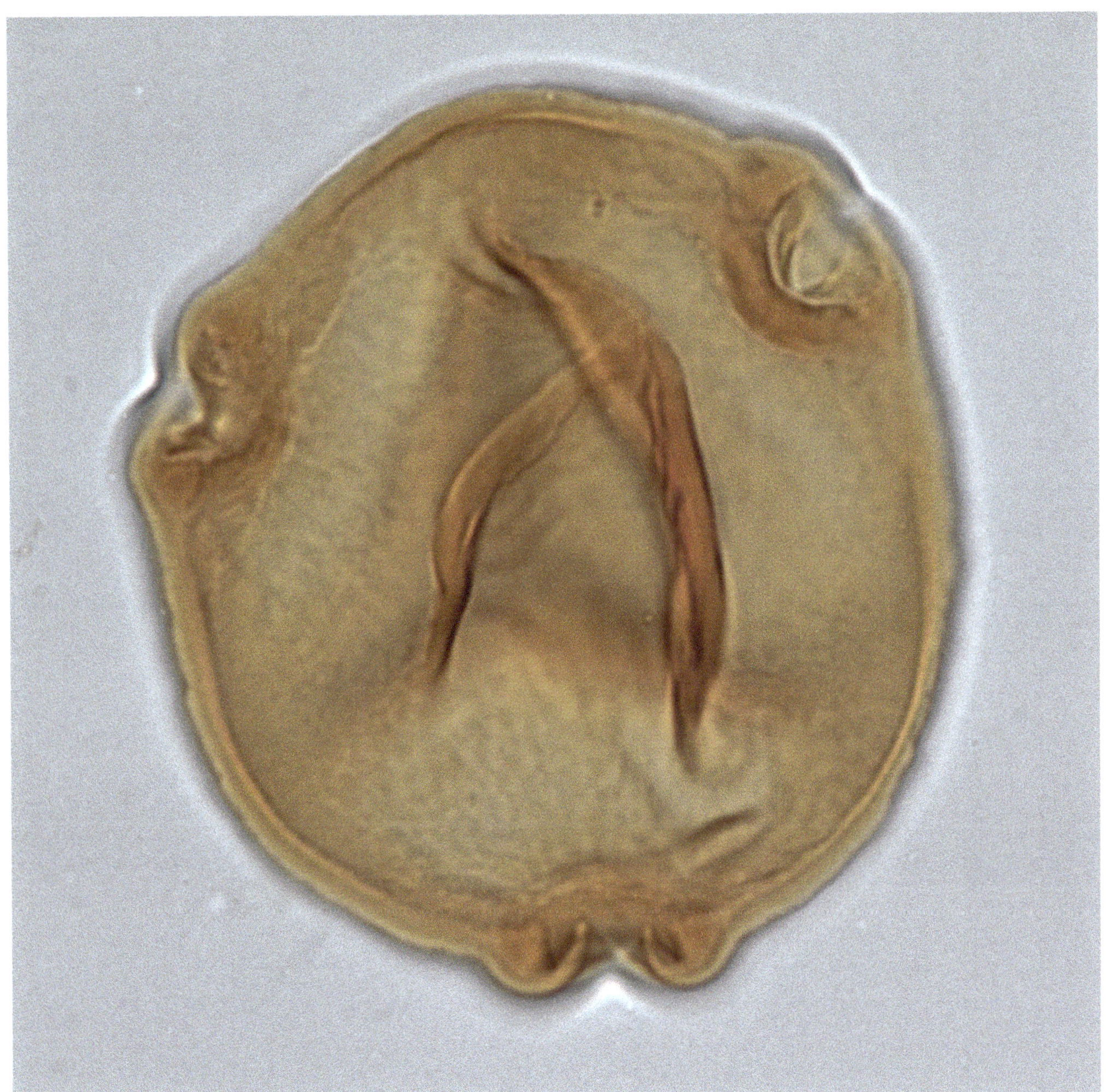

Onagraceae, possibly *Ludwigia*

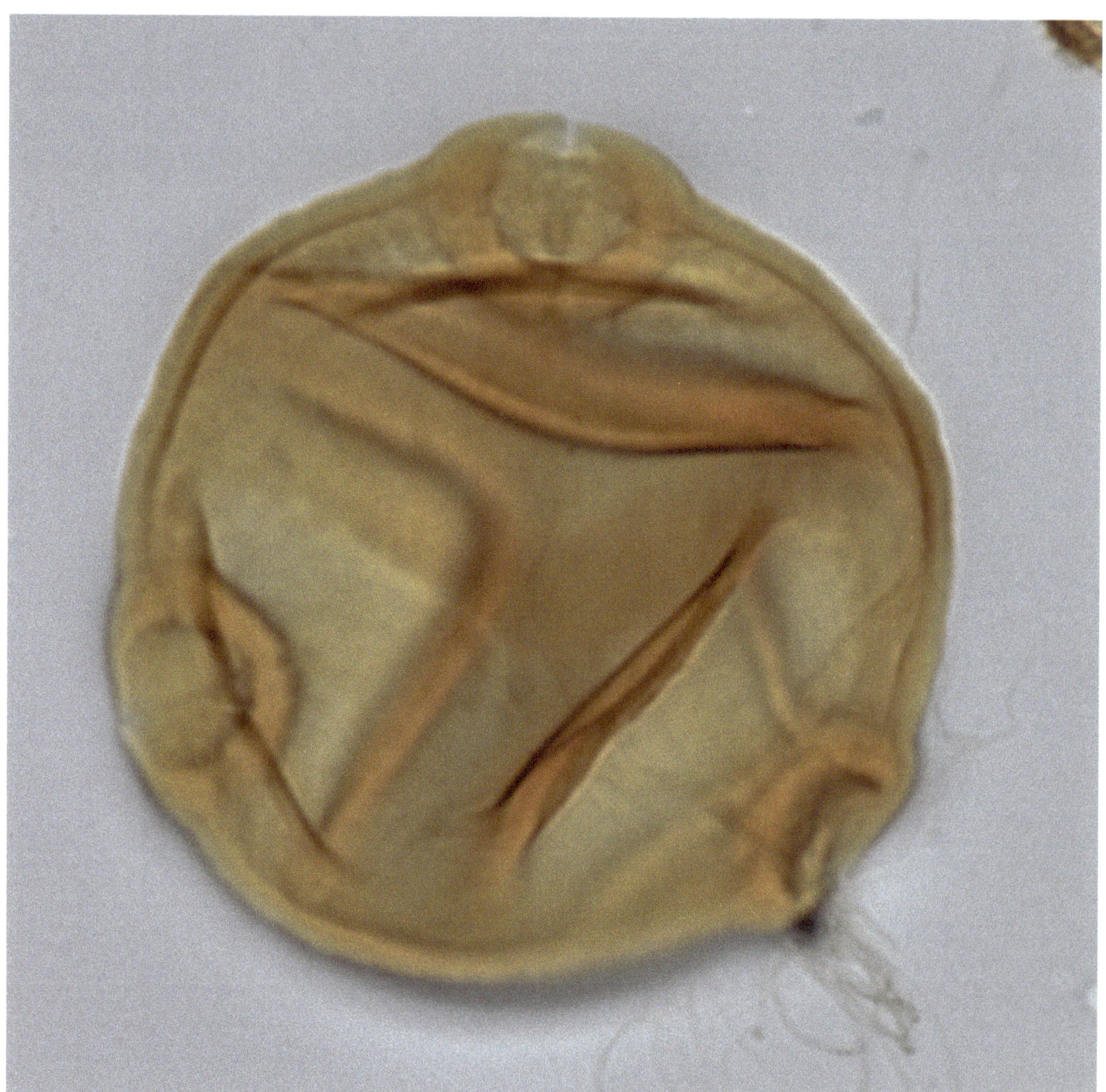

Onagraceae, possibly *Ludwigia*

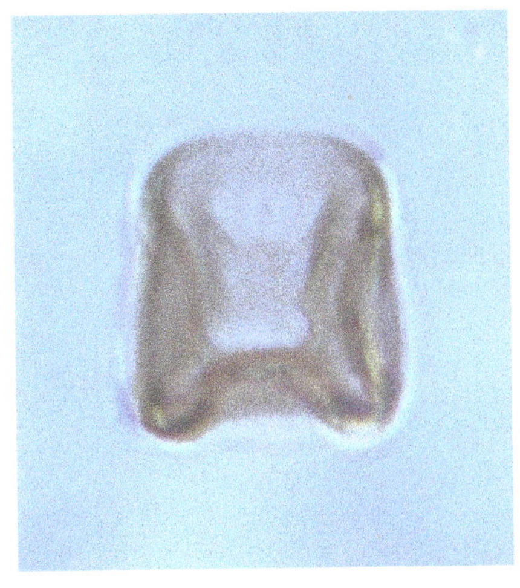
Morus

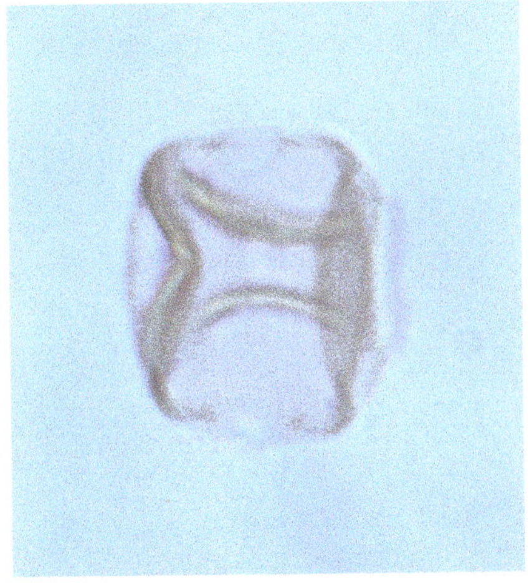
Morus

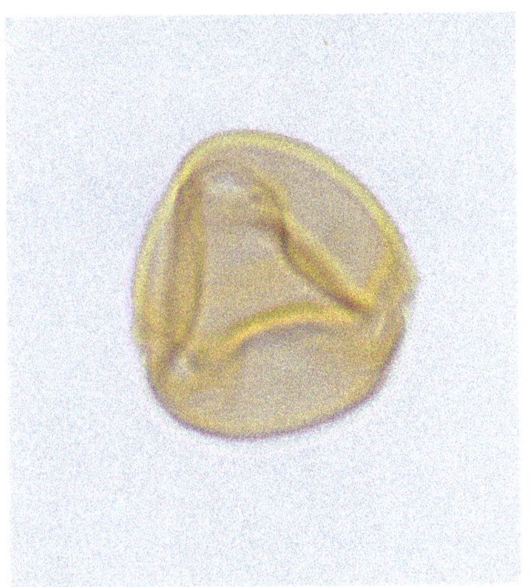
Morus

Carpinus

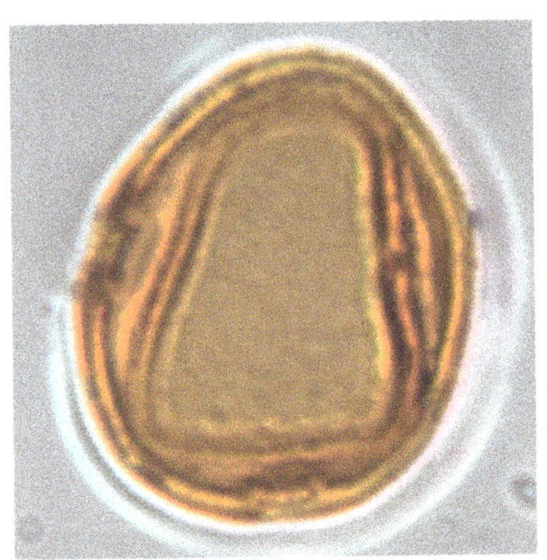
Celtis

PERIPORATE

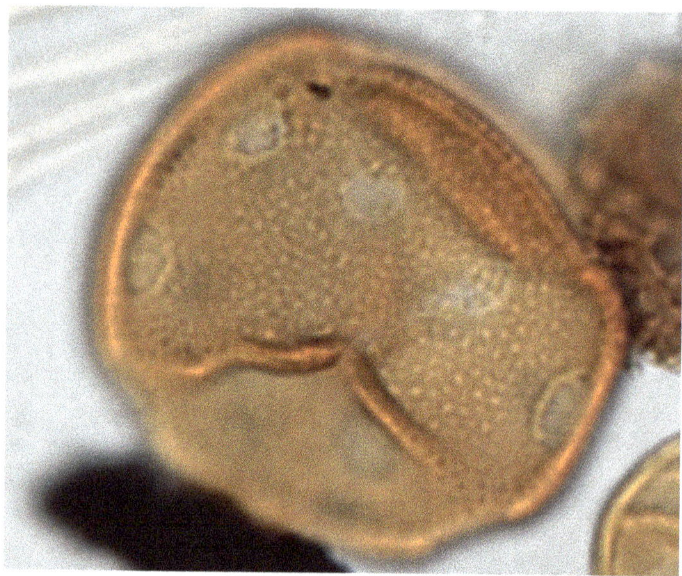

Liquidambar

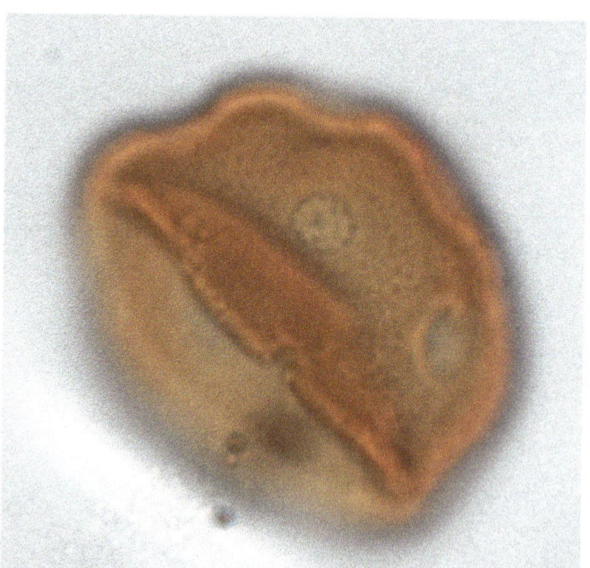

Liquidambar

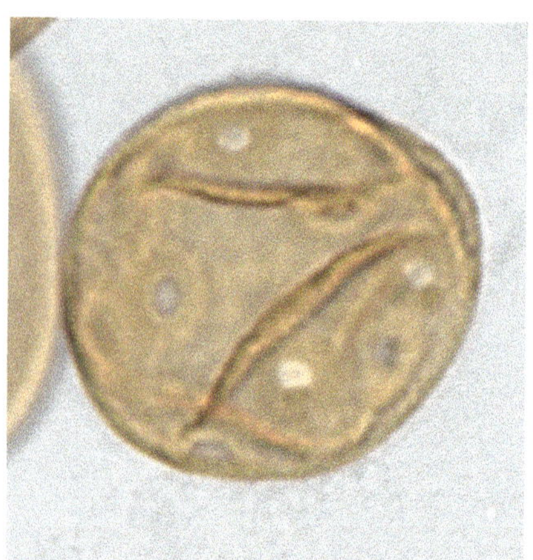

Plantago

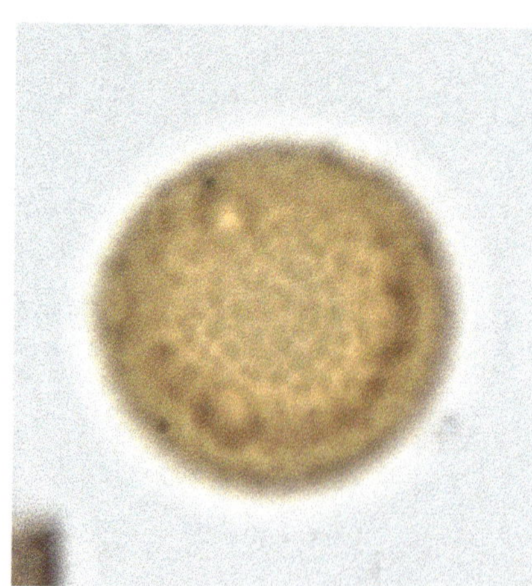

Plantago

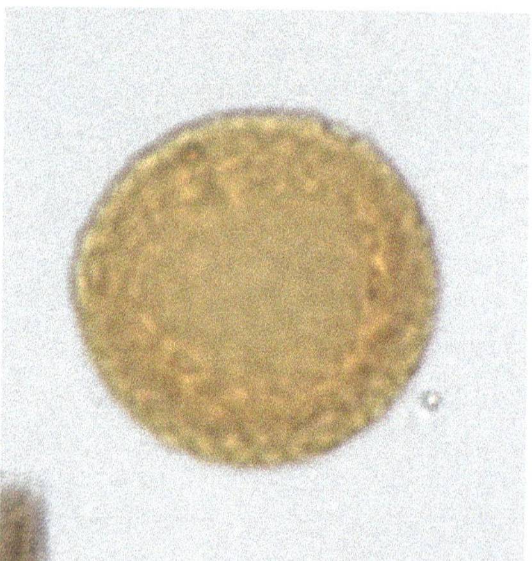

Plantago

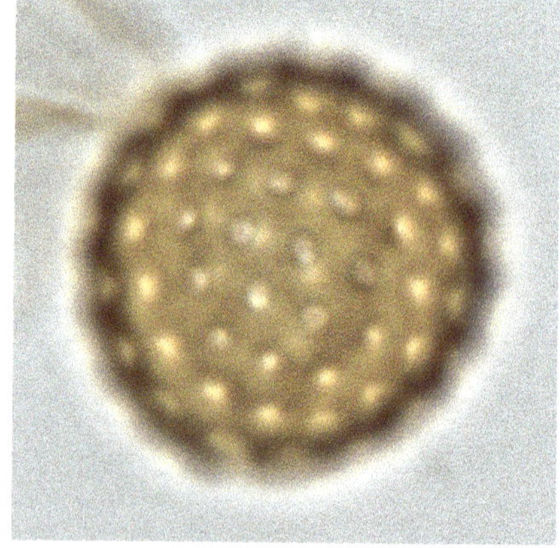

Chenopodiaceae/Amaranthaceae

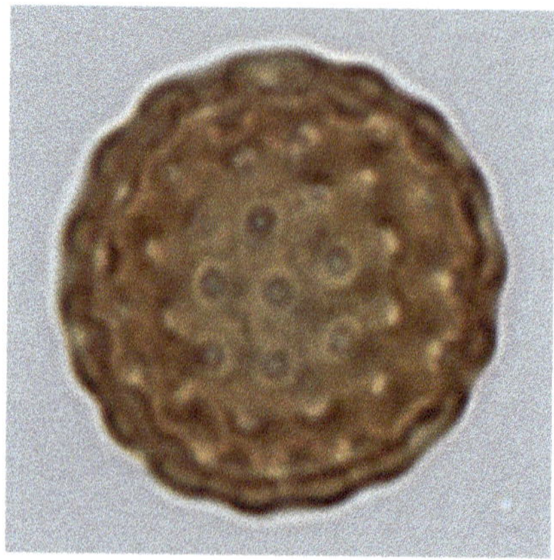

Chenopodiaceae/Amaranthaceae

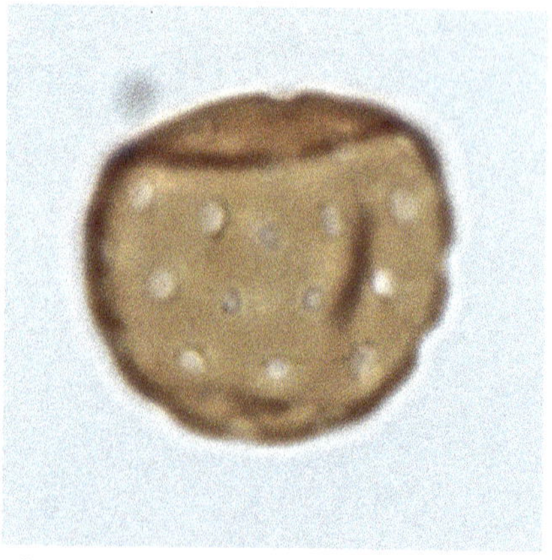

Chenopodiaceae/Amaranthaceae

Polygonum/Persicaria

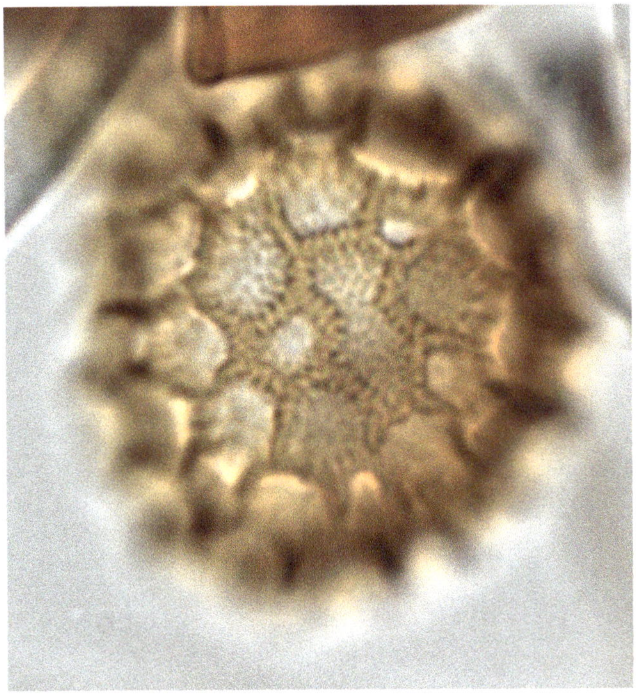

Polygonum/Persicaria

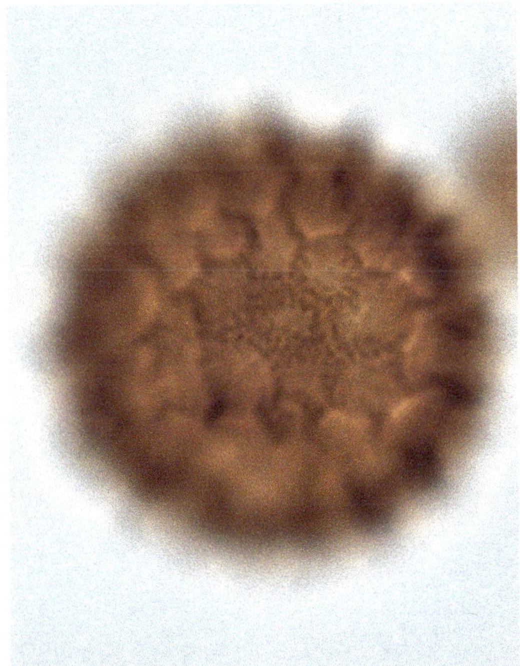

Polygonum/Persicaria

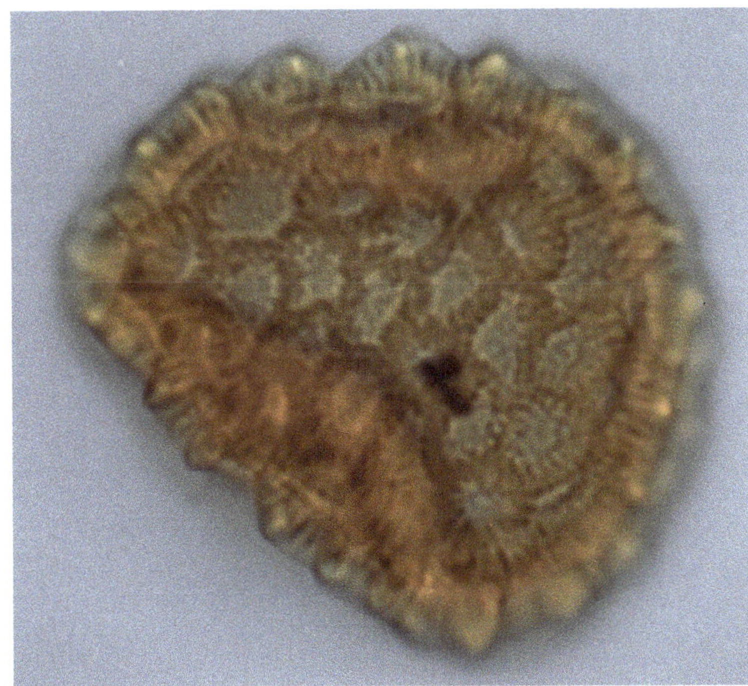

Polygonum/Persicaria

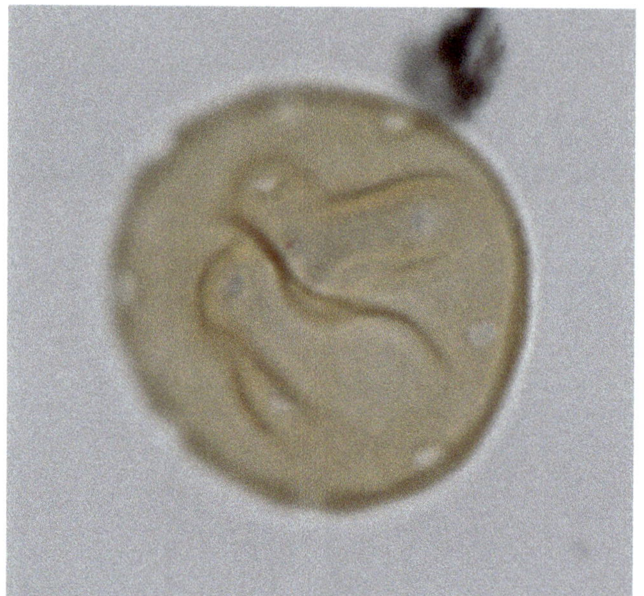

Juglans

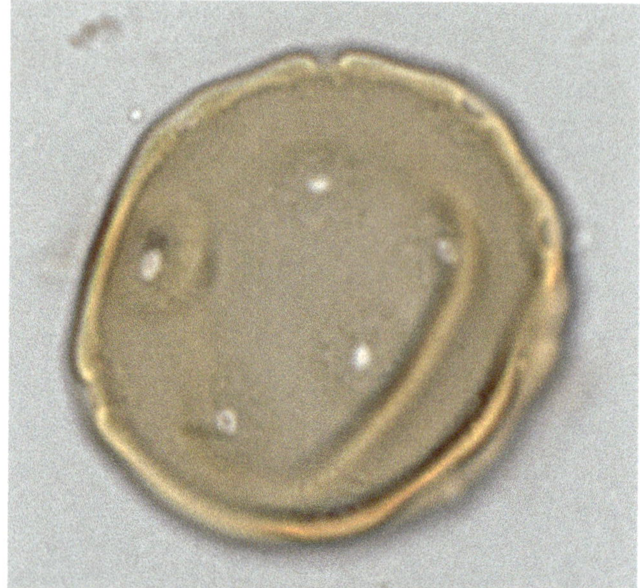

Juglans

Juglans

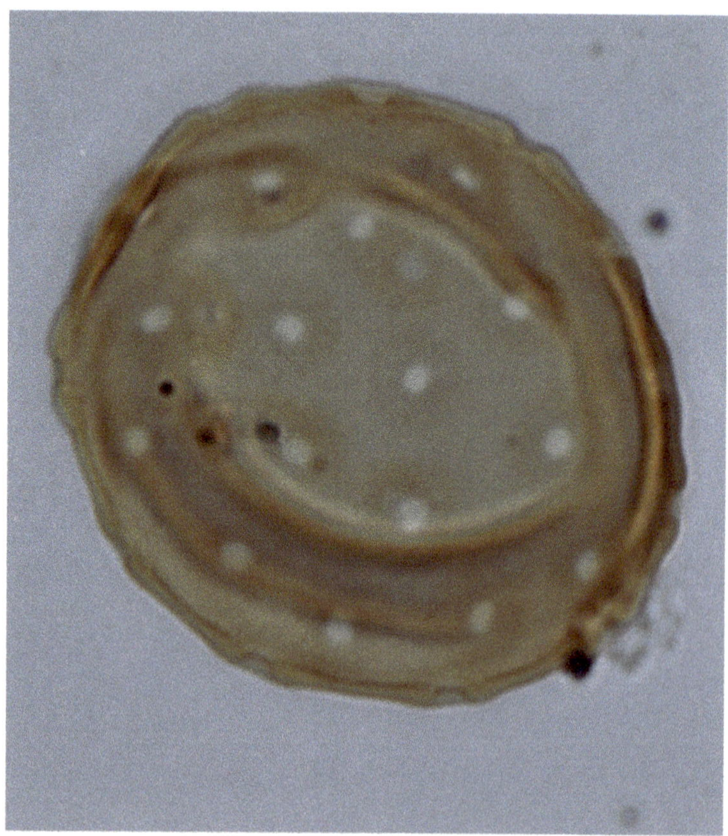

Juglans

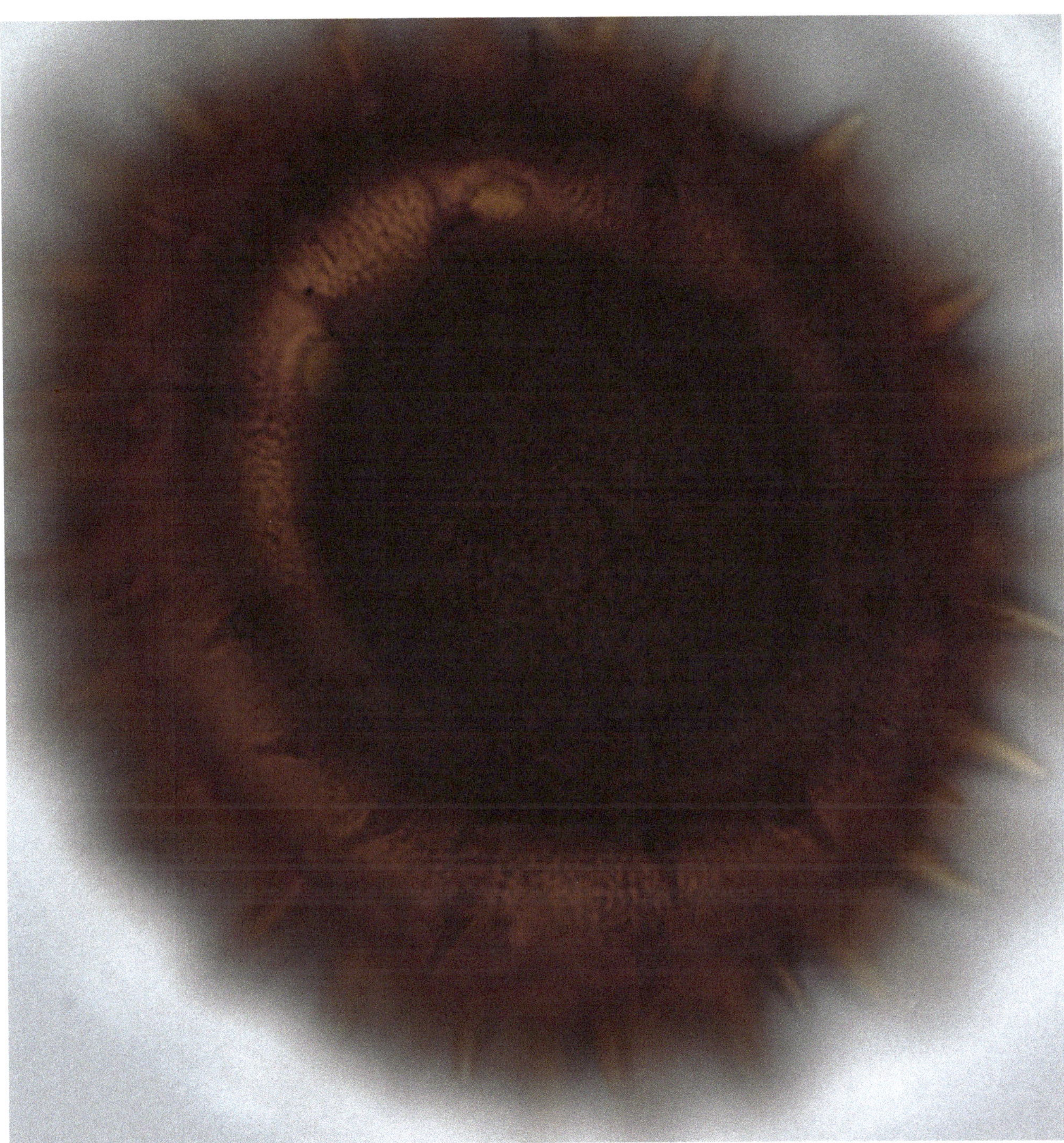

Gossypium

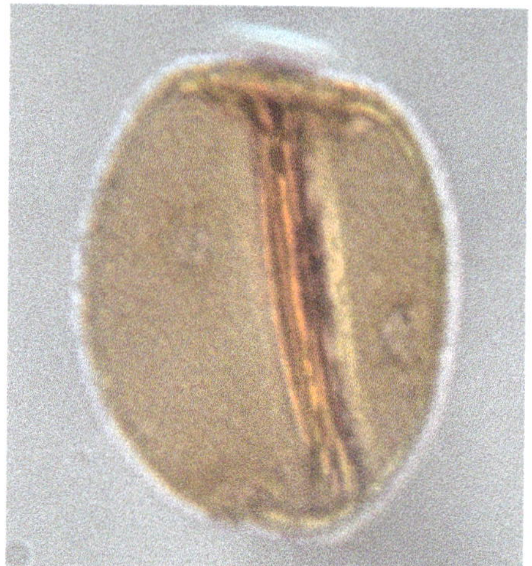

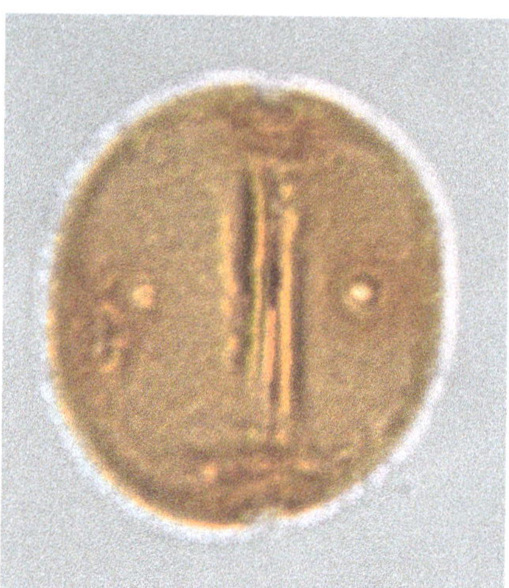

Celtis

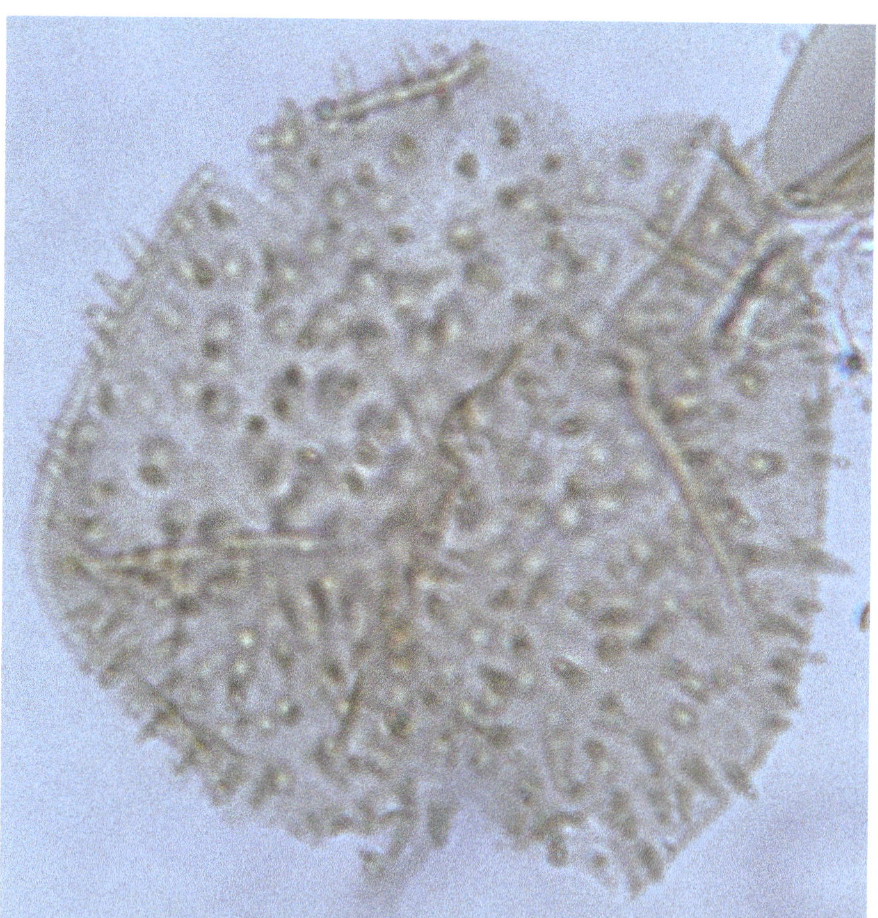

Malvaceae, possibly *Malvaviscus*

STEPHANOPORATE

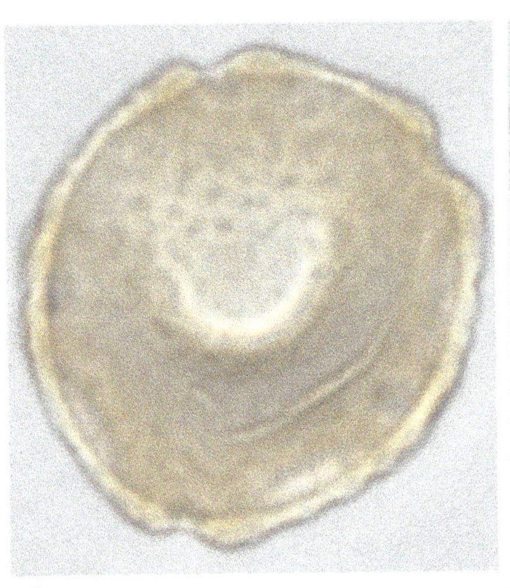

Ulmus

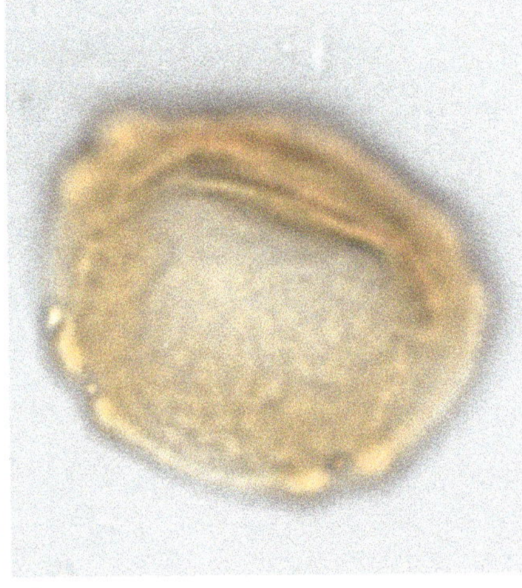

Ulmus

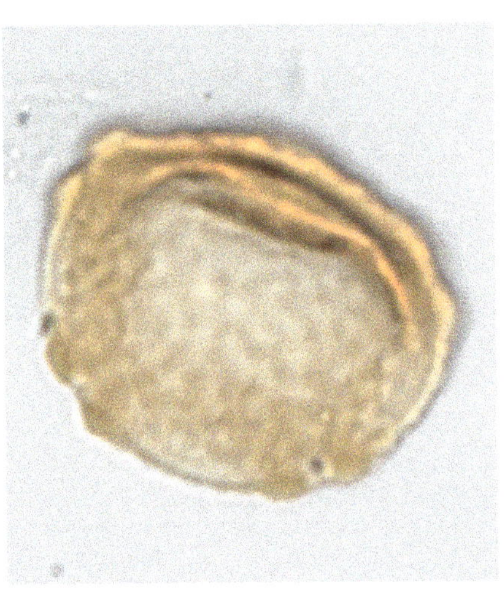

Ulmus

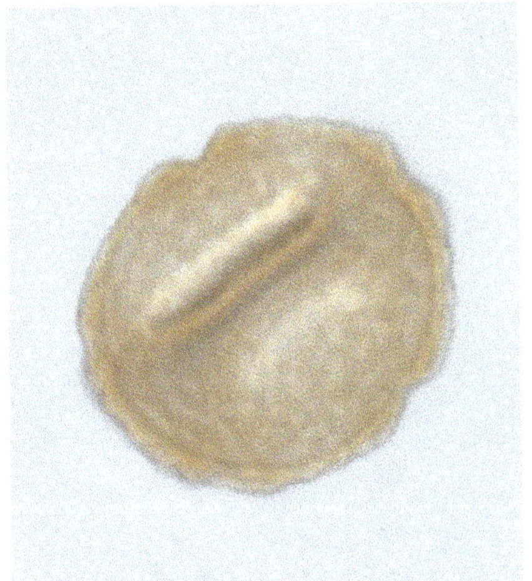

Ulmus

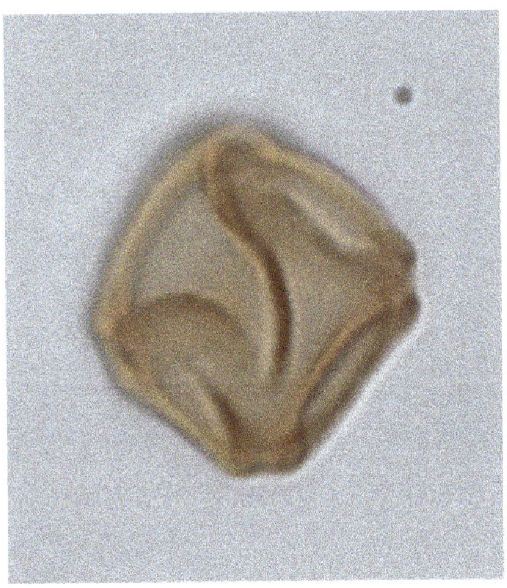

Alnus

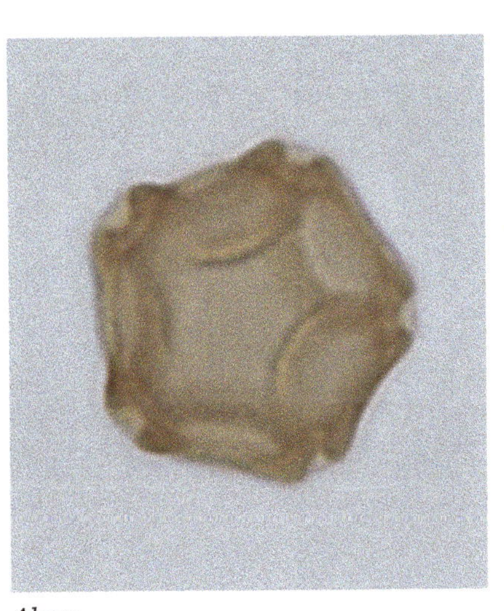

Alnus

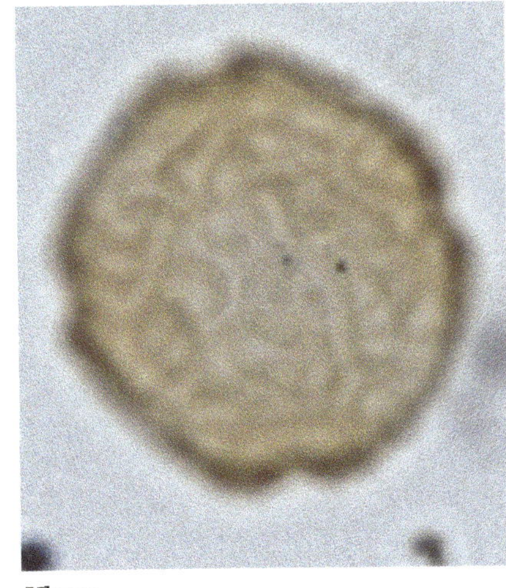

Ulmus

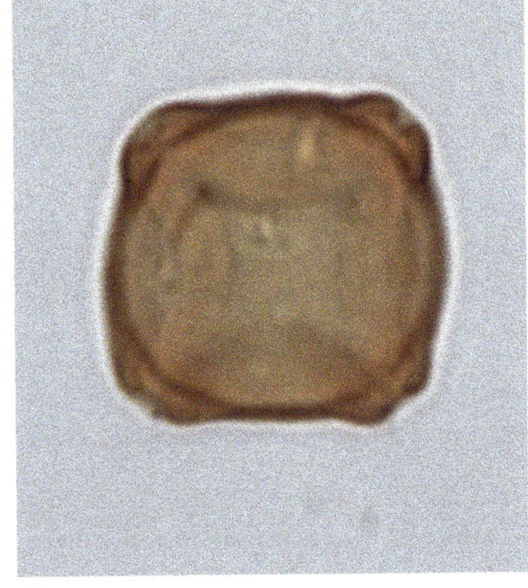

Alnus

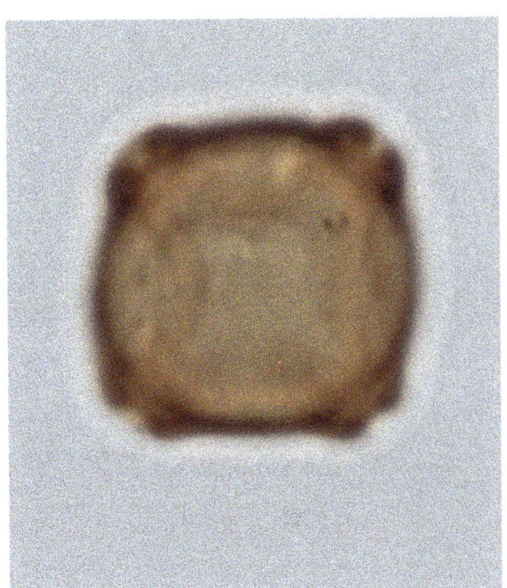

Alnus

FENESTRATE/LOPHATE

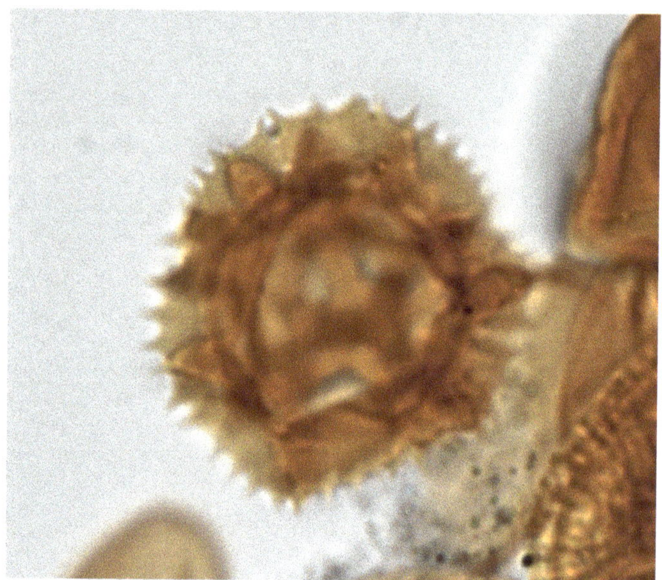

Asteraceae fenestrate

Vernonia

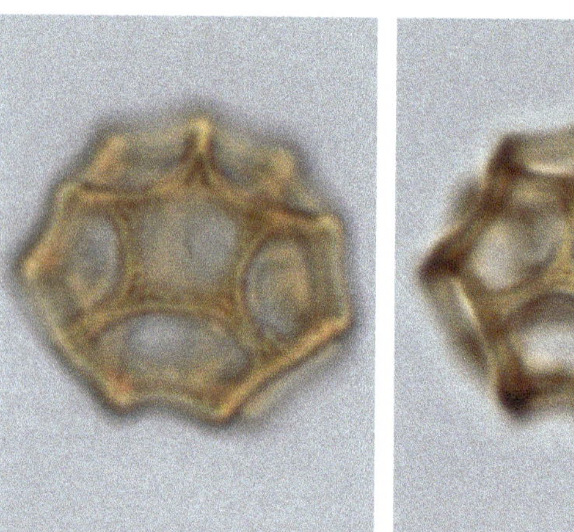

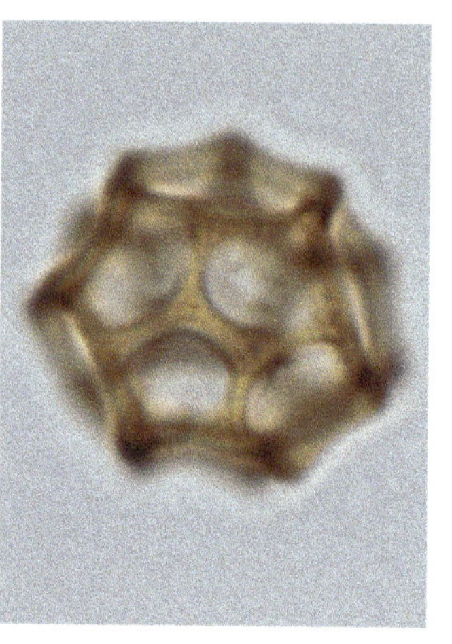

Althernanthera

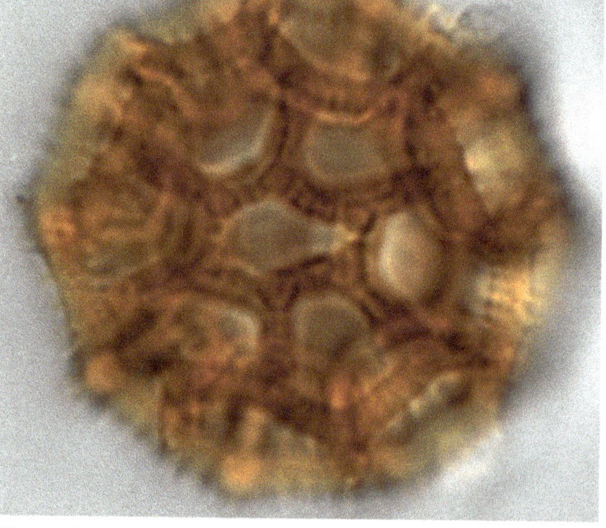

Vernonia

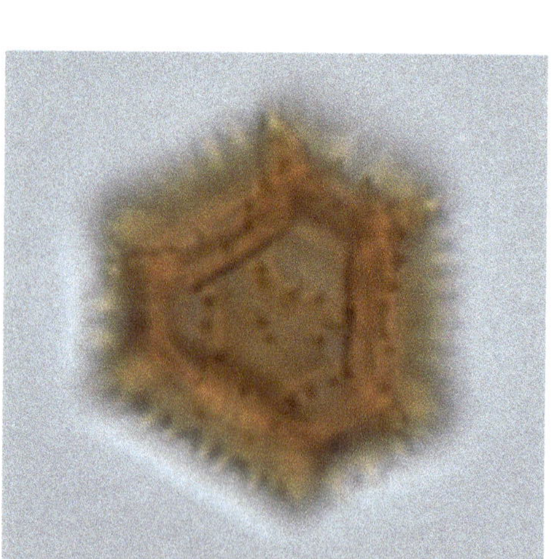

Asteraceae fenestrate

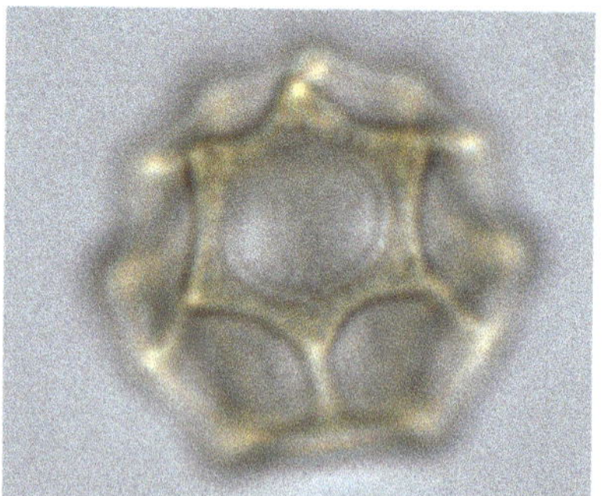

Althernantera

ECHINULATE

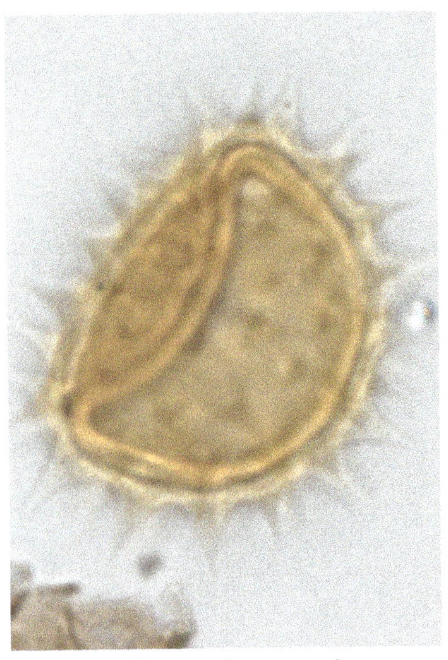

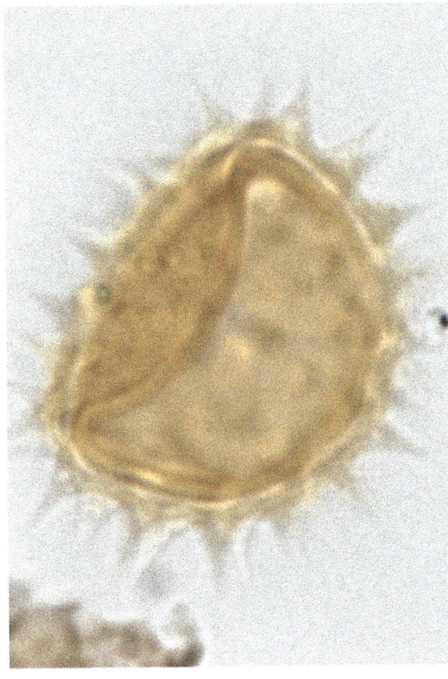

Asteraceae (e.g., sunflower type)

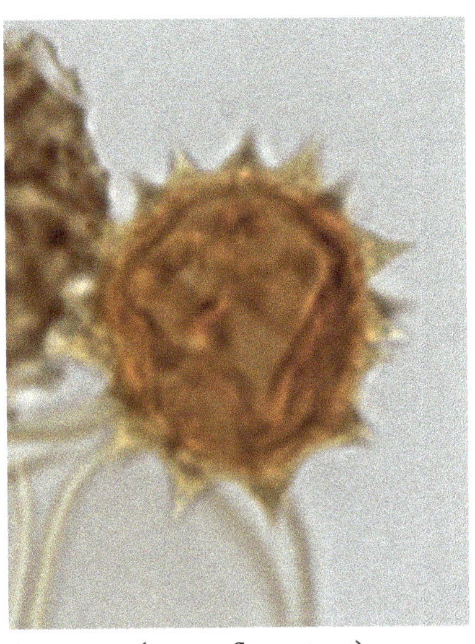

Asteraceae (e.g., sunflower type)

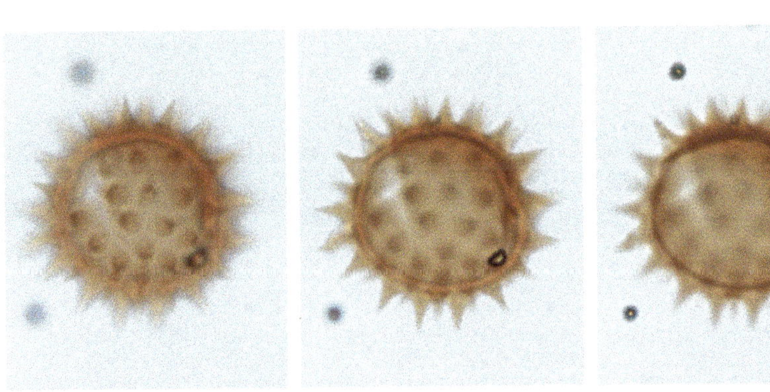

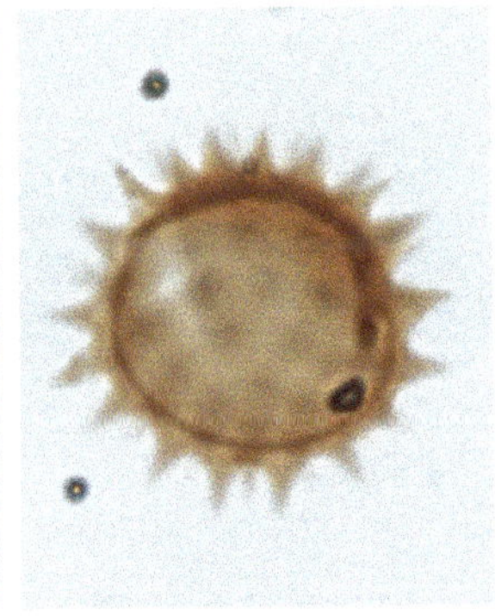

Asteraceae (e.g., sunflower type)

Asteraceae (e.g., sunflower type)

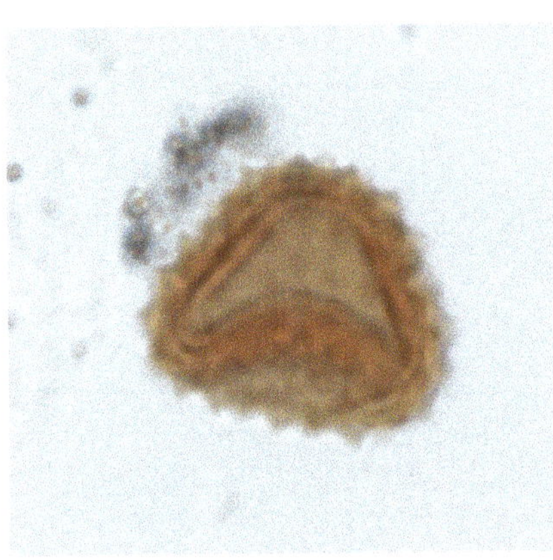

Asteraceae (e.g., *Ambrosia*)

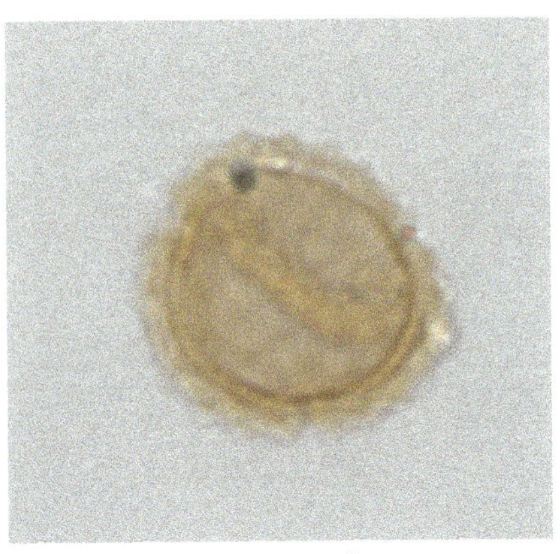

Asteraceae (e.g., *Ambrosia*)

Asteraceae (e.g., *Calendula*)

MONOCOLPATE & SULCATE

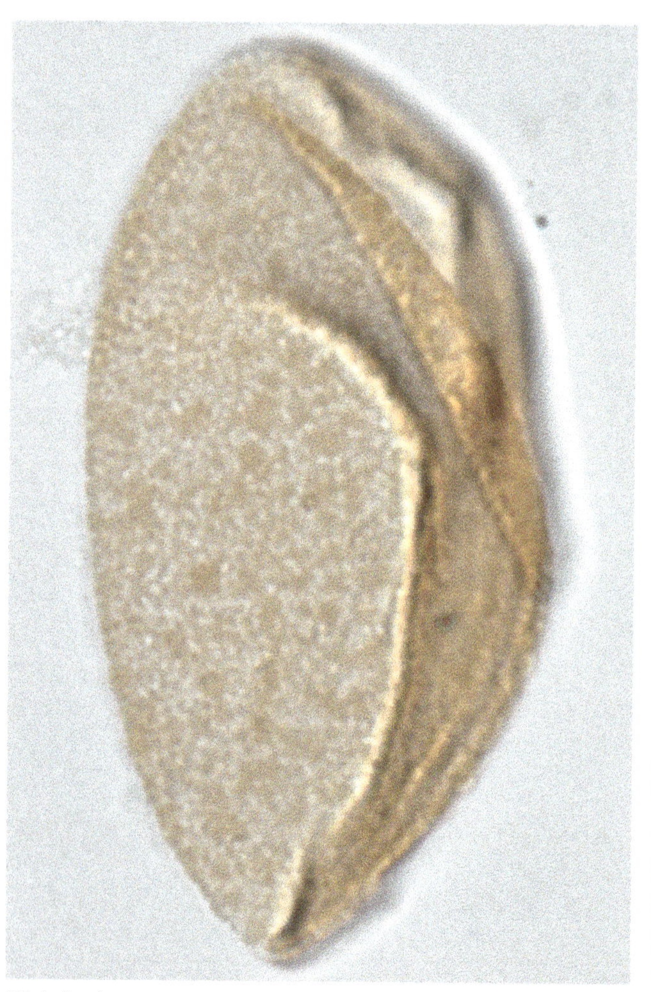

Liriodendron

Liriodendron

Tillandsia

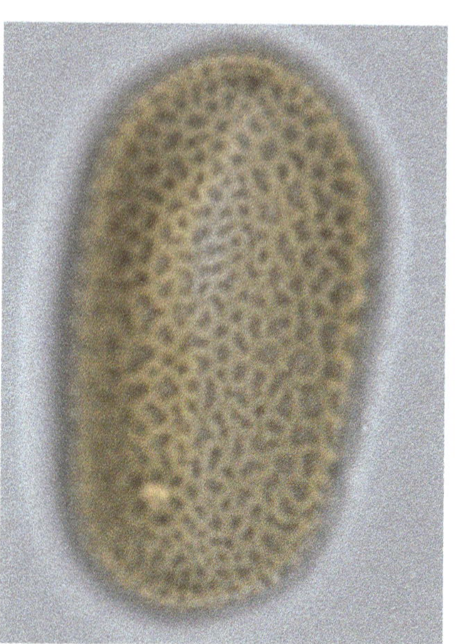

Liliaceae

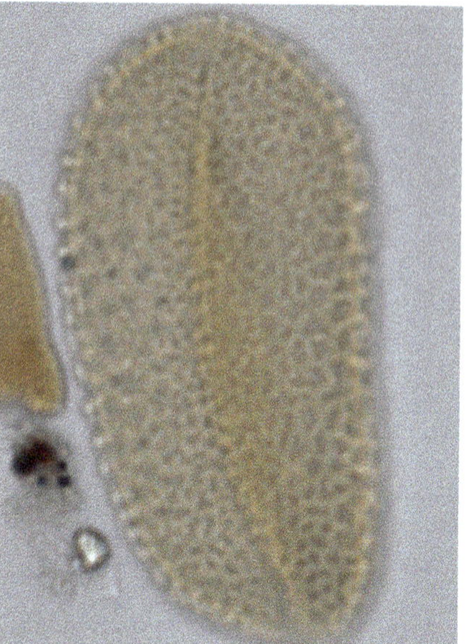

Liliaceae

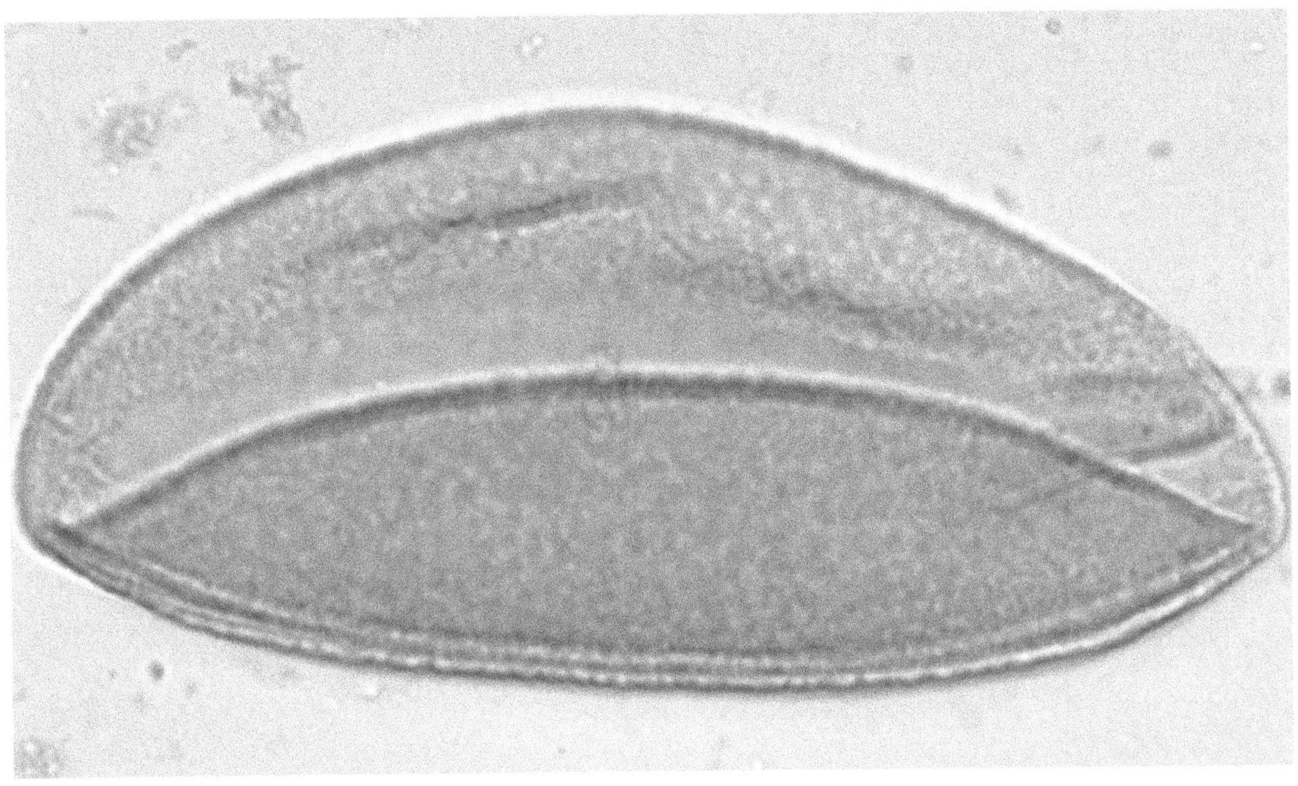

Magnolia

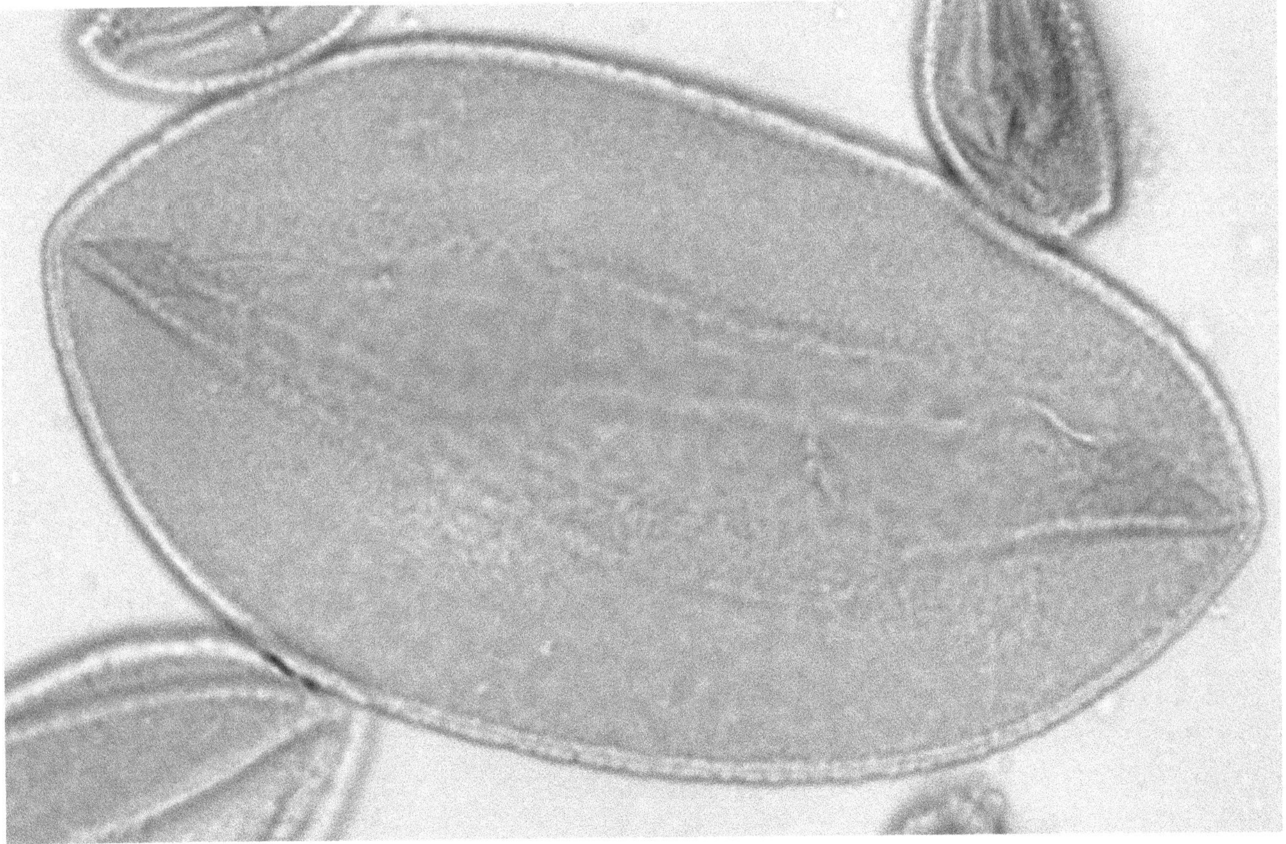

Magnolia

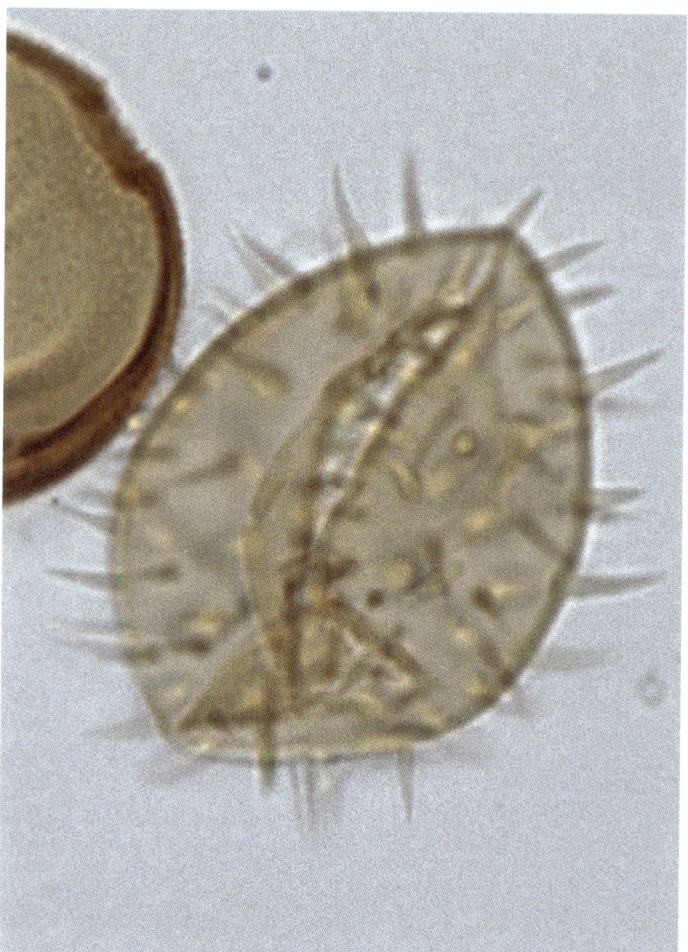

Nuphar

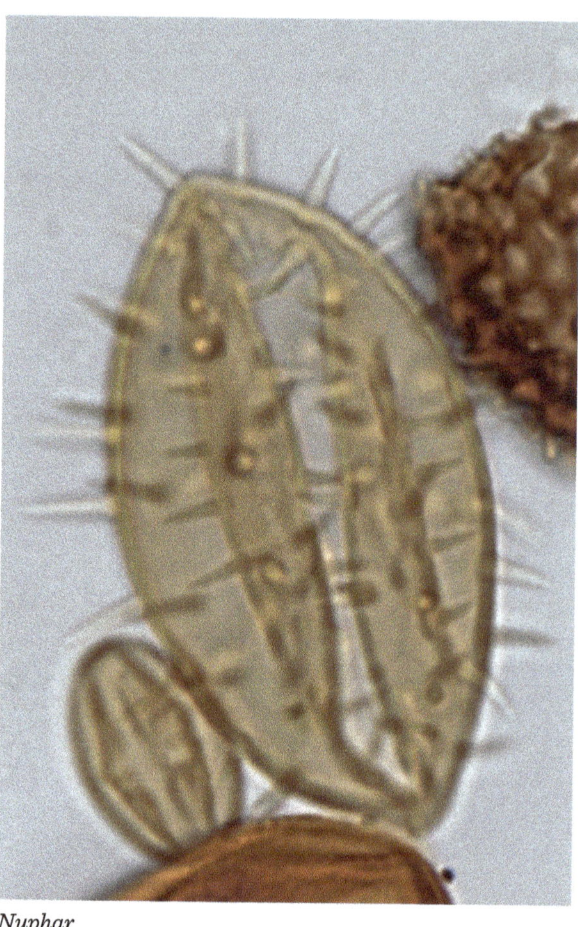

Nuphar

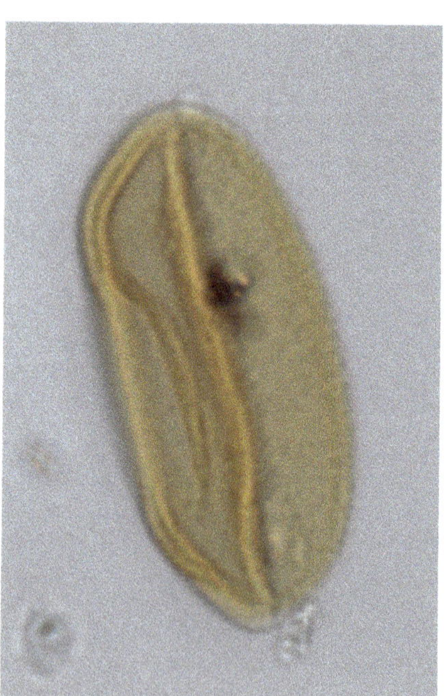

Arecaceae (*Sabal palmetto*)

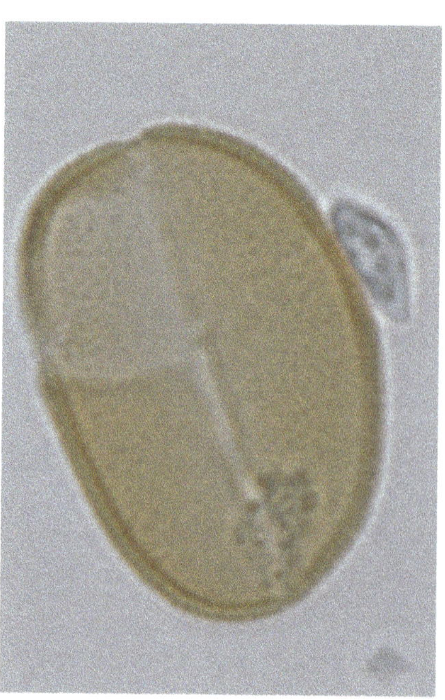

Arecaceae (*Sabal* or *Saw palmetto*)

Tradescantia

TRICOLPATE

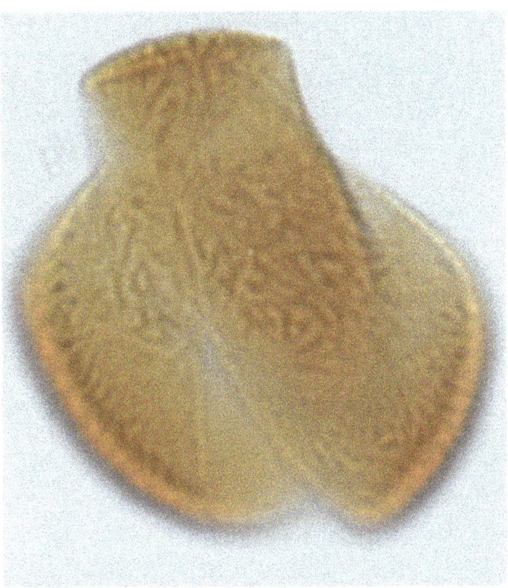

Acer

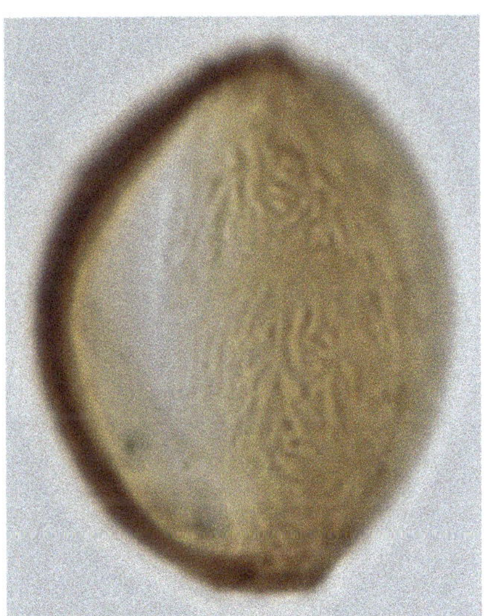

Acer

Acer

Fagopyrum

Fagopyrum

Lamium

Lamium

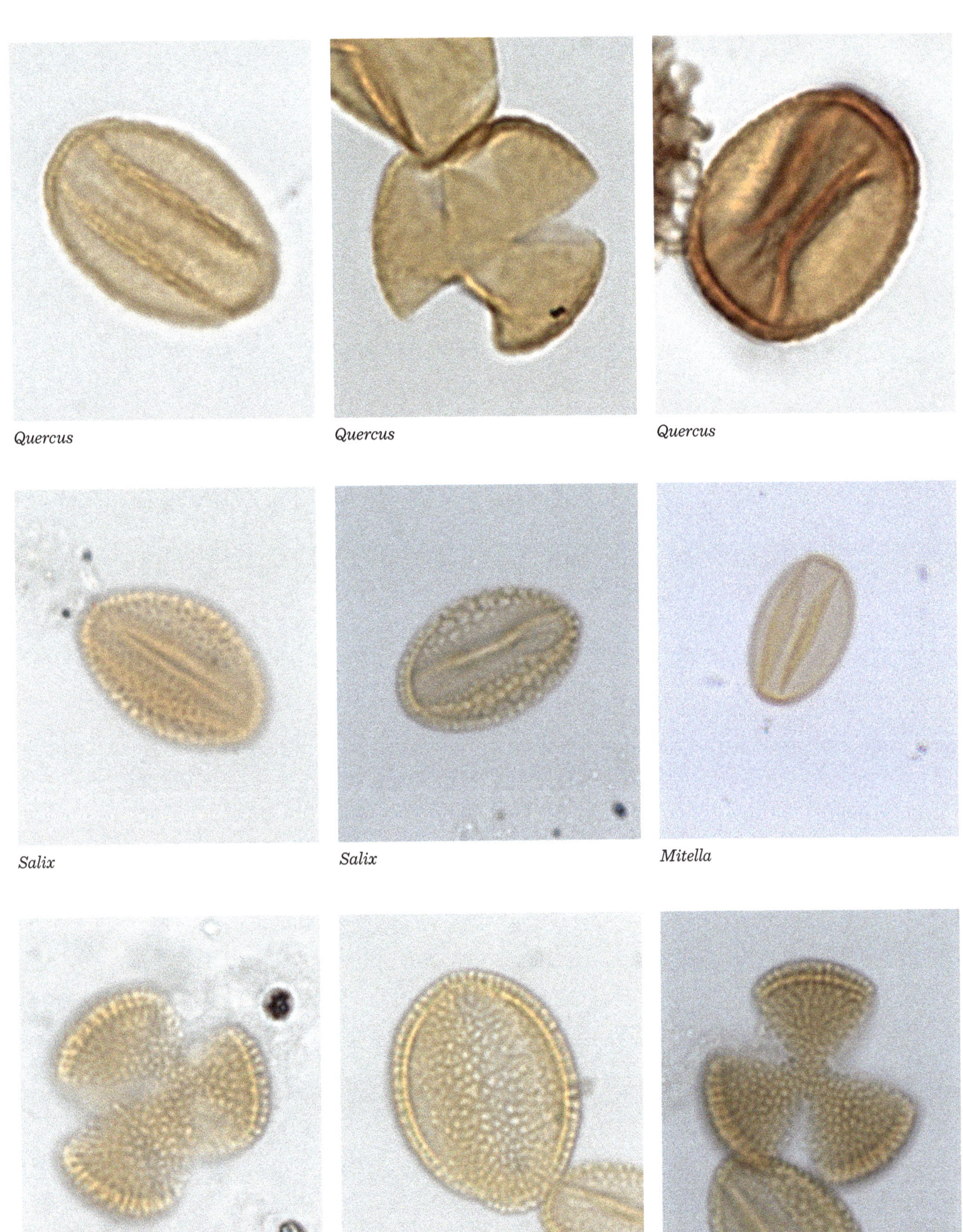

Quercus *Quercus* *Quercus*

Salix *Salix* *Mitella*

Brassica *Brassica* *Brassica*

STEPHANOCOLPATE

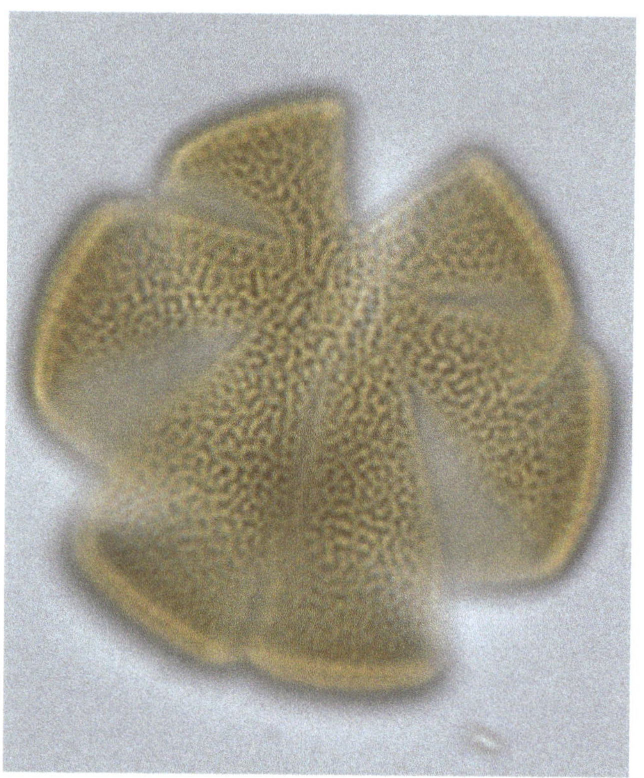

Hyssopus

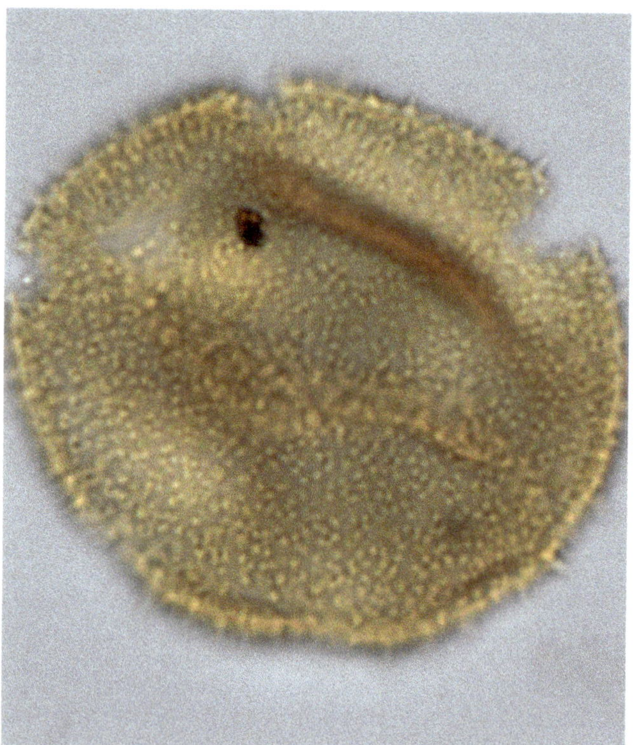

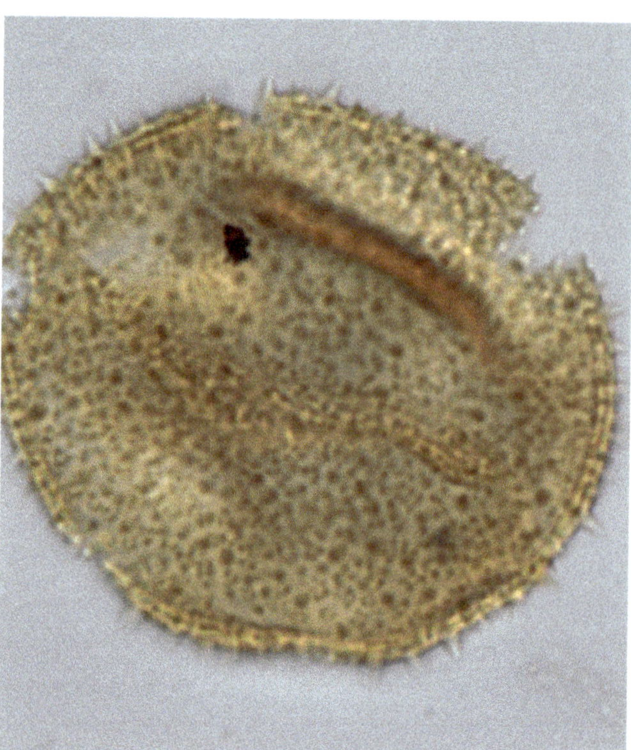

Portulaca

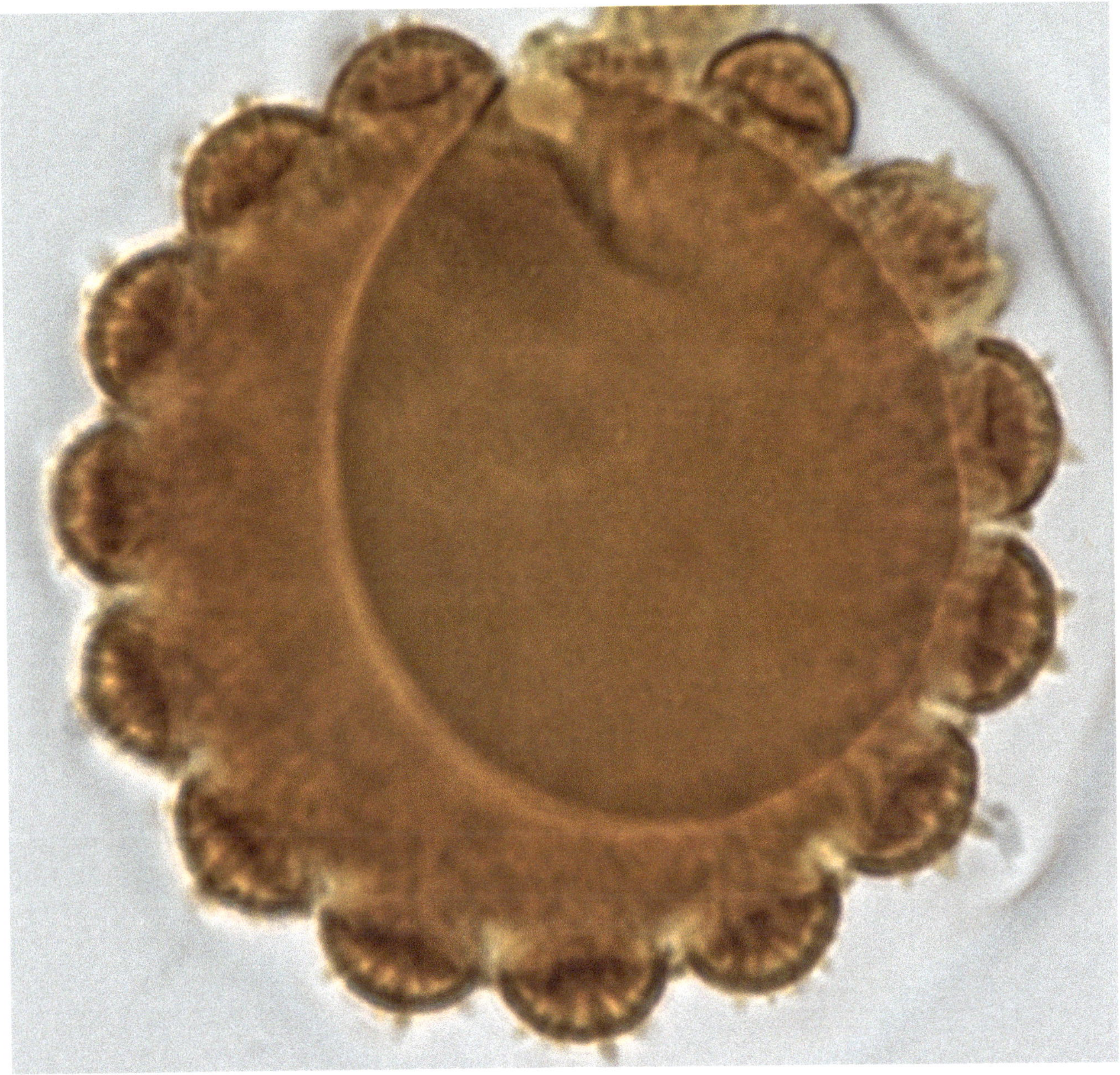

Hexasepalum terres

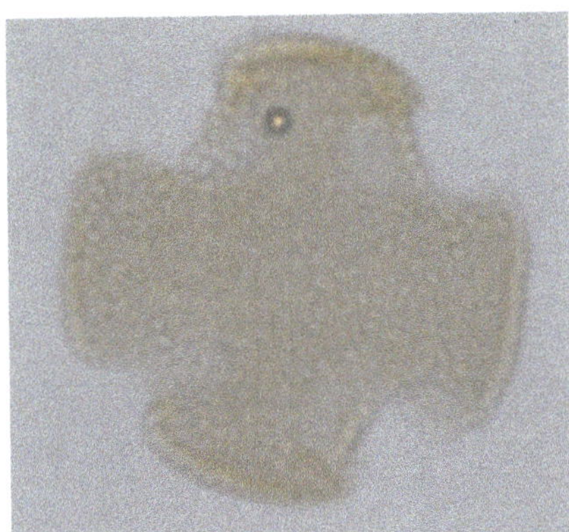

Fraxinus

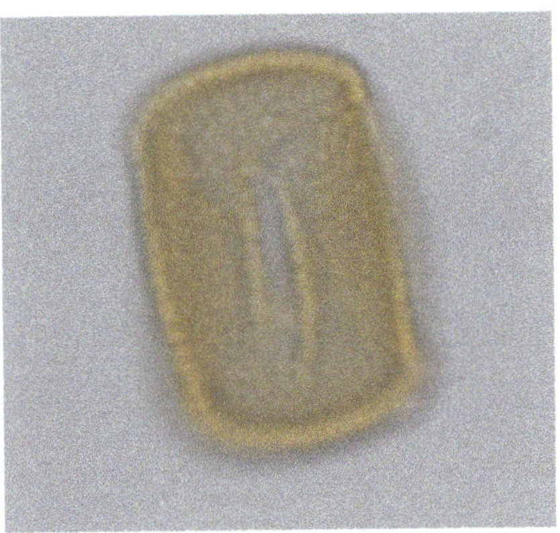

Fraxinus

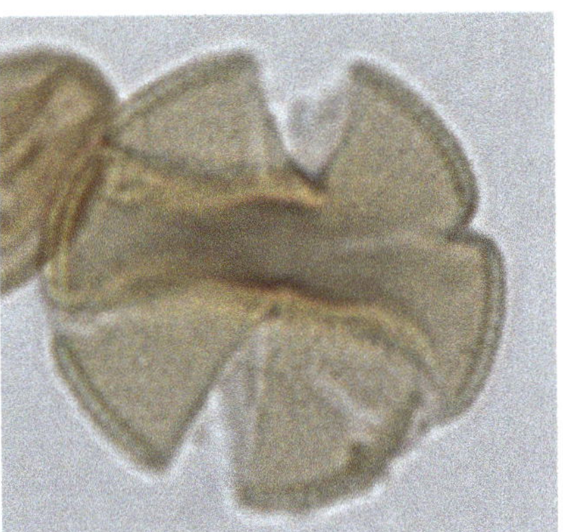

Lamiaceae

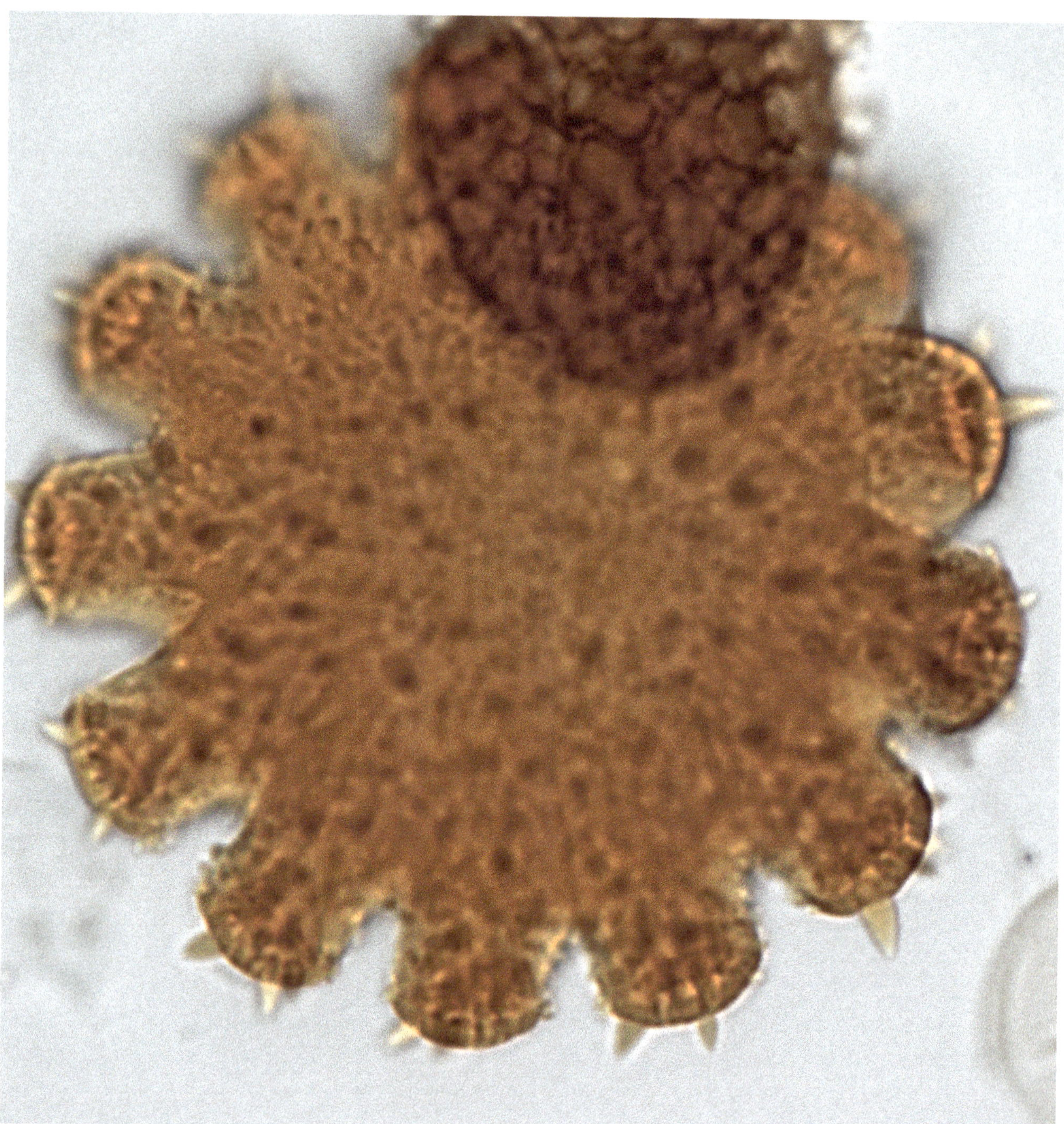

Hexasepalum teres

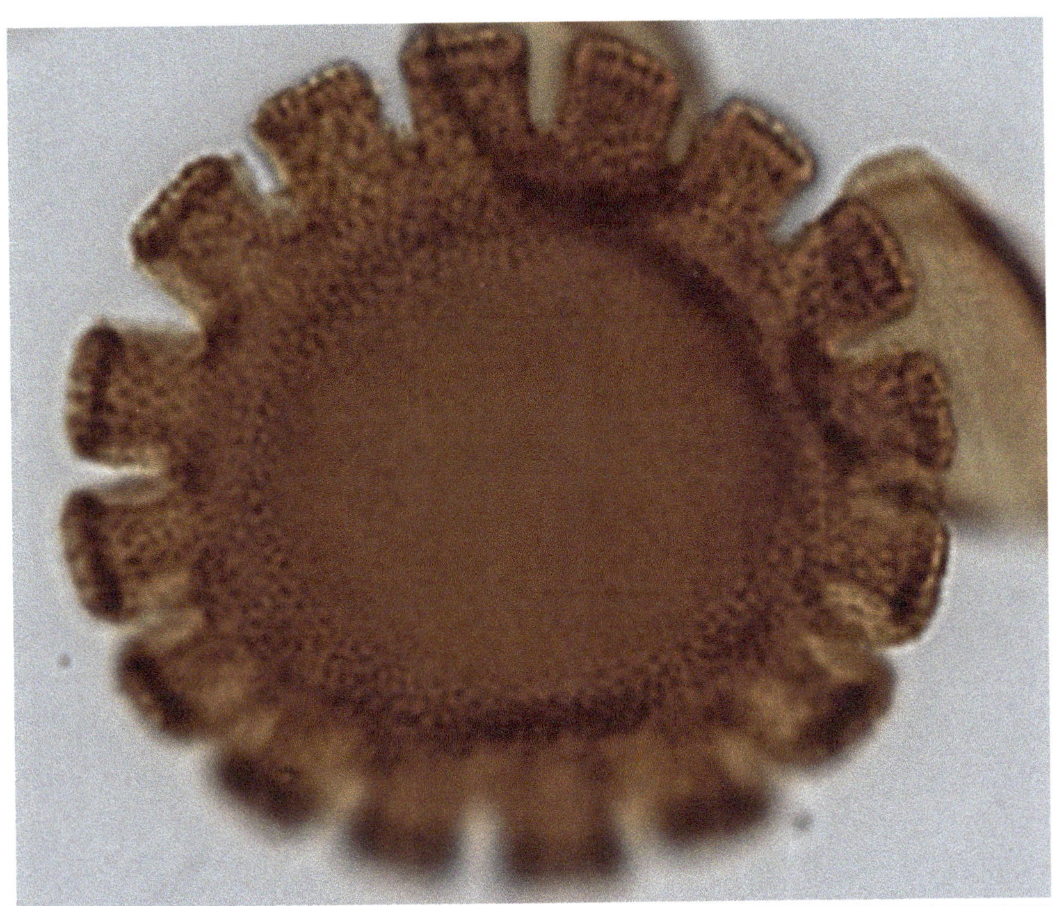

Rubiaceae, possibly *Richardia grandiflora*

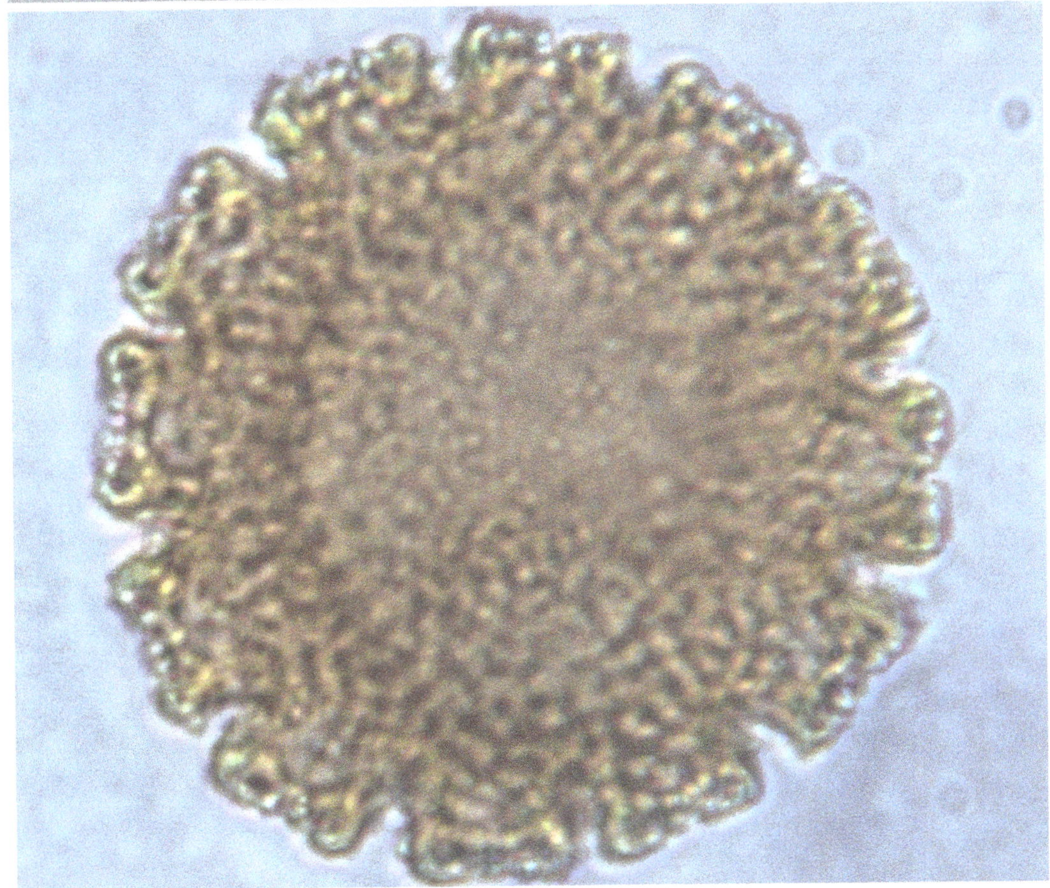

Rubiaceae, *Richardia*

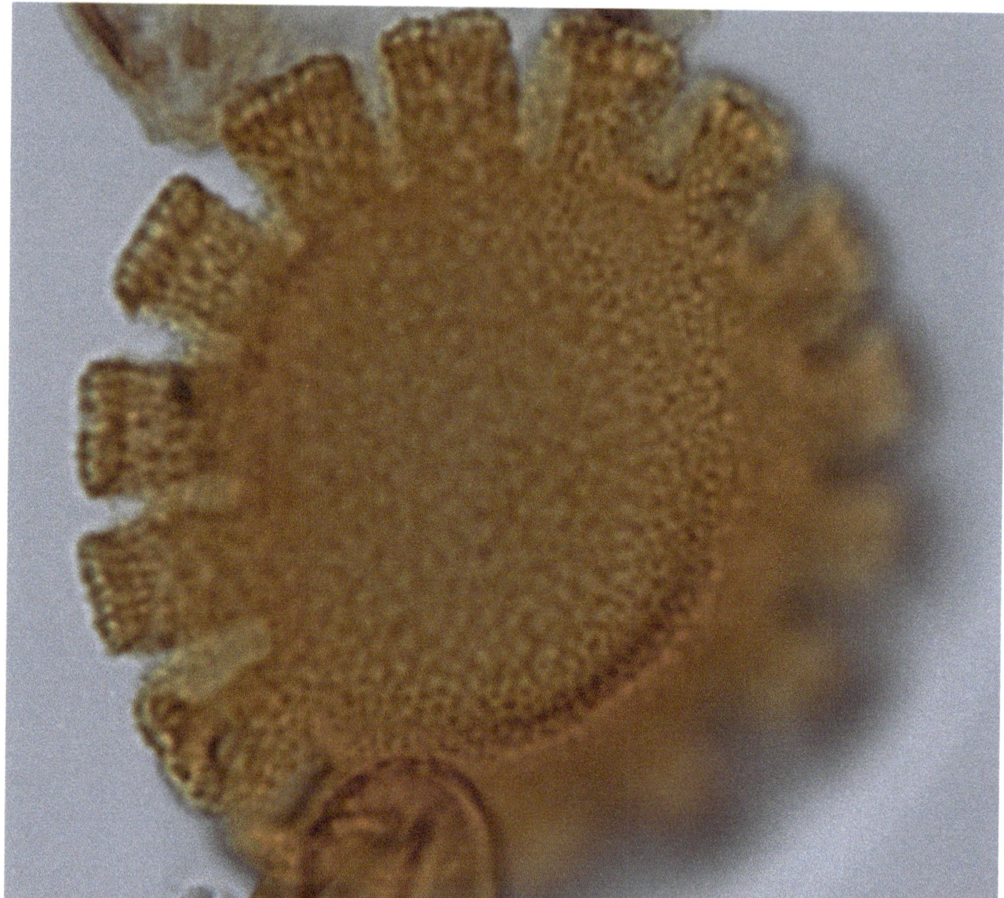

Rubiaceae, possibly *Richardia grandiflora*

Possibly another type of Rubiaceae

SYNCOLPATE

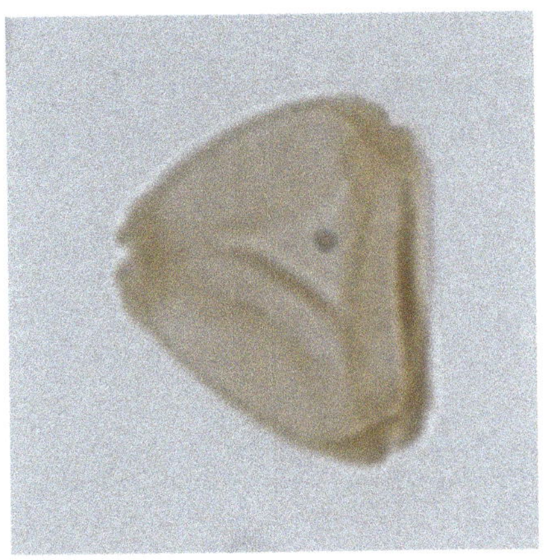

Myrtaceae

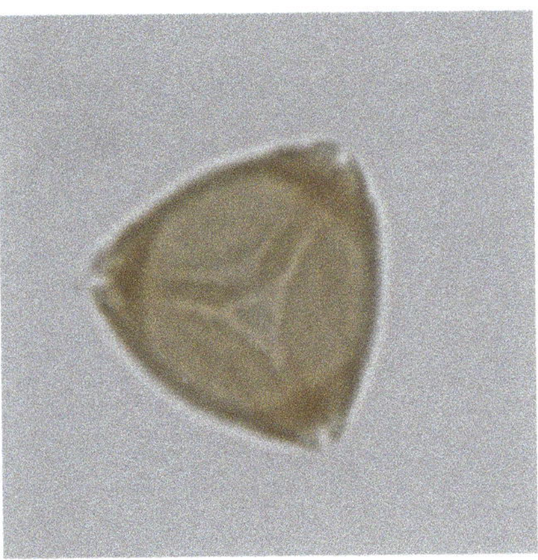

Myrtaceae

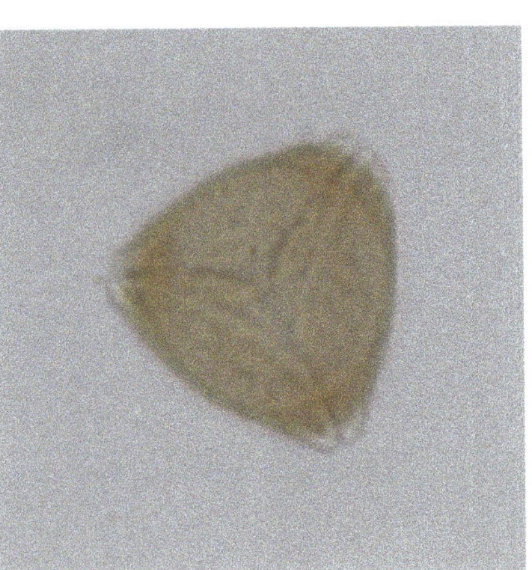

Myrtaceae

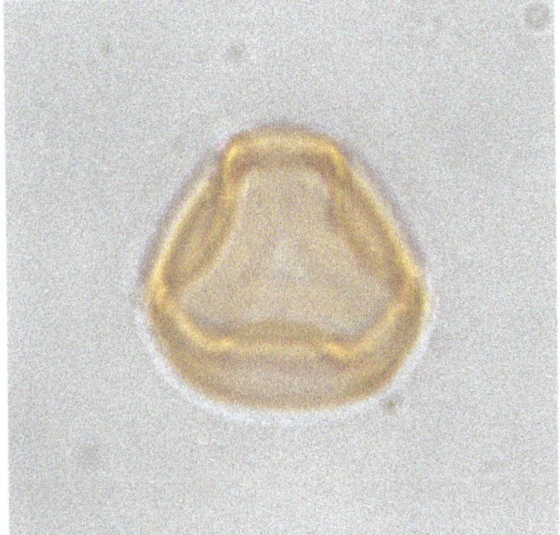

Phoaradendron

STEPHANOCOLPORATE

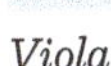

Viola

TRICOLPORATE

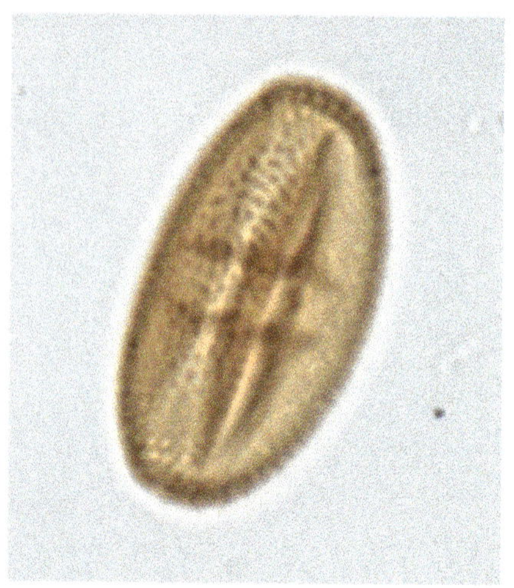
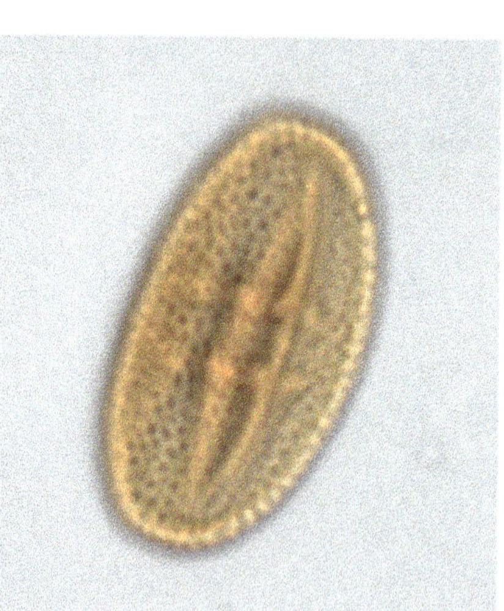

Unknown #3

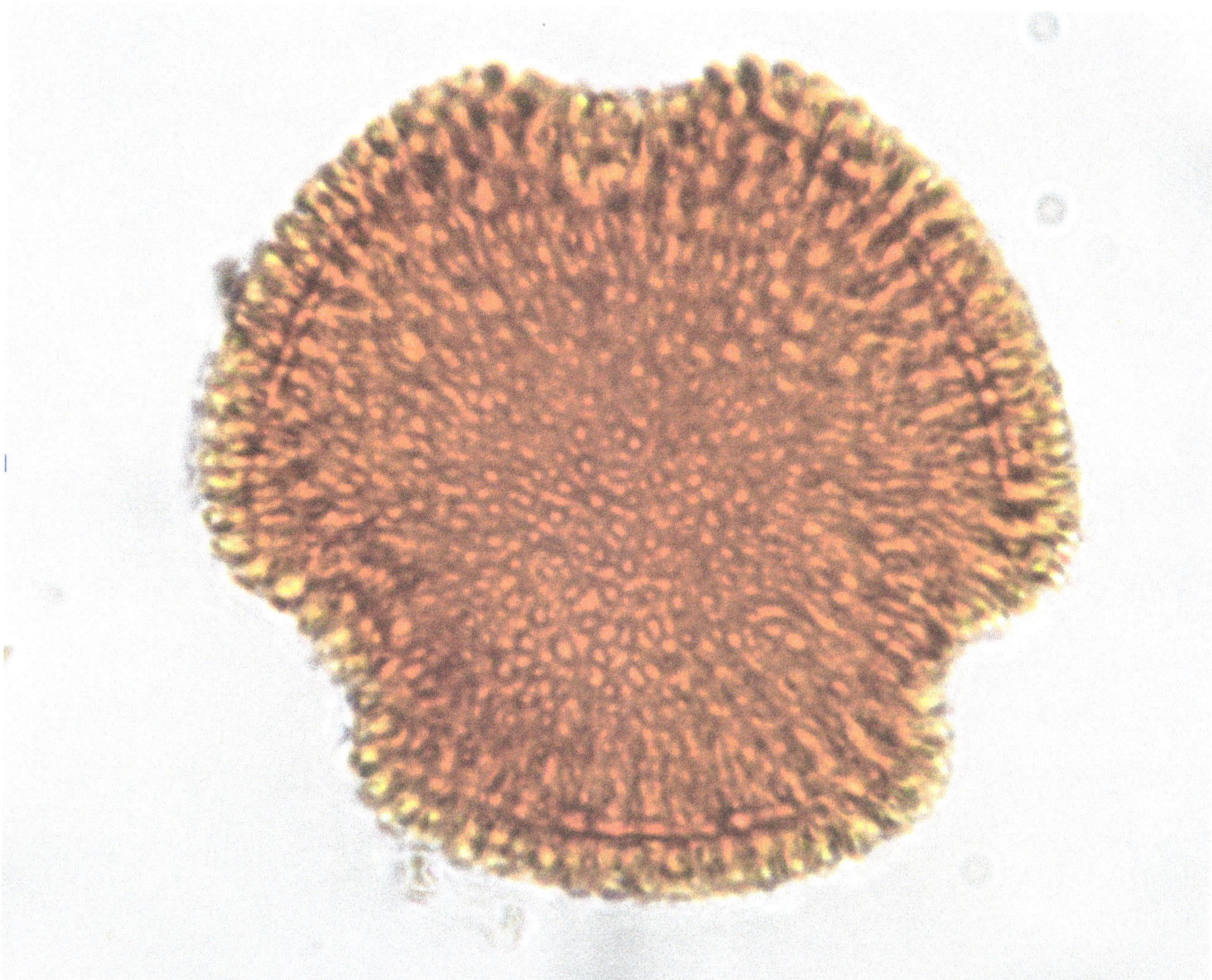

Erodium

Unknown #9

Unknown #9

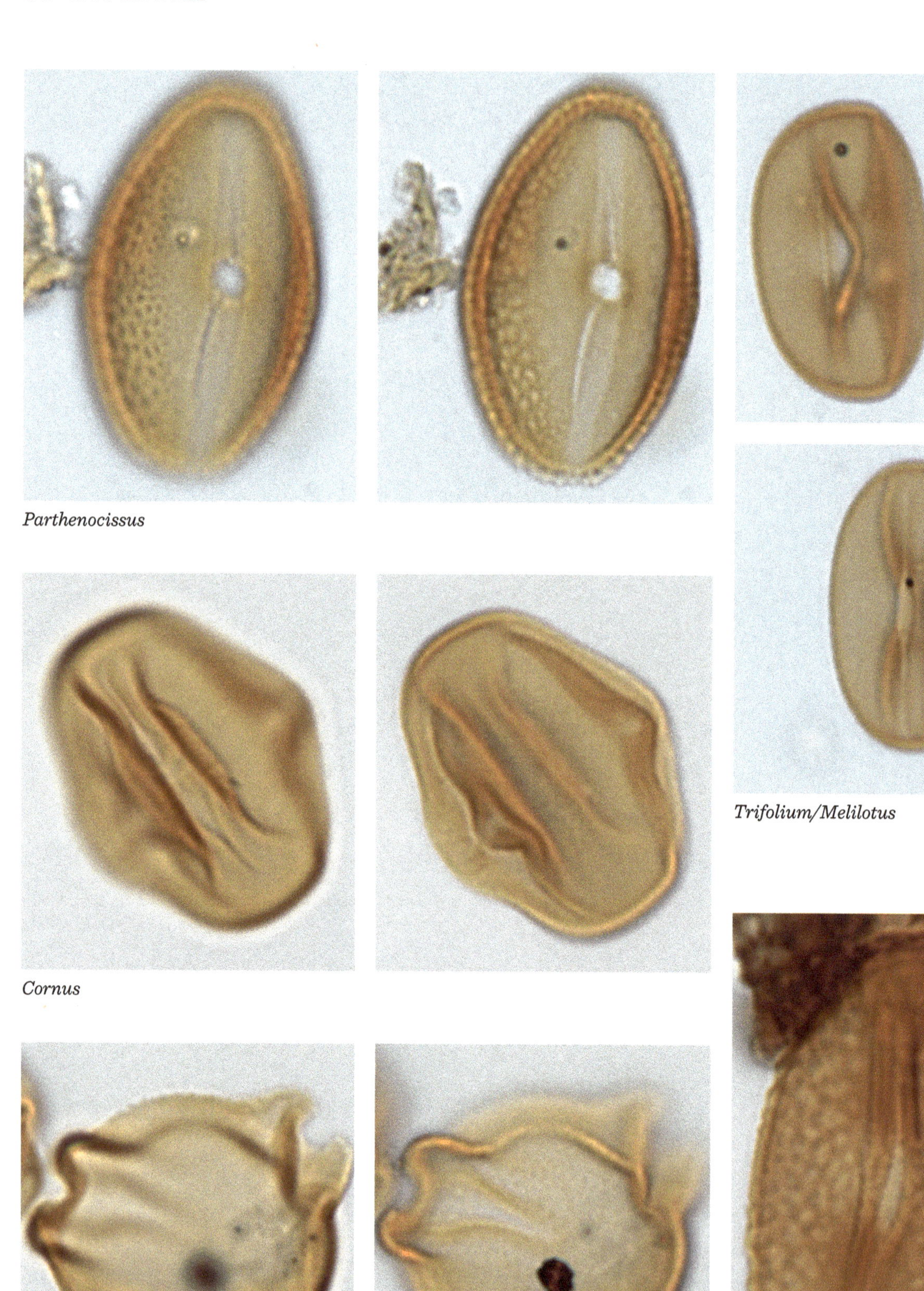

Parthenocissus

Trifolium/Melilotus

Cornus

Cornus

Trifolium

Various genera of Rosaceae

Lonicera

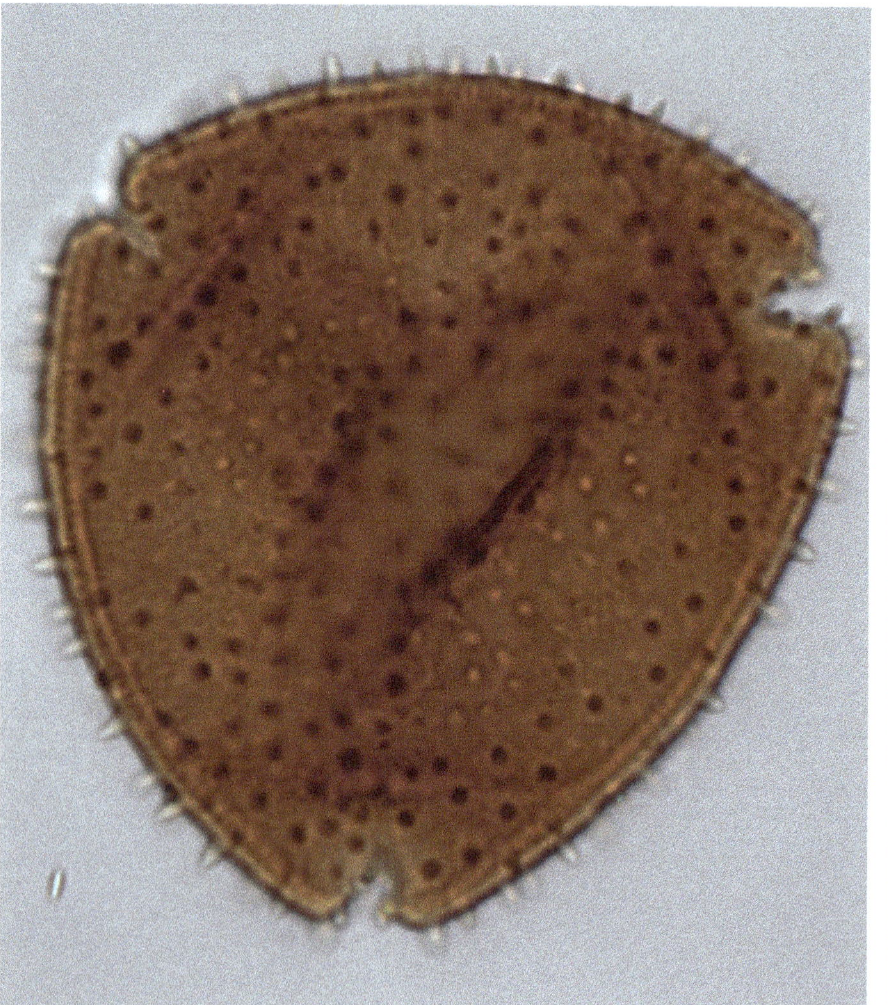

Lonicera

Camellia

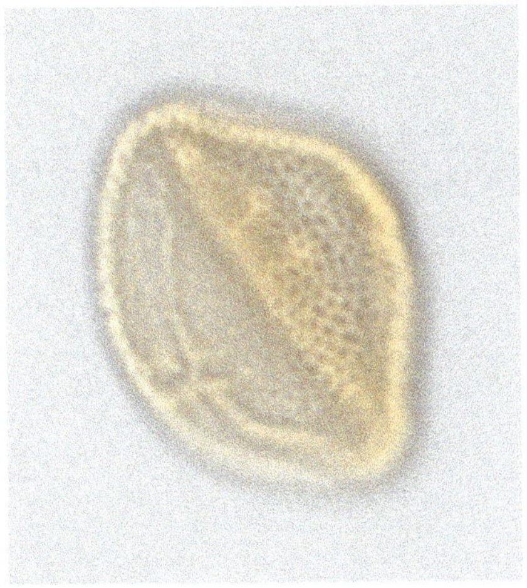
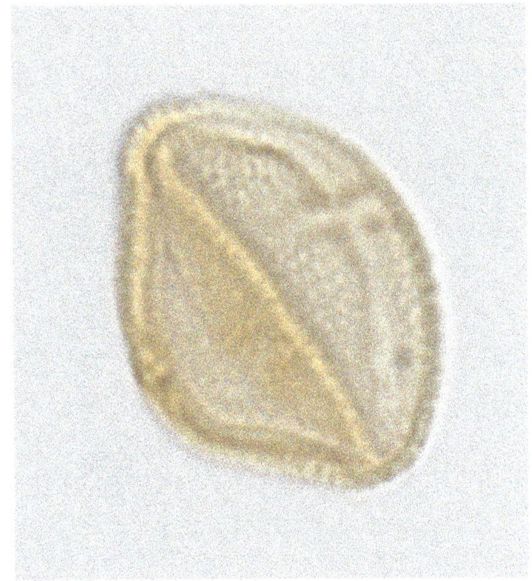
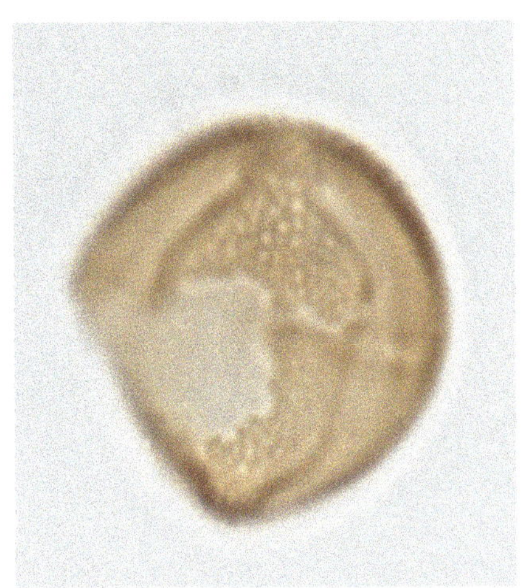

Rhus/Toxicodendron

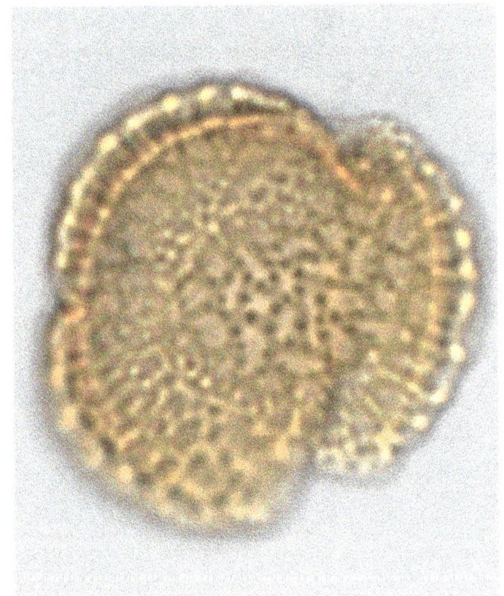
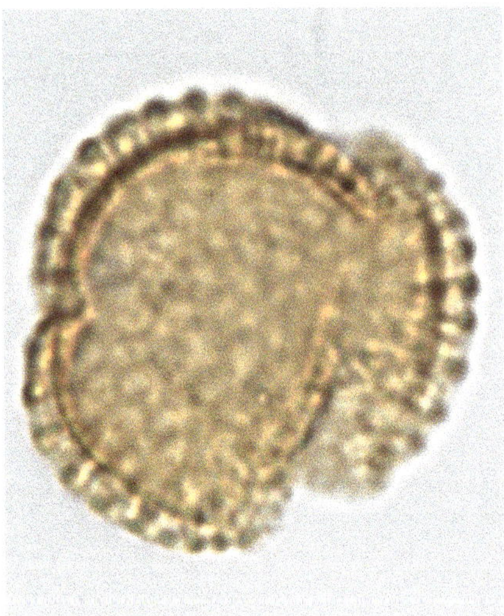
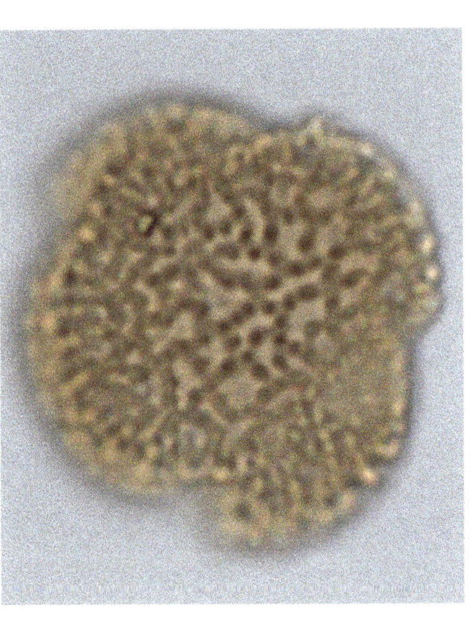

Ligustrum

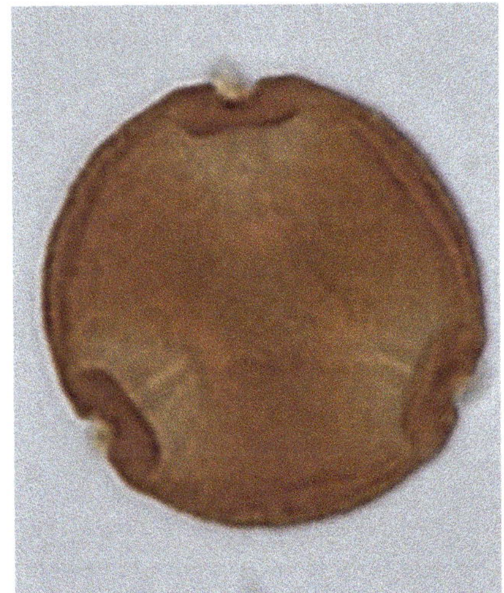

Possibly *Nyssa*

Possibly *Nyssa*

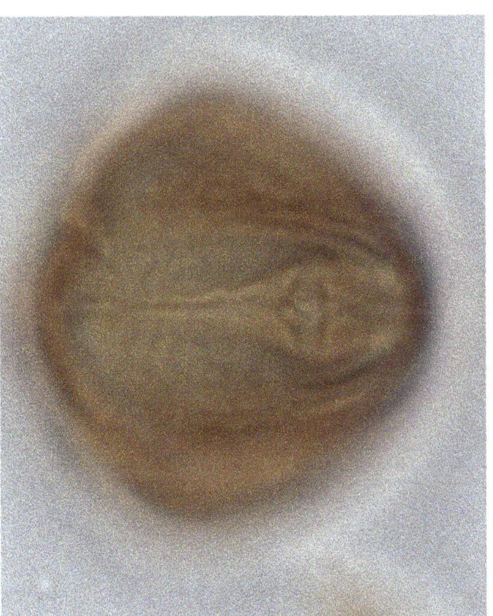

Possibly *Nyssa*

Centaurea

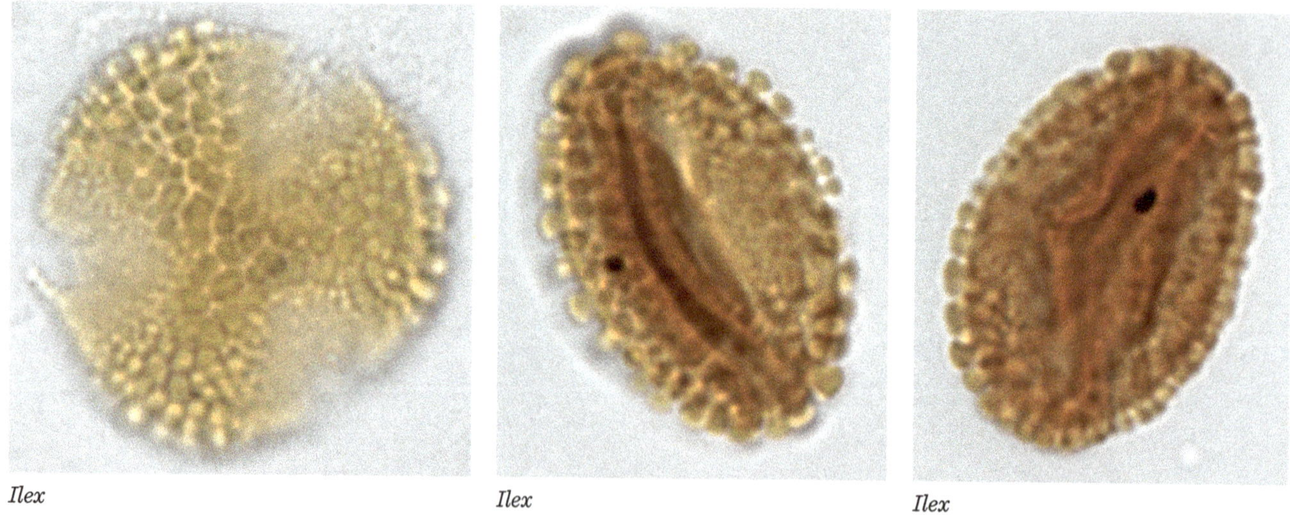

Ilex *Ilex* *Ilex*

Fagus

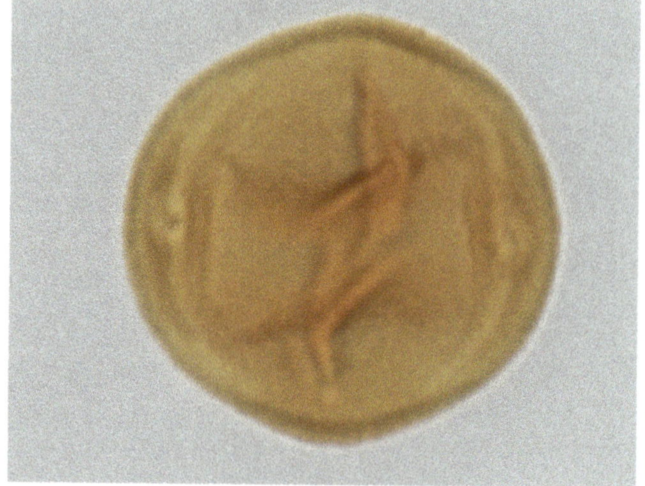

Fagus

Diospyros

Diospyros

Diospyros

Diospyros

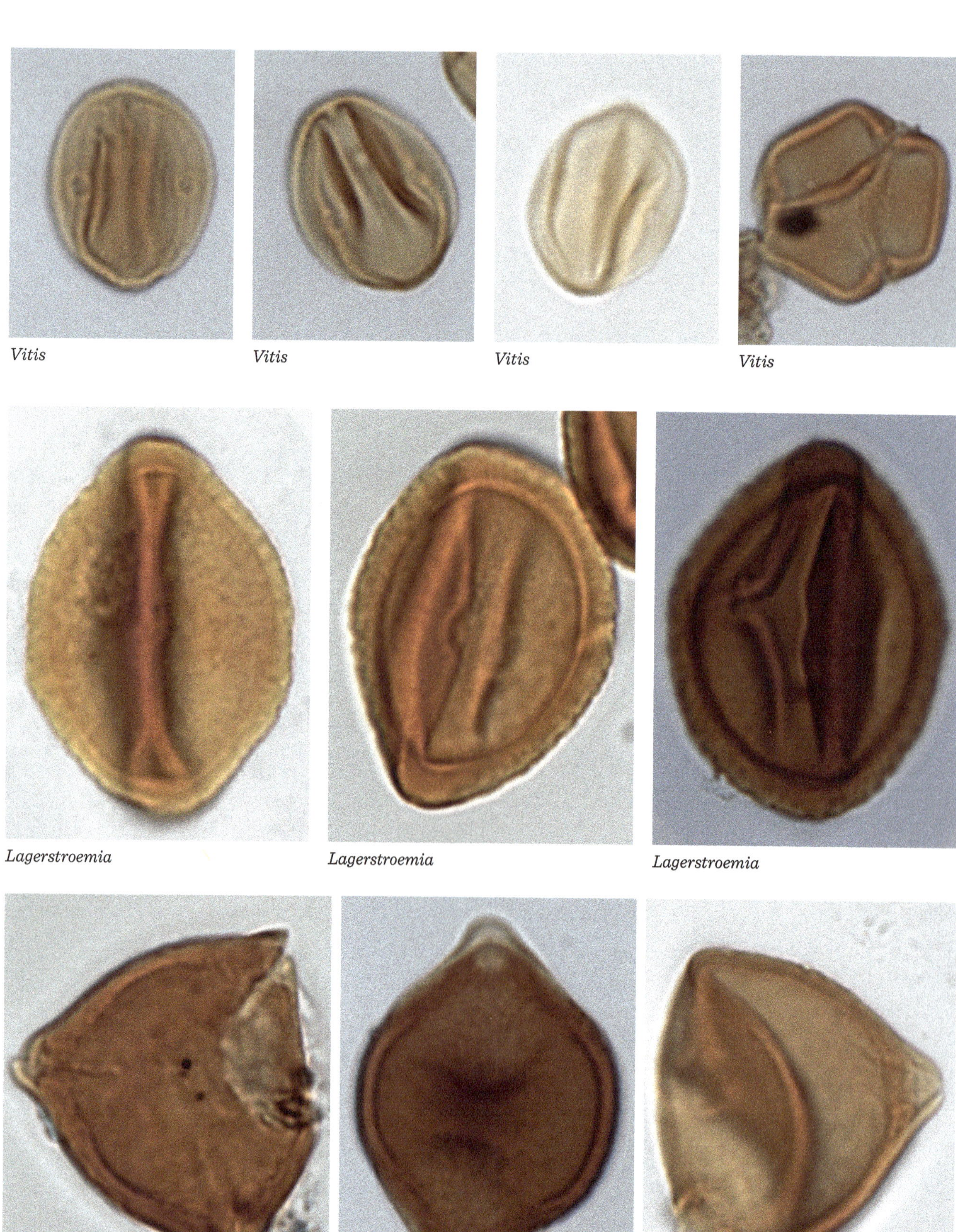

Vitis *Vitis* *Vitis* *Vitis*

Lagerstroemia *Lagerstroemia* *Lagerstroemia*

Elaeagnus

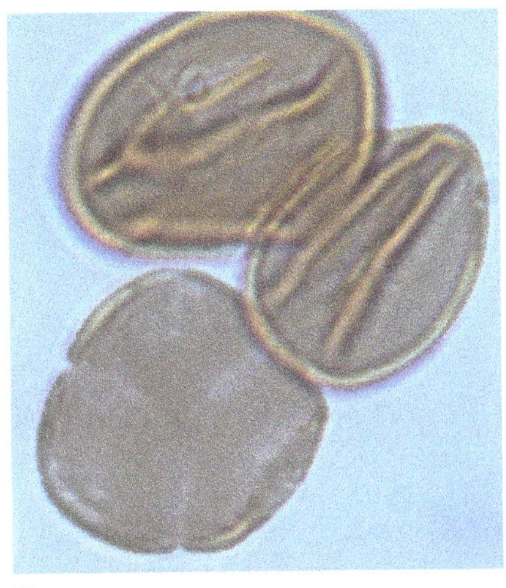
Rumex

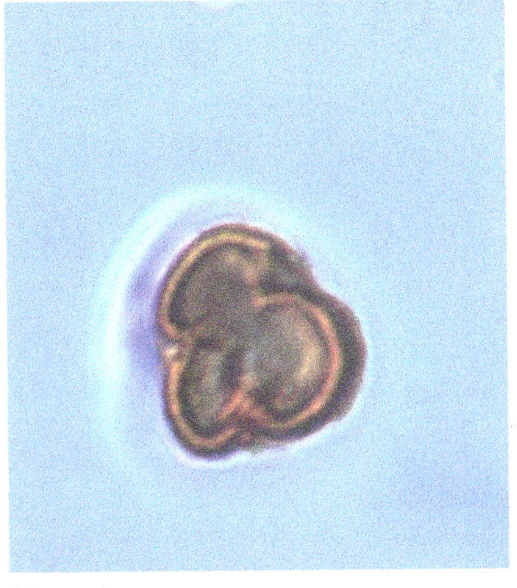
Hypercicaceae

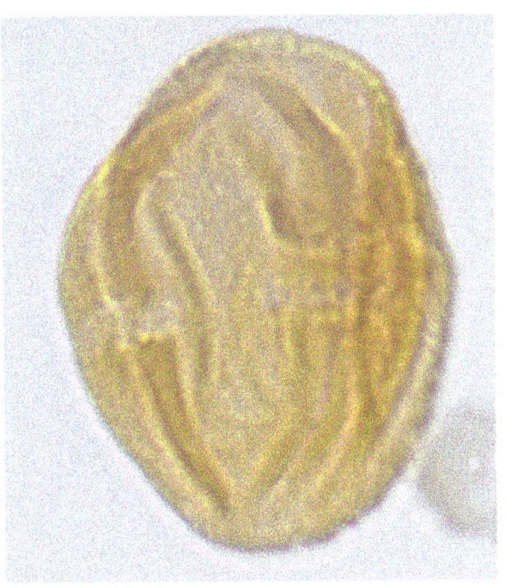
Mitchelia

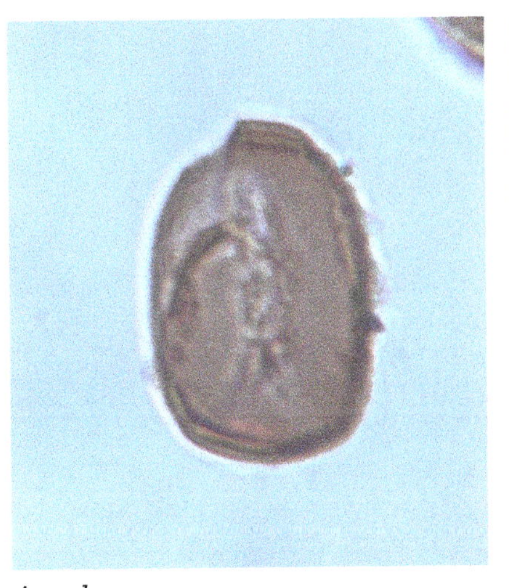
Aesculus

Castanea

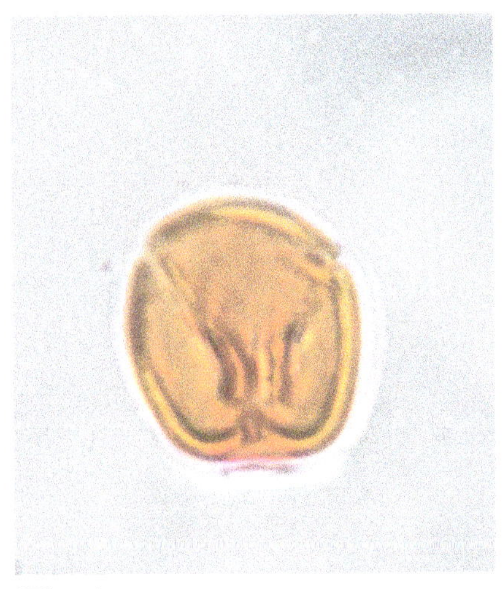
Wisteria

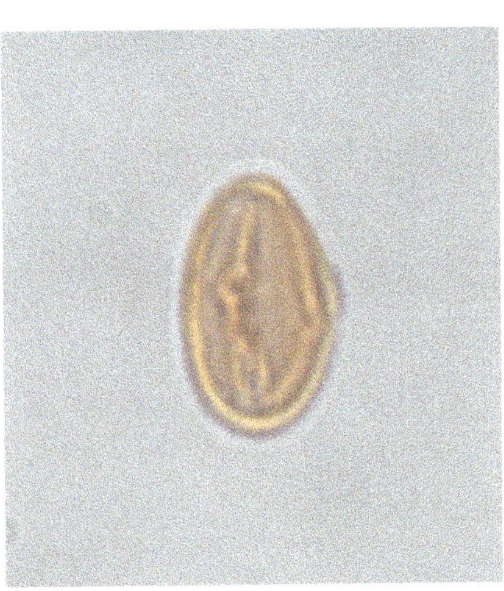
Fabaceae

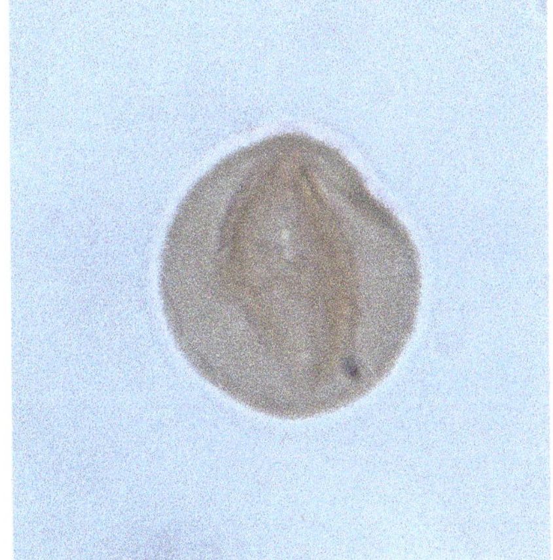
Ceanothus

Primulaceae

TETRAD & POLYAD

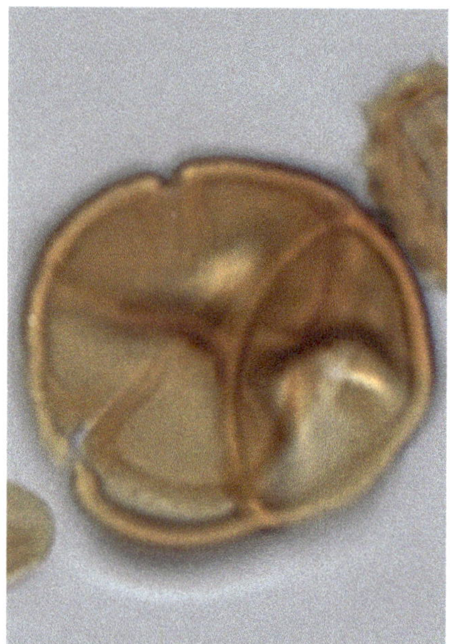

Oxydendrum

Possibly *Acacia*

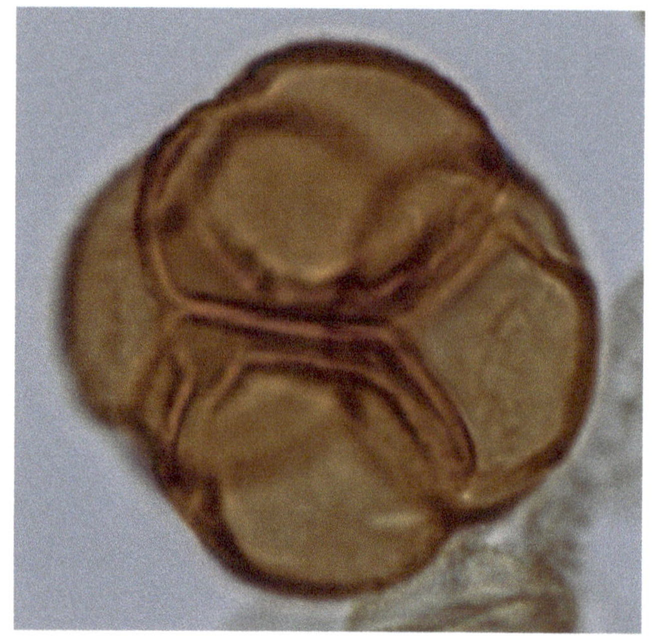

Ericaceae

TETRAD & POLYAD

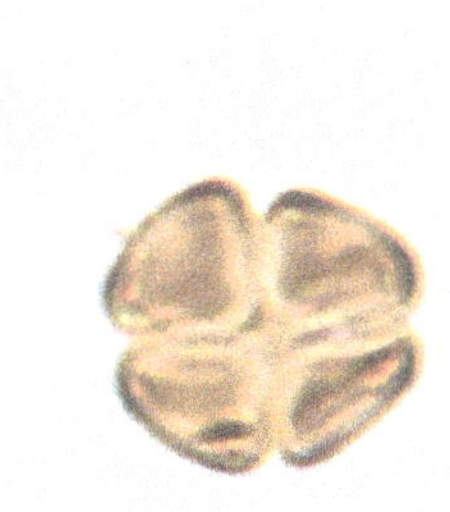

Mimosa

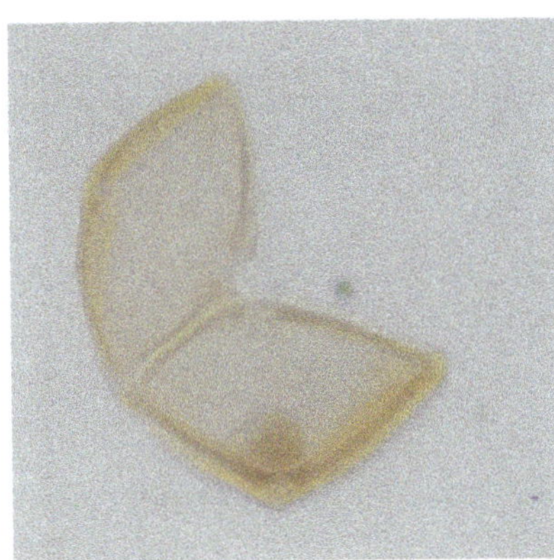

Gardenia

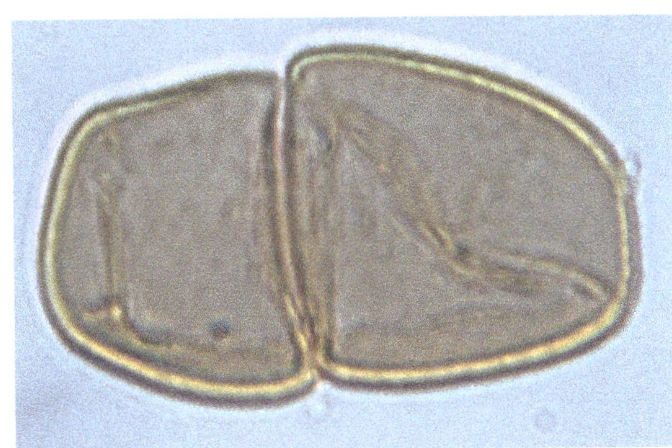

Possibly *Typha*

SPIRAPERTURATE

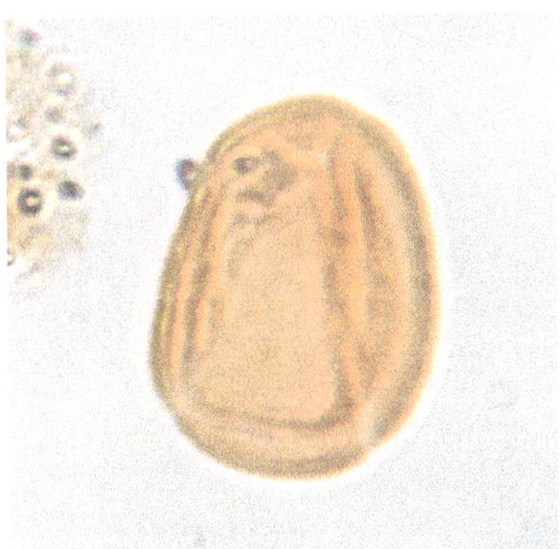

Berberis

VESICULATE

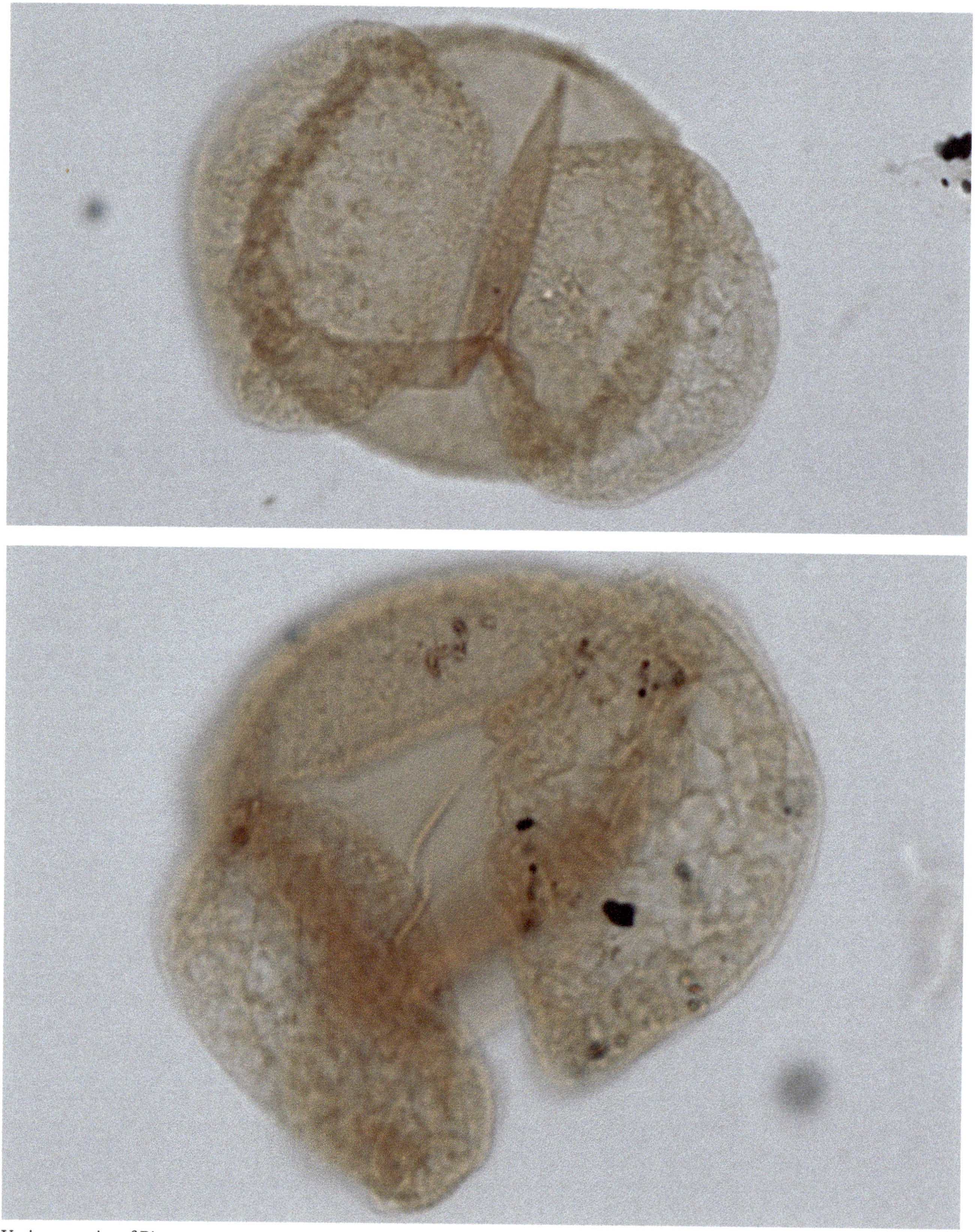

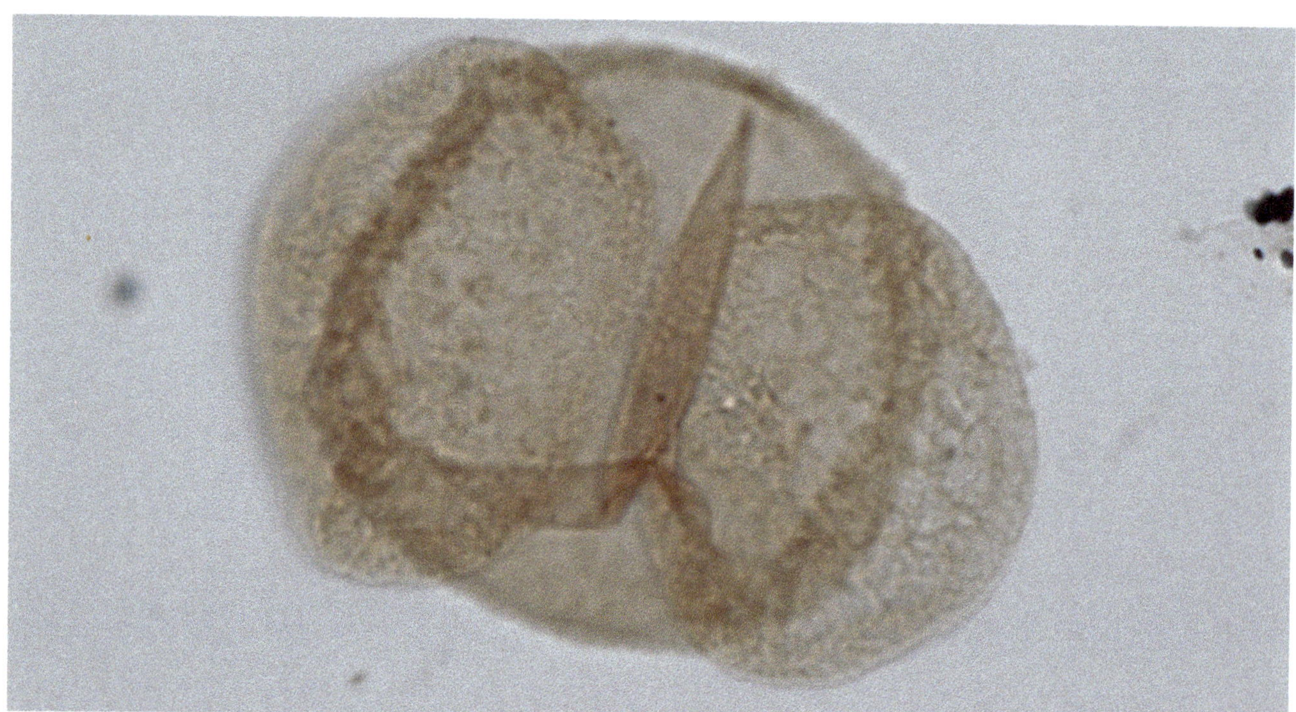

Various species of *Pinus*

Pollen Occurrences in Honey and BroodMinder Hive Weight Changes by Ecoregion

Plant species tend to be widely distributed across the United States. However, groups of plants and the other living organisms associated with them, collectively called a community, tend to be more restricted. The members of a community are able to live in a particular region because that region contains the resources and characteristics they need to survive and thrive—features like precipitation, humidity, soil type, soil moisture, soil nutrients, and temperature. Together, all these organisms (the community) and their environment define an ecosystem. These general ecosystem groupings were devised by ecologists to delineate broad-brush subdivisions of the North American landscape, referred to as Level I ecoregions (Griffith et al., 2002). Most of the eastern half of the contiguous United States falls into the Eastern Temperate Forest Level I ecoregion. Over time, and to better explain the finer-scale groupings, smaller subdivisions were made. South Carolina is broken into three Level II ecoregions: Ozark/Ouachita-Appalachian Forests, Southeastern USA Plains, and Mississippi Alluvial and Southeast USA Coastal Plains. Level II ecoregions are widely used for looking at biotic (plant, animal, etc.) patterns and managing natural resources (e.g., timber, waterways, etc.) at the national scale, but they don't capture the level of ecological information that helps us understand honey production.

Level III ecoregions do this and are intrinsically tied to the underlying geomorphology (landforms), hydrogeology (water resources), and geology (soil, sediment, and rock types). These ecoregions are most useful for beekeepers on a statewide scale, because they capture the environmental factors that influence nectar production (Trimboli, 2018). While Level IV ecoregions do a better job of capturing the nuances of varying soil types and plant communities at a local scale, they require significantly more data collection. Indeed, the USDA National Agricultural Statistics Service reports honey production at a statewide level (roughly equivalent to a Level I ecoregion) and does not capture plant contributions to honey at all. Research shows that honey production models, which are outside of the scope of this book but could be derived from the data presented, are more accurate when using Level III or Level IV ecoregion data (Trimboli, 2018). For these reasons, we chose to organize the collected data (i.e., BroodMinder hive weights and pollen diversity and abundance in honey) in terms of Level III ecoregions (Figure 4.1): Blue Ridge, Piedmont, Southeastern Plains, Mid-Atlantic Coastal Plain, and Southern Coastal Plain.

Hive weight dynamics display interesting patterns that are the result of a combination of nectar availability and honeybee colony behavior. This behavior includes both foraging and feeding behaviors, reproductive behaviors, relative moisture and humidity in the hive, and loss of community members (Meikle et al., 2018). As our focus was on nectar flow, we made a deliberate choice to base our analyses on a single daily reading (at midnight) with the goal of reducing noise generated by other bee behaviors and/or hive inspection and focusing data analysis on significant weight gains/losses within the colony (Sponsler et al., 2020). Daily hive weight is generally greatest just after sunset, when bees are all in for the night and daily honey stores have not been depleted by consumption or evaporation (Arias-Calluari et al., 2023); the weight drop that occurs between sunset and midnight seems to align with daily honey consumption and thus provides an estimate of what is being stored long term. The analyses of hive weight changes presented in this book chapter represent a fraction of a very large suite of possible analyses of hive-weight indicated behaviors within the colonies. For example, daily weight changes could be used to track the number of active foragers in the colony and identify colonies at risk of collapse, to monitor colony foraging efficiency (and thus bee lifespan, as longer flights typically mean shorter lives), to monitor the amount of food collected on each trip, and to monitor overall colony weight growth in terms of optimal colony development (Arias-Calluari

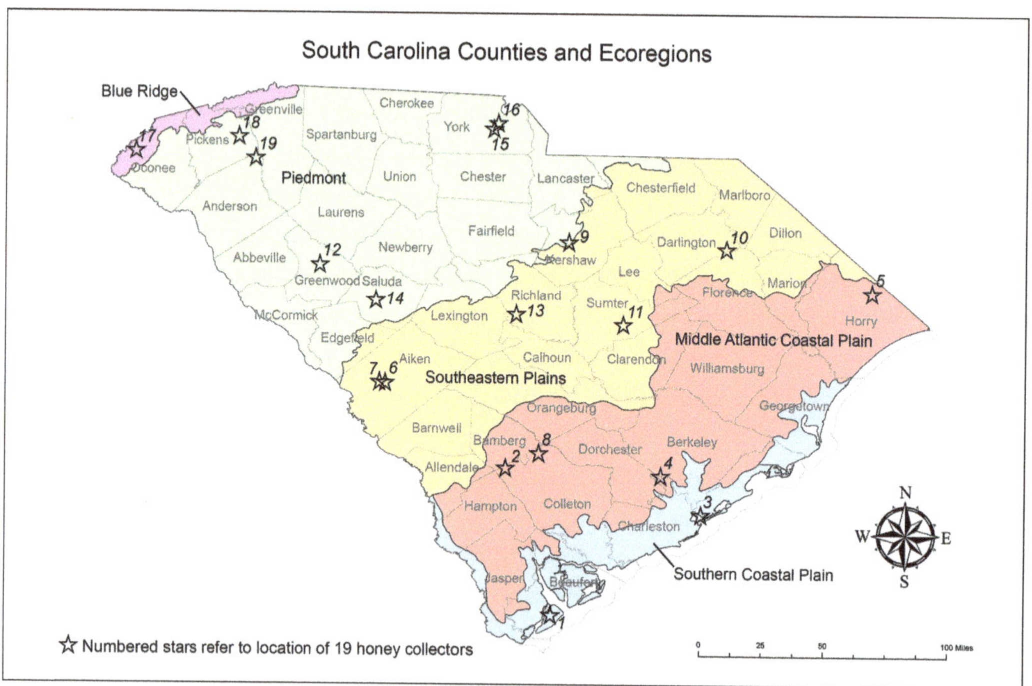

Figure 4.1. Level III ecoregions of South Carolina showing apiaries involved in the study.

et al., 2023). These analyses are, however, outside of the scope of this book. Full tables of hive weights are available from the authors upon request; indeed, we welcome the interest of bee behavioral statisticians! Here, we present the first record of identified periods of nectar flow for each ecoregion.

Before exploring the pollen in honey within each region, we need to further explain how the complex datasets were harmonized to show weekly variations in each region. While participating apiaries collected, on average, weekly samples, they did not all collect on the same day, because bees in each hive have their own foraging patterns and there are variations in nectar availability within the state, the ecoregion, and the direct surrounding of each apiary. To account for variations in sample collection day of the week, each collection date was standardized to occur between Sunday and Saturday of the same calendar week. This means that in a few cases, samples collected just a couple days apart—for example, a Friday and a Sunday—fall into two separate weeks of the year even though those samples were collected very close together. Using this standardization method, there were 53 weeks in 2022. In the very few cases that two samples were collected in the same standardized week, the data were averaged. The harmonized datasets were then analyzed in several different ways to examine variations in the pollen present in the honey (see Chapter 2). From the analyses, tables were designed to be accessible and relevant to the beekeeper. For ease of interpretation, both month and week are represented in the tables. Note that where a month changes within a given week, you'll see a transitional notation; for example, J/F denotes the end of January and beginning of February during the 6th week of the year (Table 4.1).

In this and subsequent sections, the term "plant taxa" is used repeatedly. "Taxa" (singular "taxon") is short for "taxonomic groups." All living things can be placed in taxonomic groups based on their shared characteristics and evolutionary relatedness. The most broad and inclusive taxonomic group is the domain, while the smallest, most exclusive taxonomic group is the species. In some cases, the melissopalynologists were able to identify the pollen to the species level, but this did not occur very often. One example of such pollen is from *Glycine max* (soybean) (Table 4.1). Pollen identified to the species level has a two-part name (*e.g. Glycine max*) and the name is written in italics. Much more frequently, pollen was identified to the level of the genus (plural: genera) such as *Acer* (maple) or *Pinus* (pine) (Table 4.1). There are many different species of maple, like sugar maple (*Acer saccharum*) and red maple (*Acer rubrum*). The pollen of each species looks a little different, but for the purposes of this study, a maple was simply a maple. Pollen identified to the genus level has a single name, with its first letter capitalized (note that by convention, the first letter of a species name is always lowercase); genera are also italicized. Some families (the next-largest taxonomic group after genus) have pollen that is so similar in appearance that it is very hard to tell one genus from another. In that case, the melissopalynologists only identified the pollen to the family level. As an example, the genera within the family Rosaceae look very similar. This family includes roses (*Rosa*), cherries (*Prunus*), and blackberries (*Rubus*) (Table 4.1). Within the family Asteraceae, the melissopalynologists were able to separate the genera to some extent. Family names are not written in italics but have a characteristic "-aceae" ending to their name, and the first letter is capitalized. This lesson in plant taxonomy is to make the reader aware of the limits to our analysis. In the Blue Ridge (for example), the **plant taxa** the bees are using include the species *Glycine max*, one or more species of maple (genus *Acer*), and one or more species within the Rosaceae family (Table 4.1). In summary, knowing what pollen types can confidently be assigned a species, genus, or family name is based on the morphological distinction between types, and there is a general agreement between all experienced palynologists to decide when a family is as far as one can confidently go using standard microscopic analyses.

When comparing pollen occurrences in honey to published datasets, we consulted primarily the bloom times published in the Honey Bee Net (GSFC, 2007; Ayers & Harmon, 1992) and the Honey Bee Pollen Timing Chart (Smith, 2021) published by the Home and Garden Information Center of the Clemson Cooperative Extension, and bloom times included in Beekeeping in South Carolina (Hood, 2006). While information from Honey Bee Net is older, it represents the most complete regional dataset available to date—albeit at a very broad scale divides the state into two regions: Appalachian-Ozark Upland (Region 11) and Atlantic and Gulf Coastal Plain (Region 12). Note that these region designations are not ecoregions; they reflect the types of plants previously reported from honey and bee pollen pellets. Smith (2021) is more up to date but focuses on the Piedmont and Low State regions. Thus, very little is known about the overall nuances of variations in pollen occurrences in honey in South Carolina.

We chose to discuss the Blue Ridge ecoregion first, because it is represented by only one apiary. This permits us to put both means of analyzing the data (an overview of pollen occurrence through the year and details of plant taxa present in each honey sample) side-by-side and drill into the differences we observed. As such, the discussion of the results observed in the Blue Ridge lays the foundation for understanding how to interpret the tables for other ecoregions. After this, we examine subsequent ecoregions to the east and then southeast.

In the sections below, you will see three different types of data represented in three different ways. The first, exemplified by Figure 4.2, is a hive weight gain graph; these allow us to estimate when nectar flows are occurring.

The second, exemplified by Table 4.1, is an ecoregion pollen occurrence table; these contain lists of the plant taxa whose pollen was observed in the honey samples collected. As such, it is not a "Bloom Chart," which relies upon trained botanists to make a systematic survey of the plants in flower each week. While individual beekeepers did collect bloom time information immediately surrounding their apiary at the time of sampling, these data are not representative of the entire bee foraging radius (an approximate 2-mile radius from the hive), and in many instances, the bees were incorporating pollen from nectar sources into their honey before blooms were observed by the apiarist (see below). Therefore, we do not know when the plant began blooming for certain, only when its pollen was observed in the honey. Beekeepers were instructed to collect a honey sample each week that nectar was flowing. Thus, a honey sample was not collected every week of the year. Weeks when no honey sample was collected are colored white in the ecoregion pollen occurrence table (see Table 4.1). If the plant taxon was observed in the honey sample for a particular week, that week's box is colored black; when the box is colored gray, the plant taxon was not present in a collection week. A black box in this table only indicates that the plant taxon was present (even if only represented by a single grain); it does not provide information about the quantity of that taxon's pollen that was present.

The third type of data representation in this section is an apiary pollen occurrence table for each individual honey apiary (such as Table 4.2); these tables show information about pollen quantity for each taxon. All apiary pollen occurrence tables are standardized to include the same weeks of the year; for example, even though nectar flow in the Blue Ridge didn't begin until Week 8 (mid-February), the apiary pollen occurrence table includes Weeks 4–7.

The pollen content of honey reflects the plants that contribute to the nectar the bees have foraged to generate that honey. Likewise, the proportion of pollen observed for each plant taxon provides information about the contribution that plant made to the honey. Given that premium honey varietals are typically monofloral and contain >45% of pollen from a single source (with the exception of sourwood (*Oxydendrum*) honey, which rarely contains a significant proportion of sourwood (*Oxydendrum*) pollen, because of the size of the pollen), they are of interest to beekeepers and regional honey marketers alike. We focus on monofloral varietal honeys in the following sections.

4.1 The Blue Ridge

The Blue Ridge Mountains in Oconee County, South Carolina. Image obtained from Charlotte Waters.

The southern end of the Blue Ridge Level III ecoregion occurs in extreme northwestern South Carolina, along the Georgia-North Carolina facet of the cut-diamond-shaped state boundary. This is generally the wettest part of the state, with average annual rainfall exceeding 152 cm (60 in.) (Runkle et al., 2022). Unlike most of the state—in which precipitation peaks in March and July—the Blue Ridge receives the greatest amount of precipitation in winter (although there is no true dry season). This abundance of moisture—coupled with extensive chemical and physical weathering of middle Proterozoic metamorphic rocks (amphibolite facies) and early Paleozoic igneous rocks that intruded into the older rocks—has produced a region of highly variable topography and relatively thick fine-grained loamy soils (Abella, 2003; Abella et al., 2003; Overstreet & Bell, 1965). Most of these soils are alkaline and support a distinctive flora, including extensive moist broad-leaves forests (SC DNR, 2005a). These forests are dominated by oak (*Quercus*) and pine (*Pinus*) with a blueberry (*Vaccinium*, usually counted as Ericaceae by melissopalynologists) groundcover.

Dense stands of mountain laurel (*Kalmia*, also counted as Ericaceae) and hemlock also grow and may extend onto lower north-facing slopes and other low areas, where tulip poplar (*Liriodendron*), Frasier magnolia (counted as Magnoliaceae), hickory (*Carya*), sweet birch (*Betula*), beech (*Fagus*), basswood (*Tilia*), elm (*Ulmus*), rhododendron (counted as Ericaceae), red maple (*Acer*), and walnut (*Juglans*) may also grow. In some areas, Carolina silverbell (*Halesia*) and sourwood (*Oxydendrum*) are present as mid- or understory plants, and the forest floor may contain bloodroot (*Sanguinaria*), cane (*Arundinaria*), foamflower (*Tiarella*), ginseng (*Panax*), and partridge berry (*Mitchella*). In moist soils, especially along streams and close to the Piedmont, there is also alder (*Alnus*), azalea (counted as Ericaceae), black willow (*Salix*), sweetgum (*Nyssa*), sweet pepperbush (*Clethra*), Virginia willow (*Itea*), and yellow root (*Xanthorhiza*). Dogwoods (*Cornus*), redbud (*Cercis*), sumac and poison ivy (counted as *Rhus/Toxicodendron*), blackberry, Carolina rose, and wild black cherry (all counted as *Rosaceae*), persimmon (*Diospyros*), muscadine grape (*Vitis*), winterberry (*Ilex*), and Virginia creeper (*Parthenocissus*) are all common. Especially in more disturbed or open areas within the forests, a wide variety of wildflowers grow—including many goldenrods, asters, daisies, Poor Joe, clovers, wild mustards, members of the buckwheat family, chicories, and dandelions—alongside invasive species such as Russian olive (*Eleagnus*).

Bloom timings, and thus nectar flows, vary from year to year. In general in the Blue Ridge, the bees may be active on warm days in January but do not begin foraging for nectar or pollen until February; April is generally considered the beginning of the spring nectar flow. In 2022, the monitored hive showed two small weight gains in March before falling through most of April; although, a brief period of weight gain occurred at the end of April (see Figure 4.2). While there were periodic gains and losses from April through mid-September, the general trend was toward weight loss until a significant weight gain occurred in late September.

Hood (2006) notes 13 plants as important to beekeepers in the Blue Ridge region. These are: red maple (*Acer*) from mid-March through April; yellow maple (*Acer*) from late March through April; blackberry (Rosaceae) from mid-April through June; tulip

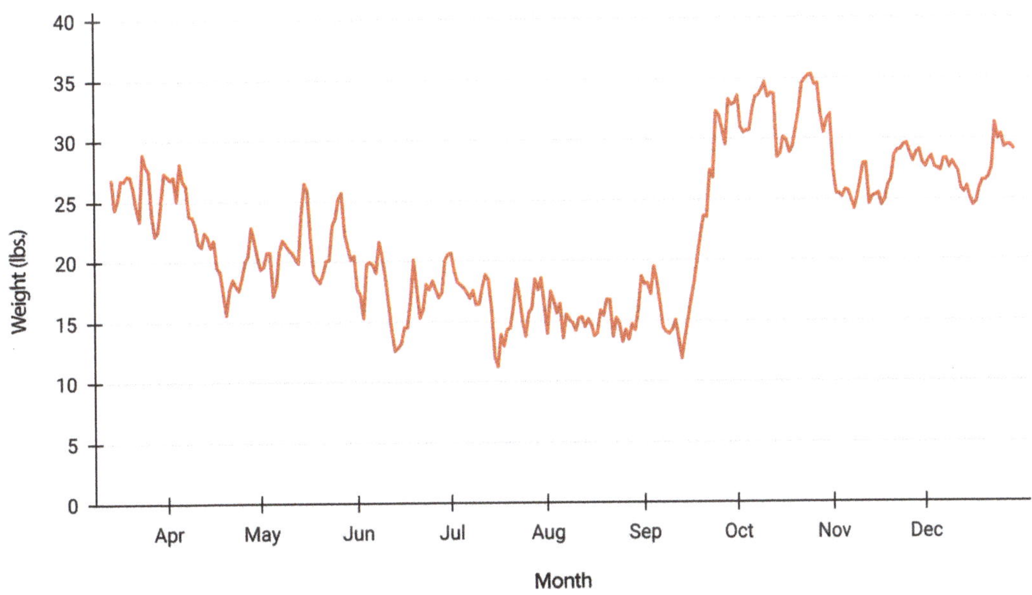

Figure 4.2. BroodMinder hive weight data collected from Apiary 17 in the Blue Ridge ecoregion during 2022.
Note: Labels indicate the first day of each listed month.

poplar (*Liriodendron*) in May; black haw (*Viburnum*) from late May through June; white sumac (*Rhus/Toxicodendron*) from mid-June through July; mountain laurel (Ericaceae) and sourwood (*Oxydendrum*) from late June through July; wild hydrangea (*Hydrangea*), rhododendron (*Ericaceae*), and goldenrod (Asteraceae [coneflower/goldenrod/sunflower family]) from late August through September; and both purple and white aster (Asteraceae [coneflower/goldenrod/sunflower family]) from mid-September through October. No other sources note plants specific to the Blue Ridge.

We now know that bees in the Blue Ridge are foraging for nectar from many more than these 13 plant taxa. Indeed, of the pollen occurring in Blue Ridge honey (Tables 4.1 and 4.2), at least 31 of the 35 taxa are known nectar producers. *Pinus*, *Quercus*, and *Poaceae* produce no nectar. It is unclear if *Ulmus* collected by regional bees is a nectar producer; most are not. Some of these plant taxa appeared in the honey for only a specific part of the year, while others appeared throughout the year. *Acer* (maple) and the Amaranthaceae (goosefoot family) were only present in spring. *Magnolia* and *Tilia* (basswood) were only present in the summer. *Ilex* (holly) and *Trifolium/Melilotus* (red/white clover) were present off and on throughout the year and produced monofloral honeys in mid-July. This makes sense with the continuous blooms seen in clovers, but not for holly, which typically blooms in the late spring. Later occurrences may be indicative of bees moving honey within the hive or moving honey from elsewhere into the hive.

One of the curious things about these observations is that sometimes, the pollen occurrence of a taxon in the honey corresponds with reported bloom times for the plant (Smith, 2021) and other times it does not. *Acer* (maple) blooms only in the spring and its pollen was only observed in the honey in the spring; however, Hood (2006) notes that red maple blooms from mid-March through April. In our study, *Acer* (maple) pollen first appearred in February and continued through the first week of April; this represents an early extension of the bloom period, likely related to the unusually warm January-early February experienced by the region in 2022 (Powell, 2022).

It is important to note that bees in the Blue Ridge were making monofloral honey throughout the year from a variety of plants (Table 4.2). Monofloral honeys are indicated with red color-coding in the apiary pollen occurrence table. The first honey produced by the hive was a polyfloral honey that contained *Trifolium/Melilotus* (red/white clover) as its most common pollen type, as well as maple (*Acer*). The first monofloral honey was *Acer* (maple) honey (*i.e.*, >45% of the pollen was *Acer*), produced in the latter half of March; this monofloral honey aligns well with increases in nectar in mid- and late March on the BroodMinder graph (Figure 4.2). Throughout April to the second week of May, monofloral Rosaceae (rose/cherry/plum/peach/blackberry family) honey was produced, corresponding to both large and small upticks in nectar production on the BroodMinder graph. The occurrence of Rosaceae pollen surged again, producing a monofloral Rosaceae honey the second week of July and an additional sample of monofloral Rosaceae honey at the August/September transition. It is highly unlikely that it is the same taxon that bloomed from April through September, particularly given that all members of this large family were grouped together. However, it is clearly an important plant family for bees in the Blue Ridge.

As noted above, the first monofloral honey was *Acer* (maple) honey. One of the major objectives of this project was to determine whether *Acer* (maple) significantly contributes to honey production in South Carolina. Our observations demonstrate that this is true in the Blue Ridge.

In addition to monofloral honeys from *Acer* and Rosaceae, other monofloral honeys were observed (Table 4.2). A significant peak in nectar production (as recorded from the BroodMinder scale [Figure 4.2]) occurred in late May, corresponding to a monofloral *Salix* (willow) honey that also contained *Oxydendrum* (sourwood) pollen. This is noteworthy because it represents a very early occurrence of sourwood pollen, which is thought to bloom in late June (Hood, 2006) and continue blooming through July. In our case, *Oxydendrum* (sourwood) pollen was not recorded again until the third week of July and continued until the 36th week, which contained the August/September transition. This extended occurrence likely represents both an earlier-than-expected and later-than-expected bloom, as *Oxydendrum* pollen is very large and unlikely to occur in moved honey. After the *Salix* honey in late May, monofloral *Trifolium/Melilotus* (red/white clover) was observed in mid-July. Unfortunately, after this mid-July sample was collected, no more samples were collected until the August/September transition; however, *Trifolium/Melilotus* (red/white clover) pollen appeared again in this later honey. It is realistic to speculate that this commonly used pollen contributed significantly to honey production throughout July and into August.

One of the interesting results from this study was that many plants/taxa contributed to the honey for extended periods of time. This means that the plants are blooming for enough time that they are available forage for bees. If we consider taxa that contributed to honey for at least three straight weeks, there were 13 such taxa in the Blue Ridge (Tables 4.1 & 4.2). This was the fewest taxa among the five ecoregions, but the Blue Ridge is the smallest of the ecoregions, and this study only sampled from a single apiary. Despite this, 13 taxa is a substantial portion (38%) of the total taxa (34) observed in this ecoregion (see Chapter 5). These 13 taxa were: *Acer* (maple), Asteraceae (ragweed/calendula/solidago family), *Glycine max* (soybean), *Ilex* (holly), *Ligustrum* (privet), *Liriodendron* (tulip poplar), *Magnolia*, *Nyssa* (tupelo/black gum), *Plantago* (plantain), *Quercus* (oak), *Rhus/Toxicodendron* (sumac/poison ivy), Rosaceae (rose/cherry/plum/peach/blackberry family), and *Trifolium/Melilotus* (red/white clover). Note that these taxa span the entire foraging season. They also include plants which do not have nectaries, such as *Quercus* (oak). Without nectaries, *Quercus* (oak) cannot contribute nectar; its pollen might have blown into other flowers on which the bees were foraging, or it could represent pollen from a pollen cell that was inadvertently broken as the honey sample was collected. Appearance of these pollen in the honey for such an extended period suggests that they are important forage for bees.

Table 4.1. Plant Taxa whose Pollen was Observed in Honey Collected from The Blue Ridge Ecoregion during 2022

Blue Ridge	Jan		J/F	Feb			F/M	Mar			M/A	Apr			
Week ▶	4	5	6	7	8	9	10	11	12	13	14	15	16	17	18
Acer (Maple)					■				■	■		■	▒		▒
Amaranthaceae (Goosefoot family)					■				▒	■		▒	▒		▒
Asteraceae (Coneflower/Golden Rod/Sunflower family)					▒				▒	▒		▒	▒		▒
Asteraceae (Dandelion/Cat's Ear/Endive family)					▒				▒	▒		▒	▒		▒
Asteraceae (Ragweed/Calendula/Solidago family)					■				■	▒		▒	▒		▒
Brassicaceae (Mustard family)					▒				▒	▒		▒	▒		▒
Cornus (Dogwood)					▒				▒	▒		▒	▒		▒
Diospyros (Persimmon)	▒				▒				▒	▒		▒	▒		▒
Elaeagnus (Autumn/Russian Olive)	▒				▒				▒	▒		■	▒		▒
Ericaceae (Heath family)					■				▒	▒		▒	▒		▒
Fagopyrum (Buckwheat)					▒				▒	▒		▒	▒		▒
Glycine max (Soybean)					▒				▒	▒		▒	▒		▒
Hexasepalum teres (Poorjoe/Rough Buttonweed)					▒				▒	▒		▒	▒		▒
Ilex (Holly)					▒				▒	▒		■	■		■
Juglans (Walnut)					■				▒	▒		▒	▒		▒
Lagerstroemia (Crepe Myrtle)					▒				▒	▒		▒	▒		▒
Ligustrum (Privet)					▒				▒	▒		▒	▒		▒
Liquidambar (Sweetgum)					■				▒	▒		▒	▒		▒
Liriodendron (Tulip Poplar)					■				▒	▒		▒	▒		▒
Lonicera (Honeysuckle)					▒				▒	▒		▒	▒		▒
Magnolia					▒				▒	▒		■	▒		■
Nyssa (Tupelo/Black Gum)					■				▒	▒		■	■		▒
Oxydendrum (Sourwood)					▒				▒	▒		▒	▒		▒
Pinus (Pine)					▒				▒	▒		▒	▒		▒
Plantago (Plantain)					■				▒	▒		▒	▒		■
Poaceae (Grass family)					■				▒	▒		▒	▒		▒
Quercus (Oak)					▒				▒	▒		▒	■		▒
Rhus/Toxicodendron (Sumac/Poison Ivy)					■				▒	▒		■	▒		▒
Rosaceae (Rose/Cherry/Plum/Peach/Blackberry family)					■				▒	■		■	■		■
Salix (Willow)					▒				▒	▒		▒	▒		▒
Tilia (Basswood)					▒				▒	▒		▒	▒		▒
Trifolium/Melilotus (Red/White Clover)					▒				▒	▒		▒	▒		▒
Ulmus (Elm)					▒				▒	▒		▒	▒		▒
Vitis (Grape)					▒				▒	▒		▒	▒		▒
Various Unknown Palynomorphs					■				▒	■		▒	▒		▒

Key: *no data collected* □ *taxon absent on this week* ▒ *taxon present on this week* ■

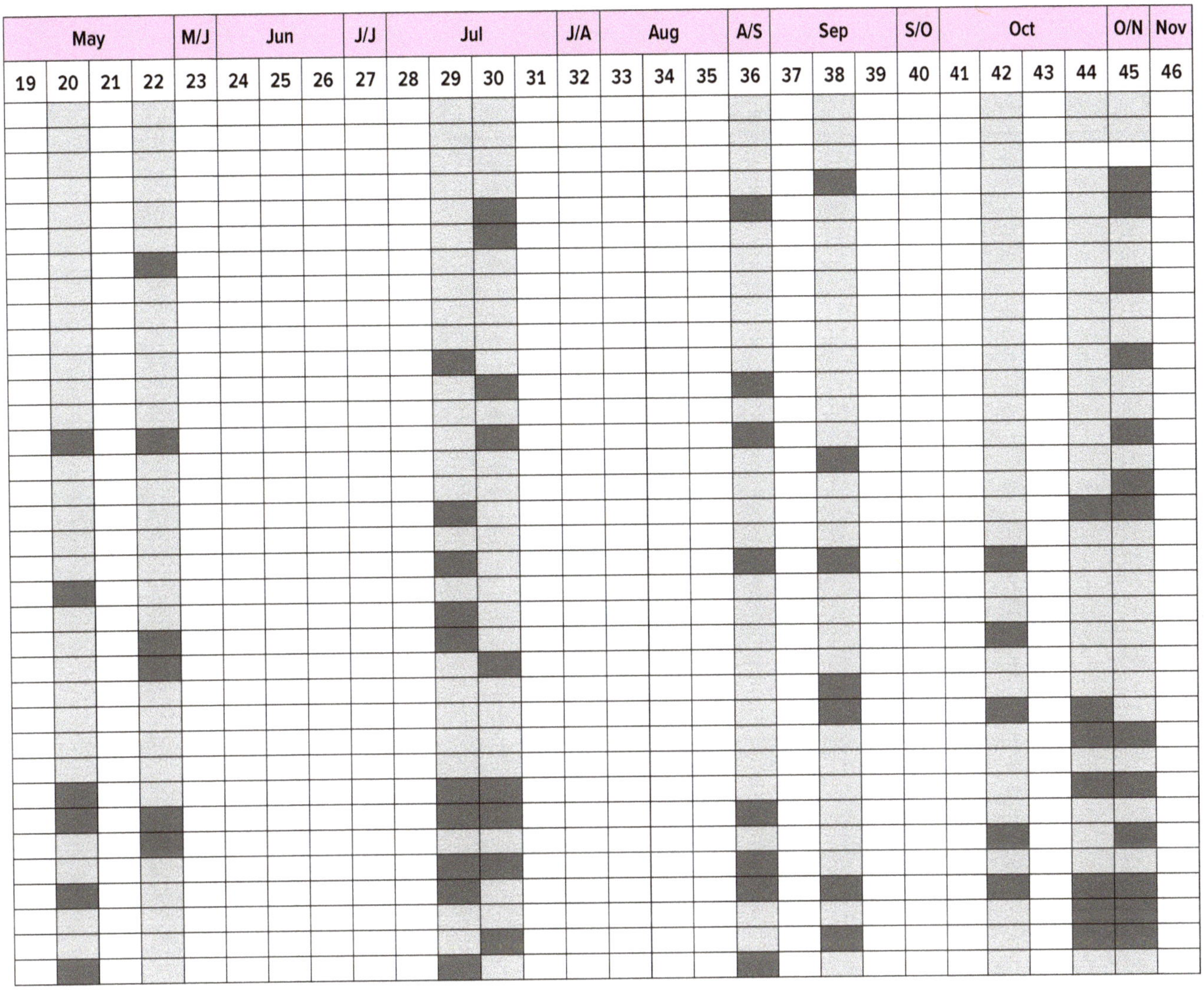

May				M/J	Jun			J/J	Jul				J/A	Aug			A/S	Sep			S/O	Oct				O/N	Nov
19	20	21	22	23	24	25	26	27	28	29	30	31	32	33	34	35	36	37	38	39	40	41	42	43	44	45	46

Table 4.2. Pollen Occurrence in Honey Collected in Apiary 17 in the Blue Ridge Ecoregion during 2022

Apiary 17	Jan		J/F	Feb				F/M	Mar					M/A	Apr						
Week ▶	4	5	6	7	8		9	10	11	12		13		14	15		16		17	18	
Type of Data ▶					%	PC				%	PC	%	PC		%	PC	%	PC		%	PC
Acer (Maple)					4.4	M				95.5	D	87.3	D		5.1	M	0.0			0.0	
Amaranthaceae (Goosefoot family)					5.9	M				0.0		2.3	L		0.0		0.0			0.0	
Asteraceae (Coneflower/Sunflower family)					0.0					0.0		0.0			0.5	L	0.0			0.0	
Asteraceae (Dandelion/Cat's Ear/Endive family)					0.0					0.0		0.0			0.0		0.0			0.9	L
Asteraceae (Ragweed/Calendula/Solidago family)					12.7	M				2.7	L	0.5	L		0.0		0.5	L		0.0	
Brassicaceae (Mustard family)					0.0					0.0		0.0			0.0		0.0			0.0	
Cornus (Dogwood)					0.0					0.0		0.0			0.0		0.0			0.0	
Diospyros (Persimmon)					1.0	L				0.0		0.0			0.0		0.0			0.0	
Elaeagnus (Autumn/Russian Olive)					0.0					0.0		0.0			1.9	L	0.0			0.0	
Ericaceae (Heath family)					4.4	M				0.0		0.0			0.0		0.0			0.0	
Fagopyrum (Buckwheat)					0.0					0.0		0.0			0.0		0.0			0.0	
Glycine max (Soybean)					0.0					0.0		0.0			0.0		0.0			0.0	
Hexasepalum teres (Poorjoe/Rough Buttonweed)					0.0					0.0		0.0			0.0		0.0			0.0	
Ilex (Holly)					0.0					0.0		0.0			5.6	M	2.8	L		6.4	M
Juglans (Walnut)					4.4	M				0.0		0.0			0.0		0.0			0.5	L
Lagerstroemia (Crepe Myrtle)					0.0					0.0		0.0			0.0		0.0			0.0	
Ligustrum (Privet)					0.0					0.0		0.0			0.0		0.0			0.9	L
Liquidambar (Sweetgum)					1.5	L				0.0		0.0			0.0		0.0			0.0	
Liriodendron (Tulip Poplar)					2.4	L				0.5	L	0.0			0.0		0.5	L		0.0	
Lonicera (Honeysuckle)					0.0					0.0		0.0			0.5	L	0.0			0.0	
Magnolia					0.0					0.0		0.0			5.1	M	0.0			1.4	L
Nyssa (Tupelo/Black Gum)					2.9	L				0.0		0.0			1.4	L	17.5	I		0.0	
Oxydendrum (Sourwood)					0.0					0.0		0.0			0.0		0.0			0.0	
Pinus (Pine)					0.0					0.0		0.0			0.5	L	0.0			0.0	
Plantago (Plantain)					2.4	L				0.9	L	0.0			0.0		0.0			2.3	L
Poaceae (Grass family)					1.5	L				0.0		0.0			0.0		0.0			0.0	
Quercus (Oak)					0.0					0.0		0.0			0.5	L	3.8	M		0.9	L
Rhus/Toxicodendron (Sumac/Poison Ivy)					15.1	M				0.0		0.0			5.6	M	0.0			0.0	
Rosaceae (Rose/Cherry/Plum/Peach/Blackberry family)					9.3	M				0.0		5.6	M		73.4	D	74.9	D		85.5	D
Salix (Willow)					0.0					0.0		0.9	L		0.0		0.0			0.0	
Tilia (Basswood)					0.0					0.0		0.0			0.0		0.0			0.0	
Trifolium/Melilotus (Red/White Clover)					20.5	I				0.0		0.0			0.0		0.0			0.5	L
Ulmus (Elm)					0.0					0.5	L	0.0			0.0		0.0			0.9	L
Vitis (Grape)					2.0	L				0.0		0.5	L		0.0		0.0			0.0	
Various Unknown Palynomorphs					9.8	M				0.0		2.8	L		0.0		0.0			0.0	
Totals ▶					100					100		100			100		100			100	

Key: *D* (predominant) >45%   I (secondary) 16-45%   M (important) 3-15%   L (minor) <3%

May						M/J	Jun			J/J	Jul						J/A	Aug			A/S		Sep				S/O	Oct						O/N		Nov
19	20		21	22		23	24	25	26	27	28	29		30		31	32	33	34	35	36		37	38		39	40	41	42		43	44		45		46
	%	PC		%	PC							%	PC	%	PC						%	PC		%	PC				%	PC		%	PC	%	PC	
	0.0			0.0								0.0		0.0							0.0			0.0					0.0			0.0		0.0		
	0.0			0.0								0.0		0.0							0.0			0.0					0.0			0.0		0.0		
	0.0			0.0								0.0		0.9	L						0.0			0.0					0.0			7.9	M	5.2	M	
	0.0			0.0								0.0		0.0							0.0			2.4	L				0.0			0.0		3.3	M	
	0.0			0.0								0.0		3.1	M						2.6	L		0.0					0.0			0.0		1.9	L	
	0.0			0.0								0.0		1.8	L						0.0			0.0					0.0			0.0		0.0		
	0.0			7.8	M							0.0		0.0							0.0			0.0					0.0			0.0		0.5	L	
	0.0			0.0								0.0		0.0							0.0			0.0					0.0			0.9	L	3.3	M	
	0.0			0.0								0.0		0.0							0.0			0.0					0.0			0.0		0.0		
	0.0			0.0								0.0		0.0							0.0			0.0					0.0			0.0		0.0		
	0.0			0.0								1.4	L	0.0							0.0			0.0					0.0			0.0		1.4	L	
	0.0			0.0								0.0		7.1	M						3.5	M		0.0					0.0			0.0		0.0		
	0.0			0.0								0.0		0.0							0.0			0.0					0.0			0.5	L	0.5	L	
	4.1	M		1.2	L							0.9	L	12.1	M						6.9	M		0.0					0.0			0.5	L	2.4	L	
	0.0			0.0								0.0		0.0							0.0			1.4	L				0.0			0.0		0.0		
	0.0			0.0								0.0		0.0							0.0			0.0					0.0			13.0	M	8.1	M	
	0.9	L		0.0								1.8	L	0.0							0.0			0.0					0.0			31.9	I	5.7	M	
	0.0			0.0								0.0		0.0							0.0			0.0					0.0			0.0		1.0	L	
	0.0			0.0								2.3	L	0.9	L						6.9	M		7.5	M				1.9	L		0.0		0.0		
	1.4	L		0.0								0.0		0.0							0.0			0.0					0.0			0.0		0.0		
	0.0			0.0								1.4	L	0.4	L						0.4	L		0.0					0.0			0.0		0.0		
	0.9	L		4.1	M							26.6	I	0.9	L						0.4	L		0.9	L				2.3	L		0.0		0.5	L	
	0.0			1.2	L							0.0		1.3	L						0.9	L		0.0					0.0			0.0		0.0		
	0.0			0.8	L							0.0		0.0							0.0			1.4	L				0.0			0.0		0.0		
	0.5	L		0.0								0.0		0.0							0.0			19.8	I				15.7	I		6.5	M	1.0	L	
	0.0			0.0								0.0		0.9	L						0.0			0.0					0.0			1.4	L	7.6	M	
	0.0			0.4	L							0.0		0.0							0.0			0.0					0.0			0.0		1.0	L	
	3.7	M		0.0								3.6	M	1.3	L						0.4	L		0.0					0.0			2.3	L	9.5	M	
	84.9	D		20.1	I							47.3	D	2.7	L						65.4	D		30.2	I				0.0			10.2	M	0.0		
	0.0			64.3	D							0.0		0.9	L						0.0			0.0					65.7	D		0.0		15.2	M	
	0.0			0.0								3.2	M	3.6	M						1.3	L		0.0					0.0			0.0		0.0		
	1.8	I		0.0								2.3	L	54.0	D						9.5	M		14.6	M				2.3	L		7.4	M	2.4	L	
	0.0			0.0								0.0		0.4							0.0			0.0					0.0			1.4	L	26.7	I	
	0.0			0.0								0.0		7.6	M						0.0			21.7	I				0.0			16.2	I	2.9	L	
	1.8	L		0.0								9.5	M	0.0							1.7	L		0.0					12.0	M		0.0		0.0		
	100			100								100		100							100			100					100			100		100		

4.2 Piedmont

Hives maintained by Brett Kahley in the Piedmont ecoregion.

The Piedmont level III ecoregion extends from the Blue Ridge Escarpment—a zone of sheared rocks produced by the Brevard fault zone, where the mountains to the northwest give way to the rolling hills then plains toward the southeast. It extends southeastward until the Atlantic Seaboard Fall Line—a long, low series of cliffs that separate the heavily weathered metamorphic terrains of the Piedmont from the Southeastern Plains; this is roughly along the trend of the Modoc fault zone. Rainfall in the Piedmont is variable; greatest precipitation of up to 152 cm (60 in.) occurs along the border with the Blue Ridge, and least precipitation of approximately 101 cm (40 in.) occurs near the Fall Line (Runkle et al., 2022). Rocks are primarily metamorphic and include slate, schist, and gneiss produced from sedimentary rocks when the small eastern Tugaloo and Cat Square landmasses and the large Carolina landmass collided and attached themselves to the North American continent during the Taconic through early Alleghenian mountain-building episodes; intrusive igneous rocks, including diabase and gabbro, also occur (Thigpen et al., 2022; Merschat et al., 2017; Griffith et al., 2002). When Pangea broke up in the Triassic period (ca. 250 million years ago), stretching and sagging along the edges of the continent permitted deposition of sedimentary rocks in river and lake basins; deposition in these places may have continued into the Jurassic period (ca. 190 million years ago). The existing low, rolling hills and reddish, clay-rich soil is the result of deep weathering and erosion of these rocks by rivers over the last 200 million years. Most soils are acidic, although those developed over igneous rocks may be alkaline (Griffith et al., 2002). This combination, along with the transition from higher hills and deeper valleys to lower hills and shallower valleys that co-occurs with a decrease in rainfall from northwest to southeast, has produced a variable landscape of farmland, managed woodlands, and urban areas (SC DNR, 2005b).

Uplands are characterized by oak-hickory (*Quercus-Carya*) forests, some with a significant pine (*Pinus*) presence. River bottoms contain hardwood-dominated woodlands that host American holly (*Ilex*), various oaks (*Quercus*), pines (*Pinus*), sweetgum (*Liquidambar*), tupelo (*Nyssa*), and bald cypress (*Taxodium*). Small streams in the Piedmont tend to have an open to dense understory and sparse to dense herbaceous groundcover, depending upon forest canopy openness. Common trees include American elm (*Ulmus*), birch (*Betula*), green ash (*Fraxinus*), hackberry (*Celtis*), red maple (*Acer*), sweetgum (*Liquidambar*), sycamore (*Platanus*), and tulip poplar (*Liriodendron*). Cove forests are generally shrub-enriched and contain a rich herbaceous flora. Common taxa include American holly (*Ilex*), basswood (*Tilia*), beech (*Fagus*), black cherry (Rosaceae), black gum (*Nyssa*), dogwood (*Cornus*), hop-hornbeam (*Ostrya*), ironwood (*Carpinus*), maples (*Acer*), oaks (*Quercus*), sourwood (*Oxydendrum*), Southern catalpa (*Catalpa*), and witch hazel (*Hamamelis*). Shrubs and understory trees throughout the Piedmont include American plum (Rosaceae), buckeye (*Aesculus*), chokecherry (*Aronia*), dogwood (*Cornus*), pawpaw (*Asminia*), redbud (*Cercis*), serviceberry (*Amalanchier*), willows (*Salix*), and Southern crabapples (Rosaceae). Grasslands and urban areas contain a wide variety of herbaceous plants and grasses. Multiple invasive taxa also grow, including Russian olive (*Eleagnus*), Japanese privet (*Ligustrum*), multiflora rose (*Rosaceae*), kudzu (*Pueraria*), Chinese and Asian wisteria (*Wisteria*), and wart-removing herb (*Murdannia*).

Hood (2006) notes 17 plants as important to beekeepers in the Piedmont. These are: red maple (*Acer*) from mid-February through March; wild plum and peaches (Rosaceae) from early March through mid-April; blackberry (Rosaceae) from April through June; tulip poplar (*Liriodendron*) from mid-April through May; crimson clover (*Trifolium/Melilotus*) from late April through May; vetch (counted as Fabaceae, undiff.) from mid-April through June; locust (*Robinia*) from late May through June; persimmon (*Diospyros*) from late May through mid-June; sumac (*Rhus/Toxicodendron*) in June; sourwood (*Oxydendrum*) from mid-June through early August; mimosa (*Mimosa*) from late June through July; cotton (*Gossypium*) and corn (*Zea mays*) from mid-July through September lespedeza (*Sericea lespedeza*) from mid-August through late September; goldenrod (Asteraceae [coneflower/goldenrod/sunflower family]) from mid-August through October; and white aster (Asteraceae [coneflower/goldenrod/sunflower family]) from mid-September through October. Fifty-three genera important to beekeepers in Region 11, which includes the Piedmont, are reported in the Honey Bee Forage Map (GSFC, 2007; Ayers & Harmon, 1992). The most significant of these for nectar production include blackberry (Rosaceae) and holly (*Ilex*) from March through June, tulip poplar (*Liriodendron*) from April through June, aster (Asteraceae [coneflower/goldenrod/sunflower family]) from June through November, sourwood (*Oxydendrum*) from June through July, and goldenrod (Asteraceae [coneflower/goldenrod/sunflower family]) from July through November. Smith notes 59 genera used by bees in the Piedmont. Notably, sourwood (*Oxydendrum*) has not been documented below 1500 ft elevation.

Hive weights were obtained for five of the six participating apiaries in the Piedmont. Those in the southwestern half of the region (Apiaries 18, 12, and 14) displayed a distinctive increase in weight in early February that lasted until mid- to late March (see Figure 4.3). Those in the northeastern half of the region (Apiaries 15, 16, and 19) did not begin collecting data until mid-March. A second increase in weight was present in all five hives in late April. Precipitous significant decreases in hive weight (vertical lines) likely represent honey harvests, which varied from hive to hive. Likewise, precipitous gains (vertical lines) likely represent the addition of supers, rather than increases due to nectar flow. That said, it is clear that nectar flow continued through mid-July in at least the southwestern half of the region, with another flow occurring late September/early October.

Bees in the Piedmont were using a much wider array of taxa in their honey than was expected. Ninety-three taxa were identified across the six apiaries participating in this study (Table 4.3). This is far more than the 17 plants noted by Hood (2006) and more than the 53 genera important to beekeepers in the entire Region 11 as reported in the Honey Bee Forage Map (GSFC, 2007; Ayers & Harmon, 1992), which includes the Piedmont.

Of the 17 plants noted by Hood (2006) as important to beekeepers in the Piedmont, all but two were encountered in this study. Only cotton (*Gossypium*) and *Sericea lespedeza* (*Lespedeza*) were absent. Pollen from the other 15 plants was observed in the honey during the time period studied. However, in almost all cases, the pollen was found in the honey well before and/or well after the times when it was expected that the bees were foraging for a given taxon. Some notable examples include maple (*Acer*), tulip poplar (*Liriodendron*), and persimmon (*Diospyros*). In this study, maple (*Acer*) pollen was observed

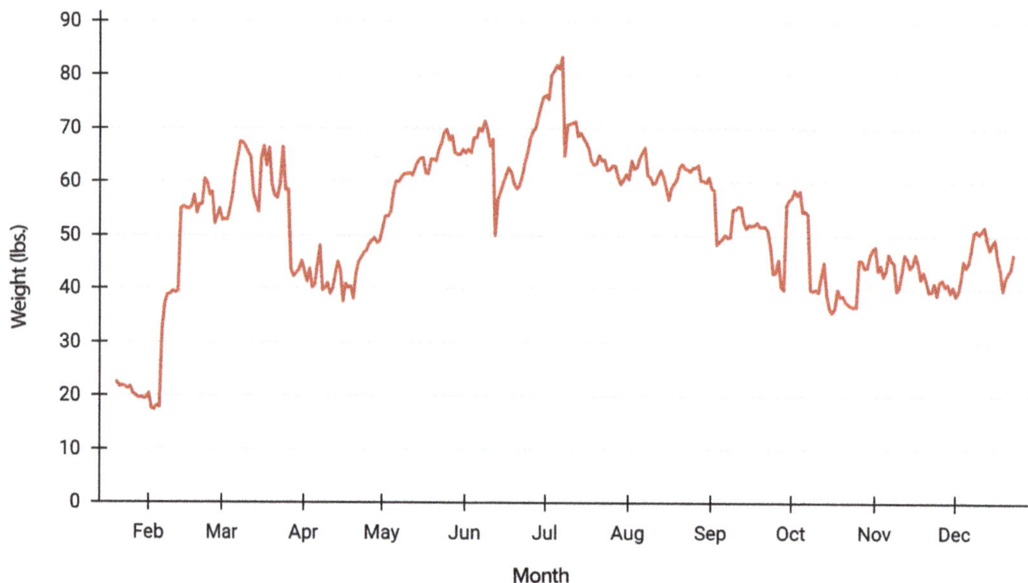

Figure 4.3. An example of BroodMinder hive weight data collected from Apiary 18 in the Piedmont ecoregion during 2022.

Note: Labels indicate the first day of each listed month.

from early February through May, whereas Hood noted red maple use from mid-February through March. Hood specifically mentions red maple (*Acer rubrum*). The discrepancy could be because all maple (*Acer*) pollen in this study was identified to the genus level, not to the species. It is doubtful that red maple was blooming from early February through May. Instead, the bees are most likely using other maple species, an interesting extension of Hood's reported pollen availability times. Tulip poplar (*Liriodendron*) was expected from mid-April through May, while it was observed in this study sporadically from mid-February through early March, then continuously from mid-April through early June and again in July. Thus, while this taxon was used by bees during the period expected, it was also used before and after the period expected. The last example for discussion is persimmon (*Diospyros*). Persimmon (*Diospyros*) was expected from late May through mid-June. In this study, it was observed in the honey almost continuously from the beginning of May through the end of July, then again in September.

There was one taxon observed in this study whose appearance in the honey samples exactly matched Hood's report: sourwood (*Oxydendrum*). Sourwood (*Oxydendrum*) was noted as being used from mid-June through early August, and this was almost exactly the period it was observed in this study. Mimosa (*Mimosa*) was expected from late June through July, but it was only observed in this study once, in mid-June. It is quite possible this shrub simply was not present near any of the apiaries participating in this study.

As noted above, one of the major objectives of this project was to determine whether *Acer* (maple) significantly contributes to honey production in South Carolina. Our observations demonstrate that *Acer* (maple) was indeed widely used by the bees in the Piedmont. All six participating apiaries (Apiaries 12, 14, 15, 16, 18, and 19) produced monofloral *Acer* (maple) honey (*i.e.*, >45% of the pollen was *Acer*), and all these samples occurred in February (Tables 4.4, 4.5, 4.6, 4.7, 4.8, 4.9). Monofloral honeys are indicated with red color-coding in the apiary pollen occurrence tables.

Acer (maple) honey was not the only monofloral honey observed in the Piedmont; our data indicate that monofloral honey was produced throughout the year from a variety of plants. Apiary 12 (Table 4.4) produced monofloral Amaranthaceae (goosefoot family) honey at the end of March, monofloral *Lagerstroemia* (crepe myrtle) honey in mid-April, monofloral *Diospyros* (persimmon) honey is mid-May, and monofloral *Lagerstroemia* (crepe myrtle) honey in early August. Apiary 15 (Table 4.6) produced monofloral Rosaceae (rose/cherry/plum/peach/blackberry family) honey in mid-March. Apiaries 16 (Table 4.7) and 18 (Table 4.8) both produced monofloral Lamiaceae (mint family) honey in mid-February. In addition to monofloral Lamiaceae (mint family) honey, Apiary 18 produced monofloral *Lagerstroemia* (crepe myrtle) honey in September. Furthermore, all six apiaries produced monofloral Rosaceae (rose/cherry/plum/peach/blackberry family) honey from March through May.

One interesting result of this study is that many plants/taxa contribute to the honey for extended periods of time. This means that the plants are blooming long enough to be foraged by bees. If we consider only taxa that contributed to honey for at least three straight weeks, there were 47 in the Piedmont (Table 4.3). This is the largest number of such taxa observed between the five ecoregions. This might not be surprising considering the range of habitats in the Piedmont compared to the other ecoregions. Furthermore, the Piedmont contained the second-most taxa present in the honey samples, with 93 total taxa present (see Chapter 5). Thus, 50.5% of the taxa appear in the honey samples for an extended period of time. These 47 taxa were: *Acer* (maple), *Alnus* (alder), *Alternanthera* (joyweed), Amaranthaceae (goosefoot family), Amaryllidaceae (amaryllis/onion family), Arecaceae (palm family), Asteraceae (coneflower/goldenrod/sunflower family), Asteraceae (dandelion/cat's ear/endive family), Asteraceae (ragweed/calendula/solidago family), *Bidens* (tickseed), Brassicaceae (mustard family), *Camellia*, *Carya* (hickory/pecan), *Castanea* (chinkapin/chestnut), *Centaurea* (knapweed), *Cornus* (dogwood), *Diospyros* (persimmon), Ericaceae (heath family), *Fraxinus* (ash), *Gelsemium* (yellow jessamine), Hypercicaceae (St. John's wort family), *Ilex* (holly), *Lagerstroemia* (crepe myrtle), Lamiaceae (mint family), *Ligustrum* (privet), *Liquidambar* (sweetgum), *Liriodendron* (tulip poplar), *Lonicera* (honeysuckle), *Magnolia*, *Nyssa* (tupelo/black gum), *Oxydendrum*

Table 4.3. Plant Taxa whose Pollen was Observed in Honey Collected from the Piedmont Ecoregion during 2022

Piedmont	Jan		J/F	Feb			F/M	Mar			M/A	Apr			
Week ▶	4	5	6	7	8	9	10	11	12	13	14	15	16	17	18
Abelia															
Acanthus (Bear's breeches)															
Acer (Maple)															
Aesculus (Buckeye)															
Alnus (Alder)															
Alternanthera (Joyweed)															
Amaranthaceae (Goosefoot family)															
Amaryllidaceae (Amaryllis/Onion family)															
Apiaceae (Carrot/Parsely/Queen Anne's Lace family)															
Arecaceae (Palm family)															
Artemisia (Wormwood/Absinthe)															
Asteraceae (Coneflower/Golden Rod/Sunflower family)															
Asteraceae (Dandelion/Cat's Ear/Endive family)															
Asteraceae (Ragweed/Calendula/Solidago family)															
Berberis (Barberry)															
Betula (Birch)															
Bidens (Tickseed)															
Brassicaceae (Mustard family)															
Buddleja (Butterfly Bush)															
Camellia															
Carex															
Carya (Hickory/Pecan)															
Caryophyllaceae (Carnation family)															
Castanea (Chinkapin/Chestnut)															
Ceanothus															
Celtis (Hackberry)															
Centaurea (Knapweed)															
Cercis (Redbud)															
Cnidoscolus/Croton (Bull Nettle/Croton)															
Cornus (Dogwood)															
Corylus/Carpinus (Hazel/Hornbeam)															
Cyperaceae (Sedge family)															
Dalea (Prairie Clover)															
Diospyros (Persimmon)															
Elaeagnus (Autumn/Russian Olive)															
Ericaceae (Heath family)															
Fabaceae (Pea family)															

(sourwood), *Parthenocissus* (Virginia creeper), *Pinus* (pine), *Plantago* (plantain), Poaceae (grass family), *Populus* (aspen/cottonwood/poplar), *Quercus* (oak), *Rhus/Toxicodendron* (sumac/poison ivy), *Robinia* (locust), Rosaceae (rose/cherry/plum/peach/blackberry family), *Salix* (willow), *Trifolium/Melilotus* (red/white clover), *Ulmus* (elm), *Veronica*, *Vicia* (vetch), *Vitis* (grape), and Unknowns #3 and #9. These taxa span the entire foraging season; the appearance of these pollens in the honey for such an extended period of time suggests that they are important forage for bees. Note that these taxa also include plants which do not have nectaries, such as *Alnus* (alder), *Fraxinus* (ash), Poaceae (grass family), *Quercus* (oak), and *Ulmus* (elm). Without nectaries, these taxa cannot contribute nectar (and are typically thought of as wind-pollinated plants); their pollen might have blown into other flowers on which the bees were foraging or could represent pollen from a pollen cell that was inadvertently broken as the honey sample was collected. No matter the precise source of the pollen, the presence of this pollen in the honey samples indicates that the pollen was in the hive and available to the bees for their nutrition.

May				M/J	Jun			J/J	Jul				J/A	Aug			A/S	Sep			S/O	Oct				O/N	Nov
19	20	21	22	23	24	25	26	27	28	29	30	31	32	33	34	35	36	37	38	39	40	41	42	43	44	45	46

(continued)

Table 4.3. (*continued*)

Piedmont	Jan		J/F	Feb			F/M	Mar			M/A	Apr			
Week ▶	4	5	6	7	8	9	10	11	12	13	14	15	16	17	18
Fagus (Beech)				□	□		□		■	□		■	□	■	□
Fraxinus (Ash)					□	■	□	■	□	□	■	■	■	■	□
Gardenia					□	□	□	□		□			□	□	□
Gelsemium (Yellow Jessamine)					□	□	□	□	□	□			□	□	□
Glycine max (Soybean)				□	□		□		□	□		□	□	□	□
Hamamelis (Witch Hazel)					□	□	□	□	□	□			□	□	□
Hexasepalum teres (Poorjoe/Rough Buttonweed)				□	□		□		□			□		□	□
Hypercicaceae (St. John's Wort family)					■	■	■	■	□	■			■	□	□
Ilex (Holly)			□	□	■	■	■	■	■	■	■	■	■	■	■
Juglans (Walnut)			□	□	□		□	□	□	□	□	□	□	□	□
Lagerstroemia (Crepe Myrtle)			■	□	■	□	■	□	■	■	■	□	□	□	□
Lamiaceae (Mint family)			□	□	■	■	■	■	□	□	■	■	■	■	■
Ligustrum (Privet)			□	□	□	□	■	□	□	■	■	■	■	□	□
Liliaceae (Lily family)					□	□	□	□		□			□	□	□
Linaria						□	□	□	■				□		
Liquidambar (Sweetgum)			□	□	□	□	□	■	■	■	■	■	■	■	■
Liriodendron (Tulip Poplar)			□	□	■	■	■	□	■	■	□	□	■	■	■
Lonicera (Honeysuckle)			□	■	■	■	□	□	□	■	□	□	□	□	■
Ludwigia					□	□	□	□		□			□	□	□
Magnolia			□	□	□	□	□	■	□	□	□	□	■	□	■
Malus (Apple/Crabapple)				□	□		□		□			□		■	□
Mimosa (Sensitive Plant)					□	□	□	□	□	□	□	□	□	□	□
Myrtaceae (Myrtle family)			□	□				□	□	□	□	□	□		□
Nerium (Oleander)					□	□	□	□		□	□	□	□	□	□
Nyssa (Tupelo/Black Gum)			□	□	□	□	□	□	□	□	□	□	■	■	□
Oxydendrum (Sourwood)				□	□		□		□		□		□	□	□
Parthenocissus (Virginia Creeper)			□	□	□	■	■	■	■	□	□	□	■	■	□
Phlox			□	□	■	□	■	□	■	□	□	□	□	□	□
Pinus (Pine)			□	□	□	□	□	□	□	■	□	■	■	■	■
Pisum sativum (Pea)							□		□	□		□	□		□
Plantago (Plantain)			■	■	■		■	■	■	□	□	□	□	■	□
Poaceae (Grass family)			□	■	■	■	■	■	■	□	□	□	□	■	■
Polygonum/Persicaria (Buckwheat/Knotweed)			■	□			□	□	□	■	□	□	□		□
Populus (Aspen/Cottonwood/Poplar)			□	□	□		□	□	□	□	■	■	□	□	□
Quercus (Oak)			□	□	□	■	■	■	■	■	□	■	■	■	■
Rhus/Toxicodendron (Sumac/Poison Ivy)			□	□	□	□	□	■	■	□	□	□	□	□	□
Ribes (Currant)					□	□	■	■	□	□	□		■	□	□
Robinia (Locust)			□	□	□	□	■	■	■	□	□	□	■	■	□
Rosaceae (Rose/Cherry/Plum/Peach/Blackberry family)			□	□	□	□	■	■	■	■	■	■	■	■	■
Rubus (Blackberry/Dewberry)					□	□	□	□		■			□	□	□
Salix (Willow)			□	□	□	■	■	□	■	□	□	■	■	■	■
Sarcocapnos (Poppy)			□	□				□	□	■	□	□	□		□
Saururus (Lizard's Tail)					□	□	□	□		□			□	□	□
Tilia (Basswood)				□	□	□	■	□	□	□	□		□	□	□
Triadica (Chinese Tallow)					□	□	□	□		□			□	□	□
Trifolium/Melilotus (Red/White Clover)			□	□	■	■	■	■	■	■	■	■	■	■	■
Ulmus (Elm)			■	□	■	■	■	■	■	■	■	■	□	■	□
Verbena					□	□	□	□		□			□	□	□
Vernonia				□	□		□		□			□		□	□
Veronica			□	□				□	□	□	□	□	□		□
Vicia (Vetch)							□		□	□		□	□		□
Viola (Violet)			■	□	□	□	□	■	□	□	■	□	□	□	□
Vitis (Grape)			□	□	□	□	□	□	□	□	□	□	■	□	□
Zea mays (Corn)				□	□	□	□	□	■	□			□	□	□
Unknown #1					□	□	□	□		□			□	■	□
Unknowns #3 and #9			■	■	■	■	■	□	■	□	■	□	■	□	□
Various Unknown Palynomorphs			■	□	□	■	■	■	■	■	■	■	■	■	■

Key: *no data collected* ☐ *taxon absent on this week* □ *taxon present on this week* ■

May				M/J	Jun			J/J	Jul				J/A	Aug			A/S	Sep			S/O	Oct				O/N	Nov
19	20	21	22	23	24	25	26	27	28	29	30	31	32	33	34	35	36	37	38	39	40	41	42	43	44	45	46

Table 4.4. Pollen Occurrence in Honey Collected in Apiary 12 in the Piedmont Ecoregion during 2022

Apiary 12	Jan		J/F		Feb				F/M	Mar						M/A		Apr						
Week ▶	4	5	6		7		8	9	10	11		12		13		14		15		16		17	18	
Type of Data ▶			%	PC	%	PC				%	PC	%	PC	%	PC	%	PC	%	PC	%	PC		%	PC
Acer (Maple)			77.1	D	100.0	D				95.9	D	10.2	M	11.5	M	17.6	I	1.8	L	0.0			0.0	
Alnus (Alder)			0.0		0.0					0.0		0.0		0.0		0.9	L	0.0		0.0			0.0	
Amaranthaceae (Goosefoot family)			2.6	L	0.0					0.0		27.6	I	62.0	D	1.8	L	0.0		0.0			0.0	
Amaryllidaceae (Amaryllis/Onion family)			0.0		0.0					0.0		0.0		0.0		0.0		0.0		12.9	M		0.0	
Asteraceae (Coneflower/Golden Rod/Sunflower family)			2.9	L	0.0					1.9	L	33.2	I	0.0		0.0		0.0		0.0			0.0	
Asteraceae (Dandelion/Cat's Ear/Endive family)			0.0		0.0					0.0		0.0		0.0		0.0		0.0		1.0	L		0.0	
Asteraceae (Ragweed/Calendula/Solidago family)			0.0		0.0					0.0		0.0		5.8	M	2.3	L	2.7	L	0.5	L		0.0	
Brassicaceae (Mustard family)			1.3	L	0.0					0.0		1.0	L	0.0		0.0		2.7	L	0.5	L		0.0	
Buddleja (Butterfly Bush)			0.0		0.0					0.0		0.0		0.0		0.0		0.0		0.0			0.0	
Camellia			1.3	L	0.0					0.0		0.0		0.0		0.0		0.0		0.0			0.0	
Carya (Hickory/Pecan)			0.0		0.0					0.0		0.0		0.0		0.0		0.0		0.0			0.0	
Caryophyllaceae (Carnation family)			0.0		0.0					0.0		0.0		0.0		0.0		0.0		0.0			0.0	
Cornus (Dogwood)			0.0		0.0					0.0		0.0		0.0		0.9	L	0.0		0.0			16.2	I
Corylus/Carpinus (Hazel/Hornbeam)			0.0		0.0					0.0		0.0		0.0		0.0		0.9	L	0.0			0.0	
Dalea (Prairie Clover)			0.0		0.0					0.0		0.0		0.0		0.0		0.0		0.0			0.0	
Diospyros (Persimmon)			0.0		0.0					0.0		0.0		0.0		0.0		0.0		0.0			0.0	
Elaeagnus (Autumn/Russian Olive)			0.0		0.0					0.0		1.3	L	3.8	M	0.0		0.0		0.0			0.0	
Ilex (Holly)			0.0		0.0					0.0		0.0		0.0		1.8	L	2.7	L	0.0			0.0	
Juglans (Walnut)			0.0		0.0					0.0		0.0		0.0		0.0		0.0		0.0			0.0	
Lagerstroemia (Crepe Myrtle)			1.3	L	0.0					0.0		0.0		2.4	L	1.4	L	0.0		0.0			0.0	
Lamiaceae (Mint family)			0.0		0.0					0.0		0.0		0.0		37.1	I	34.8	I	3.0	M		0.0	
Ligustrum (Privet)			0.0		0.0					0.0		0.0		1.0	L	3.6	M	0.0		78.2	D		0.0	
Liquidambar (Sweetgum)			0.0		0.0					0.0		0.0		0.0		0.0		0.9	L	0.0			1.3	L
Liriodendron (Tulip Poplar)			0.0		0.0					0.0		0.0		0.0		0.0		0.0		0.5			0.0	
Lonicera (Honeysuckle)			0.0		0.0					0.0		0.0		0.0		0.0		0.0		0.0			0.0	
Magnolia			0.0		0.0					0.0		0.0		0.0		0.0		0.0		0.0			0.0	
Myrtaceae (Myrtle family)			0.0		0.0					0.0		0.0		0.0		0.0		0.0		0.0			0.0	
Nyssa (Tupelo/Black Gum)			0.0		0.0					0.0		0.0		0.0		0.0		0.0		0.0			0.0	
Parthenocissus (Virginia Creeper)			0.0		0.0					0.0		0.0		0.0		0.0		0.0		0.0			0.0	
Phlox			0.0		0.0					0.0		1.0	L	0.0		0.0		0.0		0.0			0.0	
Pinus (Pine)			0.0		0.0					0.0		0.0		0.0		0.0		0.0		0.0			0.0	
Plantago (Plantain)			2.6	L	0.0					1.9	L	6.3	M	0.0		0.0		0.0		0.0			0.0	
Poaceae (Grass family)			0.0		0.0					0.0		8.9	M	0.0		0.0		0.0		0.0			0.0	
Polygonum/Persicaria (Buckwheat/Knotweed)			2.3	L	0.0					0.0		0.0		1.4	L	0.0		0.0		0.0			0.0	
Populus (Aspen/Cottonwood/Poplar)			0.0		0.0					0.0		0.0		0.0		4.1	M	2.3	L	0.0			0.0	
Quercus (Oak)			0.0		0.0					0.0		0.0		0.0		0.0		0.0		0.0			1.3	L
Rhus/Toxicodendron (Sumac/Poison Ivy)			0.0		0.0					0.0		0.0		0.0		0.0		0.0		0.0			0.0	
Robinia (Locust)			0.0		0.0					0.0		4.6	M	0.0		0.0		0.0		0.0			0.0	
Rosaceae (Rose/Cherry/Plum/Peach/Blackberry family)			0.0		0.0					0.0		0.0		7.2	M	10.0	M	34.4	I	1.5	L		36.4	I
Salix (Willow)			0.0		0.0					0.0		0.0		0.0		0.0		0.0		0.0			33.3	I
Sarcocapnos (Poppy)			0.0		0.0					0.0		0.0		0.5	L	0.0		0.0		0.0			0.0	
Trifolium/Melilotus (Red/White Clover)			0.0		0.0					0.0		0.0		2.4	L	0.5	L	4.1	M	2.0	L		11.4	M
Ulmus (Elm)			1.6	L	0.0					0.0		3.6	M	1.9	L	5.0	M	12.7	M	0.0			0.0	
Veronica			0.0		0.0					0.0		0.0		0.0		0.0		0.0		0.0			0.0	
Viola (Violet)			1.0	L	0.0					0.3	L	0.0		0.0		1.4	L	0.0		0.0			0.0	
Vitis (Grape)			0.0		0.0					0.0		0.0		0.0		0.0		0.0		0.0			0.0	
Unknowns #3 and #9			1.0	L	0.0					0.0		2.3	L	0.0		11.8	M	0.0		0.0			0.0	
Various Unknown Palynomorphs			4.9	M	0.0					0.0		0.0		0.0		0.0		0.0		0.0			0.0	
Totals ▶			100		100					100		100		100.0		100		100		100.0			100	

Key: *D* (predominant) >45% I (secondary) 16-45% M (important) 3-15% L (minor) <3%

May								M/J	Jun				J/J		Jul						J/A	Aug					A/S	Sep					S/O		Oct						O/N	Nov
19		20		21		22		23	24		25	26	27		28	29		30		31	32	33		34		35	36	37		38	39		40		41	42		43		44	45	46
%	PC	%	PC	%	PC	%	PC		%	PC			%	PC		%	PC	%	PC			%	PC	%	PC			%	PC		%	PC	%	PC		%	PC	%	PC			
0.0		0.0		0.0		0.0			0.0				0.0			0.0		0.0				0.0		1.9	L			0.0			0.0		0.0			0.0		0.0				
0.0		0.0		0.0		0.0			0.0				0.0			0.0		0.0				0.0		0.0				0.0			0.0		0.0			0.0		0.0				
0.0		0.0		0.0		0.0			0.0				3.0	M		0.0		1.3	L			10.2	M	5.3	M			4.5	M		3.7	M	2.4	L		21.3	I	0.0				
0.0		0.0		0.0		0.0			0.0				0.0			0.0		0.0				0.0		0.0				0.0			0.0		0.0			0.0		0.0				
0.0		0.0		0.0		0.0			0.0				2.5	L		0.0		0.0				0.0		0.0				0.0			0.0		3.4	M		0.0		0.0				
0.0		0.0		0.0		0.0			0.0				0.0			0.0		0.0				0.0		0.0				0.4	L		3.2	M	0.0			0.0		0.0				
0.0		0.0		0.0		0.0			0.0				3.5	M		0.9	L	1.3	L			0.0		3.4	M			1.2	L		0.5	L	9.7	M		2.8	L	6.0	M			
0.0		0.0		0.0		0.0			0.0				5.0	M		0.5	L	0.0				0.0		0.0				0.0			0.0		0.0			0.0		0.0				
0.0		0.0		0.0		0.0			0.0				0.0			0.0		0.0				0.0		0.5	L						2.3	L	0.0			0.0		0.0				
0.0		0.0		0.0		0.0			0.0				0.0			0.0		0.0				0.0		0.0				0.0			0.0		0.0			0.0		0.0				
0.0		0.0		0.6	L	0.4	L		0.0				0.0			0.0		0.0				0.0		0.0				0.0			0.0		0.0			0.0		0.0				
0.0		0.9	L	0.0		0.0			0.0				0.0			0.0		0.0				0.0		0.0				0.0			0.0		0.0			0.0		0.0				
0.0		3.7	M	4.9	M	0.0			0.0				0.0			0.0		0.0				0.0		0.0				0.0			0.0		0.0			0.0		16.4	I			
0.9	L	0.0		0.0		0.0			0.0				0.0			0.0		0.0				0.0		0.0				0.0			0.0		0.0			0.0		0.0				
0.0		0.0		0.0		0.0			0.0				0.0			0.0		0.0				0.9	L	0.0				0.0			0.0		0.0			0.0		0.0				
0.0		2.1	L	80.9	D	11.1	M		12.9	M			3.0	M		2.4	L	0.4	L			0.0		0.0				20.6	I		0.5	L	0.5	L		21.7	I	0.0				
0.0		0.0		0.0		0.0			0.0				0.0			0.9	L	0.0				0.0		0.0				0.0			0.0		10.2	M		0.0		0.0				
0.0		0.4	L	0.0		0.0			0.0				0.0			0.0		0.0				0.0		0.5	L			1.6	L		0.5	L	0.0			0.4	L	0.0				
0.0		0.0		0.0		0.0			0.0				8.0	M		0.0		0.0				0.0		0.0				0.0			0.0		0.0			0.0		0.0				
0.0		0.0		0.0		0.0			0.5	L			2.5	L		5.2	M	2.2	L			67.6	D	17.5	I			5.7	M		19.8	I	14.6	M		0.8	L	1.9	L			
0.0		26.6	I	0.0		0.0			0.0				0.0			1.4	L	3.9	M			0.0		0.0				0.0			0.0		0.0			0.0		0.0				
0.0		0.0		0.0		0.0			2.9	L			0.5	L		0.5	L	3.9	M			0.0		0.0				2.4	L		0.0		0.0			0.0		0.0				
0.0		0.0		0.0		0.0			0.0				0.0			0.0		0.0				0.0		0.0				0.0			0.0		0.0			0.0		0.0				
0.0		0.0		11.0	M	9.8	M		0.5	L			0.0			0.5	L	0.0				0.0		0.0				0.0			0.0		0.0			6.4	M	0.0				
0.0		0.0		0.0		0.0			0.0				0.0			0.0		0.0				0.0		0.0				8.5	M		0.0		0.0			0.0		0.0				
0.0		2.8	L	0.0		9.8	M		1.0	L			0.5	L		0.0		3.0	M			0.0		0.5	L			1.2	L		2.3	L	0.5	L		6.4	M	0.0				
0.0		0.0		0.0		0.0			0.0				0.0			0.0		0.0				0.0		0.0				0.0			0.0		0.0			0.0		1.1	L			
0.0		3.9	M	0.0		0.0			0.0				0.0			0.5	L	1.7	L			0.0		0.0				0.4	L		0.0		0.0			0.0		0.0				
0.0		0.0		0.0		0.0			0.0				0.0			1.4	L	0.0				0.0		1.9	L			0.0			2.8	L	0.5	L		0.0		0.0				
0.0		0.0		0.0		0.0			0.0				0.0			0.0		0.0				0.0		0.0				0.0			0.0		1.0	L		0.0		0.0				
0.0		0.0		0.0		0.0			0.0				0.0			0.0		0.0				0.0		1.5	L			0.0			0.0		0.0			0.0		0.0				
0.0		0.0		0.0		0.0			0.0				0.0			0.0		0.0				0.0		0.0				5.7	M		0.0		3.9	M		0.0		0.0				
0.0		0.0		1.9	L	0.0			0.0				0.0			0.9	L	2.2	L			18.7	I	19.9	I			4.0	M		12.0	M	17.0	I		0.0		2.2	L			
0.0		0.0		0.0		0.0			0.0				0.0			0.0		0.0				0.4	L	0.0				0.0			0.0		0.0			0.0		0.0				
0.0		3.9	M	0.0		0.0			0.0				0.0			0.0		0.0				0.0		0.0				0.0			0.0		0.0			0.0		0.4	L			
0.0		8.1		0.0		0.0			42.4	I			2.0	L		0.0		8.3	M			0.0		0.0				0.0			0.0		0.0			0.0		0.0				
0.0		0.0		0.0		2.0	L		0.0				0.0			0.0		0.0				0.0		6.8	M			10.1	M		26.7	I	14.1	M		26.5	I	1.1	L			
0.0		0.0		0.0		0.0			0.0				0.0			0.0		0.0				0.0		0.0				0.0			0.0		0.0			0.0		0.0				
64.7	D	19.8	I	0.0		4.9	M		5.2	M			59.5	D		24.2	I	44.3	I			0.0		3.9	M			10.9	M		3.7	M	2.4	L		0.0		0.7	L			
5.1	M	21.5		0.6	L	1.2	L		0.5	L			1.5	L		45.5	D	0.4	L			0.0		16.5	I			2.8	L		3.2	M	2.4	L		0.4	L	0.0				
0.0		0.0		0.0		0.0			0.0				0.0			0.0		0.0				0.0		0.0				0.0			0.0		0.0			0.0		0.0				
7.2	M	5.9	M	0.0		47.5	D		18.1	I			8.5	M		13.3	M	16.1	I			1.3	L	4.9	M			10.5	M		11.1	M	10.2	M		13.3	M	0.0				
0.0		0.0		0.0		0.0			0.0				0.0			0.0		0.4	L			0.0		10.7	M			2.8	L		0.0		2.9	L		0.0		0.0				
0.0		0.0		0.0		0.0			0.0				0.0			0.0		0.0				0.0		2.9	L			0.8	L		0.5	L	0.0			0.0		0.0				
0.0		0.0		0.0		0.0			0.0				0.0			0.0		0.0				0.0		0.0				0.0			0.0		0.0			0.0		0.0				
0.0		0.0		0.0		0.0			0.0				0.0			1.9	L	0.0				0.9	L	1.5	L			5.7	M		4.1	M	0.0			0.0		0.0				
22.1	I	0.0		0.0		0.0			0.0				0.0			0.0		1.3	L			0.0		0.0				0.0			3.2	M	4.4	M		0.0		0.0				
0.0		0.4	L	0.0		13.1	M		16.2	I			0.0			0.0		9.1	M			0.0		0.0				0.0			0.0		0.0			0.0		70.1	D			
100		100		100		100			100				100			100.0		100				100		100				100			100		100			100		100				

Table 4.5. Pollen Occurrence in Honey Collected in Apiary 14 in the Piedmont Ecoregion during 2022

Apiary 14	Jan		J/F	Feb					F/M		Mar					M/A	Apr						
Week ▶	4	5	6	7	8		9		10		11		12	13		14	15	16		17		18	
Type of Data ▶					%	PC	%	PC	%	PC	%	PC		%	PC			%	PC	%	PC	%	PC
Abelia					0.0		0.0		0.0		0.0			0.0				0.0		0.0		0.0	
Acer (Maple)					51.5	D	47.1	D	47.9	D	16.2	I		52.5	D			5.7	M	0.0		0.0	
Alnus (Alder)					8.2	M	1.6	L	4.2	M	0.7	L		0.0				0.0		0.0		0.0	
Alternanthera (Joyweed)					0.0		0.0		0.0		0.0			0.0				0.0		0.0		0.0	
Amaranthaceae (Goosefoot family)					0.0		0.0		0.0		0.0			1.6	L			0.0		0.0		0.0	
Amaryllidaceae (Amaryllis/Onion family)					0.0		0.0		0.0		0.0			0.0				0.0		0.0		0.0	
Apiaceae (Carrot/Parsely/Queen Anne's Lace family)					0.0		0.0		0.0		0.0			0.0				0.0		0.0		0.0	
Asteraceae (Coneflower/Golden Rod/Sunflower family)					0.0		1.6	L	4.5	M	2.0	L		1.9	L			0.3	L	0.0		0.0	
Asteraceae (Dandelion/Cat's Ear/Endive family)					0.5	L	0.3	L	1.5	L	0.3	L		0.0				1.0	L	0.3	L	0.0	
Asteraceae (Ragweed/Calendula/Solidago family)					0.0		0.0		0.0		0.0			0.0				0.3	L	0.0		0.0	
Bidens (Tickseed)					0.0		0.0		0.0		0.0			0.0				0.0		0.0		0.0	
Brassicaceae (Mustard family)					2.1	L	4.0	M	1.0	L	0.7	L		0.0				1.1	L	0.0		0.0	
Buddleja (Butterfly Bush)					0.0		0.0		0.0		0.0			0.0				0.0		0.0		0.0	
Camellia					2.6	L	0.0		1.5	L	2.4	L		2.8	L			0.8	L	0.0		0.0	
Carya (Hickory/Pecan)					0.0		0.0		0.0		0.0			0.0				0.3	L	0.0		0.0	
Castanea (Chinkapin/Chestnut)					0.0		0.0		0.0		0.0			4.7	M			0.0		0.0		0.0	
Ceanothus					0.0		0.0		0.0		0.0			0.0				0.0		0.0		0.0	
Celtis (Hackberry)					0.0		0.0		0.0		0.0			0.0				0.0		0.0		0.0	
Cercis (Redbud)					0.0		0.0		0.0		0.0			0.0				10.6	M	1.6	L	0.0	
Cnidoscolus/Croton (Bull Nettle/Croton)					0.0		0.0		0.0		0.0			0.0				0.0		0.0		0.0	
Cornus (Dogwood)					0.0		0.0		0.0		0.0			0.0				3.4	M	1.6	L	0.0	
Cyperaceae (Sedge family)					0.0		0.0		0.0		0.0			0.0				0.3	L	0.0		0.0	
Diospyros (Persimmon)					0.0		0.0		0.0		0.0			0.0				0.0		0.0		0.0	
Elaeagnus (Autumn/Russian Olive)					0.0		0.0		0.0		0.0			0.0				0.0		0.0		0.0	
Ericaceae (Heath family)					0.0		0.0		0.0		0.0			3.1	M			2.3	L	0.0		0.0	
Fraxinus (Ash)					0.0		0.0		0.0		0.7	L		0.0				0.3	L	0.0		0.0	
Gardenia					0.0		0.0		0.0		0.0			0.0				0.0		0.0		0.0	
Gelsemium (Yellow Jessamine)					0.0		0.0		0.0		0.0			0.0				0.0		0.0		0.0	
Hamamelis (Witch Hazel)					0.0		0.0		0.0		0.0			0.0				0.0		0.0		0.0	
Hypercicaceae (St. John's Wort family)					0.5	L	1.1	L	1.5	L	1.7	L		0.6	L			0.8	L	0.0		0.0	
Ilex (Holly)					0.2	L	0.3		0.0		0.3	L		1.3	L			0.0		0.0		0.0	
Lagerstroemia (Crepe Myrtle)					0.2		0.0		0.0		0.0			0.0				0.0		0.0		0.0	
Lamiaceae (Mint family)					31.2	I	39.2	I	24.6	I	18.6	I		0.0				5.9	M	29.7	I	6.7	M
Ligustrum (Privet)					0.0		0.0		0.0		0.0			0.0				0.0		0.0		0.0	
Liliaceae (Lily family)					0.0		0.0		0.0		0.0			0.0				0.0		0.0		0.0	
Liquidambar (Sweetgum)					0.0		0.0		0.0		0.0			1.9	L			5.7	M	37.7	I	0.0	
Liriodendron (Tulip Poplar)					0.5	L	0.3	L	0.0		0.0			0.3	L			0.0		0.0		0.4	L
Lonicera (Honeysuckle)					1.4	L	0.3	L	0.0		0.0			0.3	L			0.0		0.0		0.0	
Ludwigia					0.0		0.0		0.0		0.0			0.0				0.0		0.0		0.0	
Magnolia					0.0		0.0		0.0		0.7	L		0.0				0.0		0.0		0.0	
Mimosa (Sensitive Plant)					0.0		0.0		0.0		0.0			0.0				0.0		0.0		0.0	
Nerium (Oleander)					0.0		0.0		0.0		0.0			0.0				0.0		0.0		0.0	
Nyssa (Tupelo/Black Gum)					0.0		0.0		0.0		0.0			0.0				1.3	L	0.0		0.0	
Parthenocissus (Virginia Creeper)					0.0		0.0		0.0		0.3	L		0.0				0.0		1.6	L	0.0	
Phlox					0.2	L	0.0		0.2	L	0.0			0.0				0.0		0.0		0.0	
Pinus (Pine)					0.0		0.0		0.0		0.0			2.2	L			2.8	L	0.3	L	0.0	
Poaceae (Grass family)					0.5	L	0.0		0.0		0.0			0.0				0.0		0.0		0.0	
Quercus (Oak)					0.0		0.3	L	0.0		0.0			0.6	L			2.1	L	0.6	L	0.4	L
Rhus/Toxicodendron (Sumac/Poison Ivy)					0.0		0.0		0.0		0.7	L		0.0				0.0		0.0		0.0	
Ribes (Currant)					0.0		0.0		3.0	M	0.3	L		0.0				0.8	L	0.0		0.0	
Robinia (Locust)					0.0		0.0		4.0	M	2.7	L		0.0				4.7	M	0.3	L	0.0	
Rosaceae (Rose/Cherry/Plum/Peach/Blackberry family)					0.0		0.0		0.2	L	3.4	M		8.1	M			0.0		0.0		52.4	D
Rubus (Blackberry/Dewberry)					0.0		0.0		0.0		0.0			0.3	L			0.0		0.0		0.0	
Salix (Willow)					0.0		0.0		0.0		0.0			0.0				37.7	I	0.0		0.0	
Saururus (Lizard's Tail)					0.0		0.0		0.0		0.0			0.0				0.0		0.0		0.0	
Tilia (Basswood)					0.0		0.0		0.2	L	0.0			0.0				0.0		0.0		0.0	

May								M/J		Jun					J/J		Jul				
19		20		21		22		23		24		25		26	27		28	29		30	31
%	PC	%	PC	%	PC	%	PC	%	PC	%	PC	%	PC		%	PC		%	PC		
0.0		0.0		0.0		0.0		0.0		0.0		0.0			0.0			0.5	L		
0.0		0.0		0.0		0.0		0.0		0.3	L	0.0			0.0			0.0			
0.0		0.0		0.0		0.0		0.0		0.0		0.0			0.0			0.0			
0.0		0.4	L	0.0		4.9	M	0.3	L	0.3	L	0.0			0.0			0.0			
0.0		0.0		0.0		0.0		0.0		0.0		0.0			0.0			0.9	L		
0.0		0.0		0.0		0.0		0.0		0.0		0.0			0.0			0.0			
0.0		0.0		0.0		0.0		0.0		0.0		1.4	L		0.0			0.0			
0.0		0.4	L	0.3	L	0.0		0.3	L	0.3	L	0.0			0.0			0.0			
0.0		0.4	L	0.0		0.0		0.0		3.1	M	0.9	L		0.0			0.0			
0.0		0.0		0.0		0.0		0.0		0.0		0.0			0.0			0.0			
0.0		0.0		0.0		0.0		0.0		0.0		0.0			0.0			0.0			
0.0		0.0		0.0		0.0		0.9	L	0.0		0.0			1.3	L		2.3	L		
0.0		0.0		0.0		0.0		0.0		0.0		0.0			0.0			0.0			
0.0		0.0		0.0		0.0		0.0		0.0		0.0			0.0			0.5	L		
0.0		0.4	L	0.0		0.0		0.0		0.0		0.0			0.0			0.0			
0.5	L	0.0		0.0		0.0		0.0		0.0		0.0			0.0			0.0			
0.0		0.0		0.0		0.0		0.0		0.0		0.0			0.0			0.0			
0.0		0.0		0.0		0.0		0.0		0.0		0.0			0.0			0.0			
0.0		0.0		0.0		0.0		0.0		0.0		0.0			0.0			0.0			
0.0		0.0		0.0		0.0		0.0		0.0		0.0			0.0			0.0			
1.0	L	0.0		0.0		0.0		0.0		0.0		0.0			1.3	L		0.0			
0.0		0.0		0.0		0.0		0.0		0.0		0.0			0.0			0.0			
0.0		0.0		0.0		0.3	L	0.0		0.0		1.7	L		0.0			0.0			
0.0		0.0		0.0		0.0		0.0		0.0		0.0			0.0			0.0			
0.0		0.0		0.0		0.0		0.0		0.0		0.0			0.3	L		0.0			
0.0		0.0		0.0		0.0		0.0		0.0		0.0			0.0			0.0			
0.0		0.0		0.0		0.0		0.0		0.0		0.3	L		0.0			0.0			
0.0		0.0		0.0		0.0		0.0		0.0		0.3	L		2.0	L		0.0			
0.0		0.0		0.0		0.0		0.0		0.0		0.0			0.0			0.0			
0.5	L	0.8	L	0.0		0.0		1.6	L	1.1	L	14.9	M		2.0	L		15.6	I		
0.0		0.4	L	0.3	L	0.0		0.0		0.0		0.0			0.0			0.0			
0.0		0.0		0.0		0.3	L	0.3	L	4.3	M	0.3	L		0.8	L		1.8	L		
4.9	M	1.6	L	1.4	L	1.0	L	0.0		0.0		0.0			0.0			0.5	L		
0.0		0.4	L	1.1	L	0.3	L	0.0		4.3	M	0.6	L		0.5	L		0.5	L		
0.0		0.0		0.0		0.0		0.0		0.0		0.6	L		0.0			0.0			
0.0		0.0		0.0		0.0		0.0		0.0		0.0			0.0			0.0			
0.5	L	0.0		0.3	L	0.0		0.3	L	0.0		0.0			0.3	L		0.0			
0.0		0.0		0.0		0.0		0.0		0.0		0.0			0.0			0.0			
0.0		0.0		0.0		0.0		0.0		0.6	L	0.0			0.0			0.0			
0.0		0.0		0.0		0.0		0.6		1.4	L	1.1	L		0.0			0.5	L		
0.0		0.0		0.0		0.0		0.0		0.0		0.3	L		0.0			0.0			
0.0		0.0		0.0		0.0		0.0		0.6	L	0.0			0.0			0.0			
0.0		0.0		0.3	L	29.7	I	0.9	L	0.0		0.0			0.0			0.9	L		
0.0		0.0		0.8	L	0.0		0.0		0.0		0.0			1.3	L		0.0			
0.0		0.0		0.0		0.0		0.0		0.0		0.0			0.0			0.0			
0.0		0.0		0.0		0.0		1.3	L	0.0		0.0			0.0			0.0			
0.5	L	0.4	L	0.0		0.0		0.0		0.0		0.6	L		1.3	L		0.0			
0.0		0.0		0.8	L	0.3	L	0.0		0.0		0.0			0.0			0.0			
0.0		0.0		0.0		0.0		0.0		0.0		0.0			0.0			0.0			
0.0		0.0		0.3	L	0.0		0.0		0.0		0.0			0.0			0.0			
0.0		0.0		0.0		0.0		0.0		0.0		0.0			0.3	L		0.0			
46.1	D	63.0	D	55.2	D	22.5	I	31.1	I	25.1	I	19.5	I		36.5	I		25.7	I		
0.0		0.0		0.0		0.0		0.0		0.0		0.0			0.0			0.0			
0.0		0.0		0.0		0.0		0.0		0.0		0.0			0.0			0.0			
0.0		0.0		0.0		0.0		0.0		0.9	L	0.0			0.0			0.0			
0.0		0.0		0.0		0.0		0.0		0.0		0.3	L		0.0			0.0			

J/A		Aug				A/S		Sep				S/O	Oct						O/N	Nov
32		33	34		35	36		37	38		39	40	41		42	43	44		45	46
%	PC		%	PC		%	PC		%	PC			%	PC			%	PC		
0.0			0.0			0.0			0.0				0.0				0.0			
0.0			0.0			0.0			0.7	L			0.0				0.0			
0.0			0.0			0.0			0.0				0.0				0.0			
0.0			0.0			0.0			0.0				0.0				0.0			
1.8	L		7.5	M		5.8	M		0.0				5.1	M			4.1	M		
0.3	L		2.2	L		2.3	L		0.0				0.7	L			0.3	L		
0.0			0.0			2.3	L		0.0				0.0				0.0			
0.0			6.5	M		1.2	L		18.3	I			10.2	M			9.8	M		
0.0			0.9	L		1.5	L		8.7	M			1.0	L			3.8	M		
0.0			0.0			0.0			0.0				0.0				0.0			
0.0			0.0			0.0			0.0				1.0	L			0.0			
2.8	L		0.3	L		0.0			0.0				0.7	L			0.3	L		
0.5	L		16.1	I		8.1	M		3.0	M			1.0	L			1.9	L		
0.0			0.0			0.0			0.0				0.0				0.0			
0.0			0.0			0.0			0.0				0.3	L			0.0			
0.0			0.0			0.0			0.0				0.0				0.0			
0.5	L		6.8	M		0.0			0.0				0.0				0.0			
0.0			0.0			0.0			0.0				3.4	M			1.6	L		
0.0			0.0			0.0			0.0				0.0				0.0			
0.0			0.0			0.0			1.3	L			0.3	L			0.0			
0.0			0.0			0.0			0.0				0.3	L			0.0			
0.0			0.0			0.0			0.0				0.0				0.0			
0.8	L		0.0			0.4	L		0.3	L			0.0				0.0			
0.0			1.2	L		0.0			0.0				0.0				0.0			
0.0			0.0			0.0			0.0				0.0				0.0			
0.0			0.0			0.0			0.0				0.0				0.0			
0.0			0.0			0.0			0.0				0.0				0.0			
3.6	M		0.3	L		0.0			0.0				0.0				0.0			
0.0			0.0			0.0			0.0				0.0				0.6	L		
3.9	M		0.6	L		0.4	L		0.0				0.0				0.0			
0.0			0.0			0.0			0.0				0.3	L			0.0			
7.0	M		3.1	M		3.9	M		2.7	L			5.5	M			3.2	M		
1.0	L		3.1	M		1.2	L		0.3	L			0.0				0.0			
0.3	L		0.0			0.0			0.0				0.0				0.0			
0.0			0.0			0.0			0.0				0.0				0.0			
0.0			0.0			0.0			0.0				0.0				0.0			
0.3	L		0.0			0.0			0.0				0.0				0.0			
0.0			0.0			0.0			0.0				0.0				0.0			
0.0			0.0			0.0			0.0				0.0				0.0			
0.0			0.0			0.0			0.0				0.0				0.0			
0.0			0.0			0.0			0.0				0.0				0.0			
0.0			0.0			0.0			0.0				0.0				0.0			
0.0			0.0			0.0			0.0				1.4	L			0.0			
0.0			0.0			0.0			0.0				0.3	L			0.0			
0.0			0.0			0.0			0.0				0.0				0.0			
0.0			0.0			0.0			0.0				0.0				0.0			
0.5	L		3.4	M		3.5	M		5.0	M			11.6	M			13.9	M		
0.0			0.0			0.0			0.0				0.0				0.0			
1.8	L		1.2	L		4.6	M		4.7	M			8.2	M			10.1	M		
0.0			0.0			0.0			0.0				0.0				0.0			
0.3	L		0.0			0.0			0.0				0.0				0.0			
16.0	I		3.4	M		13.5	M		15.0	M			9.9	M			6.6	M		
0.0			0.0			0.0			0.0				0.0				0.0			
0.0			0.0			0.0			0.0				0.0				0.0			
0.0			0.0			0.0			0.0				0.0				0.0			
0.0			0.0			0.0			0.0				0.0				0.0			

(continued)

Table 4.5. (*continued*)

Apiary 14	Jan		J/F	Feb						F/M		Mar					M/A	Apr					
Week ▶	4	5	6	7	8		9		10		11		12	13		14	15	16		17		18	
Type of Data ▶					%	PC	%	PC	%	PC	%	PC		%	PC			%	PC	%	PC	%	PC
Triadica (Chinese Tallow)					0.0		0.0		0.0		0.0			0.0				0.0		0.0		0.0	
Trifolium/Melilotus (Red/White Clover)					0.0		0.3	L	0.5	L	2.0	L		11.3	M			2.3	L	24.4	I	40.1	I
Ulmus (Elm)					0.5	L	3.7	M	5.2	M	0.7	L		0.3	L			0.0		0.0		0.0	
Verbena					0.0		0.0		0.0		0.0			0.0				0.0		0.0		0.0	
Viola (violet)					0.0		0.0		0.0		0.0			0.0				0.0		0.0		0.0	
Vitis (Grape)					0.0		0.0		0.0		0.0			0.0				0.0		0.0		0.0	
Zea mays (Corn)					0.0		0.0		0.0		0.0			0.0				0.0		0.0		0.0	
Unknown #1					0.0		0.0		0.0		0.0			0.0				0.0		1.9	L	0.0	
Unknowns #3 and #9					0.0		0.3	L	0.0		0.0			0.0				0.5	L	0.0		0.0	
Various Unknown Palynomorphs					0.0		0.0		0.0		45.6	D		6.3	M			9.3	M	0.0		0.0	
Totals ▶					100		100		100		100			100				100		100		100	

Key: *D* (predominant) >45% I (secondary) 16-45% M (important) 3-15% L (minor) <3%

Table 4.6. Pollen Occurrence in Honey Collected in Apiary 15 in the Piedmont Ecoregion during 2022

Apiary 15	Jan		J/F	Feb			F/M	Mar					M/A		Apr						
Week ▶	4	5	6	7	8	9	10	11	12		13		14		15		16		17		18
Type of Data ▶									%	PC	%	PC	%	PC	%	PC	%	PC	%	PC	
Acer (Maple)									18.5	I			5.9	M	0.9	L			5.2	M	
Arecaceae (Palm family)									0.0				0.0		6.9	M			0.9	L	
Asteraceae (Dandelion/Cat's Ear/Endive family)									3.4	M			0.0		2.8	L			0.5	L	
Cornus (Dogwood)									0.0				0.0		28.6	I			20.9	I	
Fraxinus (Ash)									0.0				2.0	L	1.4	L			1.9	L	
Ilex (Holly)									6.8	M			0.0		2.3	L			0.0		
Liquidambar (Sweetgum)									0.0				1.5	L	2.3	L			1.9	L	
Pinus (Pine)									0.0				0.0		0.0				0.9	L	
Quercus (Oak)									0.0				0.0		8.8	M			0.0		
Rosaceae (Rose/Cherry/Plum/Peach/Blackberry family)									71.2	D			17.7	I	17.5	I			51.2	D	
Trifolium/Melilotus (Red/White Clover)									0.0				3.0	M	0.0				0.0		
Various Unknown Palynomorphs									0.0				70.0	D	28.6	I			16.6	I	
Totals ▶									100				100		100				100		

Key: *D* (predominant) >45% I (secondary) 16-45% M (important) 3-15% L (minor) <3%

May				M/J	Jun			J/J	Jul				J/A	Aug			A/S	Sep			S/O	Oct				O/N	Nov
19	20	21	22	23	24	25	26	27	28	29	30	31	32	33	34	35	36	37	38	39	40	41	42	43	44	45	46
% PC	% PC	% PC	% PC	% PC	% PC	% PC		% PC		% PC			% PC		% PC		% PC		% PC			% PC			% PC		
0.0	0.0	0.0	0.0	0.0	0.0	0.9 L		0.0		0.0			0.0		0.0		0.0		0.0			0.0			0.0		
37.7 I	25.1 I	38.2 I	37.3 I	54.1 D	56.3 D	40.4 I		40.1 I		41.3 I			28.4 I		12.7 M		17.4 I		19.0 I			4.4 M			0.9 L		
0.0	1.2 L	0.0	0.0	0.0	0.3 L	0.0		0.0		0.0			0.0		0.0		0.0		0.0			0.0			0.0		
0.0	0.0	0.0	0.0	0.0	0.0	0.0		0.0		0.5 L			0.0		6.2 M		9.7 M		7.7 M			8.9 M			13.3 M		
0.0	0.0	0.0	0.0	0.0	0.0	0.0		0.0		0.0			3.4 M		0.6 L		1.2 L		0.7 L			2.0 L			2.5 L		
8.3 M	5.3 M	0.0	2.0 L	4.1 M	0.3 L	2.9 L		7.9 M		0.5 L			0.0		0.0		0.0		0.7 L			0.0			0.0		
0.0	0.0	0.0	0.0	0.0	0.9 L	1.1 L		0.0		0.0			1.0 L		0.9 L		0.8 L		0.0			0.0			0.0		
0.0	0.0	0.0	0.0	0.0	0.0	0.0		0.0		0.0			0.0		0.0		0.0		0.0			0.0			0.0		
0.0	0.0	0.0	0.0	0.0	0.0	0.0		0.0		0.5 L			3.4 M		0.3 L		0.0		0.3 L			0.0			0.0		
0.0	0.0	0.8 L	1.3 L	4.1 L	0.0	12.0 M		4.3 M		7.8 M			22.5 I		22.4 I		22.4 I		11.7 M			23.2 I			26.9 I		
100	100	100	100	100.0	100	100		100		100			100		100		100		100			100			100		

May				M/J	Jun			J/J	Jul				J/A	Aug			A/S	Sep			S/O	Oct				O/N	Nov
19	20	21	22	23	24	25	26	27	28	29	30	31	32	33	34	35	36	37	38	39	40	41	42	43	44	45	46
% PC		% PC																									
6.7 M		47.4 D																									
2.9 L		0.0																									
0.0		1.4 L																									
1.4 L		0.5 L																									
1.4 L		0.0																									
0.0		0.0																									
0.0		0.0																									
0.0		0.0																									
0.0		0.0																									
80.3 D		47.9 D																									
0.0		0.0																									
7.2 M		2.8 L																									
100		100																									

Table 4.7. Pollen Occurrence in Honey Collected in Apiary 16 in the Piedmont Ecoregion during 2022

Apiary 16	Jan		J/F	Feb			F/M		Mar				M/A	Apr						
Week ▶	4	5	6	7	8	9	10		11	12		13	14	15		16		17		18
Type of Data ▶							%	PC		%	PC			%	PC	%	PC	%	PC	
Acer (Maple)							66.2	D		59.8	D			0.9	L			0.0		
Amaranthaceae (Goosefoot family)							0.0			0.0				0.0				0.0		
Asteraceae (Coneflower/Golden Rod/Sunflower family)							16.9	I		10.5	M			0.0				0.0		
Asteraceae (Dandelion/Cat's Ear/Endive family)							0.5	L		1.0	L			0.0				0.0		
Cornus (Dogwood)							0.0			0.0				17.8	I			46.9	D	
Diospyros (Persimmon)							0.0			0.0				0.0				0.0		
Fagus (Beech)							0.0			2.9	L			3.2	M			0.0		
Fraxinus (Ash)							0.0			0.0				0.0				8.1	M	
Ilex (Holly)							0.0			0.0				0.0				0.5	L	
Lagerstroemia (Crepe Myrtle)							4.1	M		4.3	M			0.0				0.0		
Ligustrum (Privet)							0.0			0.0				0.0				0.0		
Liquidambar (Sweetgum)							0.0			0.0				10.5	M			2.4	L	
Liriodendron (Tulip Poplar)							0.0			0.0				0.0				5.3	M	
Lonicera (Honeysuckle)							0.0			0.0				0.0				0.0		
Magnolia							0.0			0.0				0.0				0.0		
Nyssa (Tupelo/Black Gum)							0.0			0.0				0.0				2.4	L	
Pinus (Pine)							0.0			0.0				0.5	L			1.9	L	
Quercus (Oak)							0.0			0.0				0.0				19.6	I	
Rhus/Toxicodendron (Sumac/Poison Ivy)							0.0			1.4	L			0.0				0.0		
Rosaceae (Rose/Cherry/Plum/Peach/Blackberry family)							0.9	L		12.9	M			5.5	M			1.4	L	
Salix (Willow)							0.0			0.0				0.5	L			8.1	M	
Trifolium/Melilotus (Red/White Clover)							0.0			1.0	L			0.0				0.0		
Ulmus (Elm)							5.0	M		4.3	M			0.0				1.4	L	
Various Unknown Palynomorphs							6.4	M		1.9	L			61.2	D			1.9	L	
Totals ▶							100			100				100				100		

Key: *D* (predominant) >45% I (secondary) 16-45% M (important) 3-15% L (minor) <3%

May						M/J		Jun					J/J	Jul				J/A	Aug			A/S	Sep			S/O	Oct				O/N	Nov
19		20	21		22	23		24		25		26	27	28	29	30	31	32	33	34	35	36	37	38	39	40	41	42	43	44	45	46
%	PC		%	PC		%	PC	%	PC	%	PC																					
0.0			1.0	L		0.0				0.0																						
0.0			0.0			1.3	L			0.0																						
0.0			0.0			0.0				0.0																						
0.0			0.0			0.0				0.0																						
0.0			1.9	L		0.0				0.0																						
0.0			1.0	L		1.8	L			2.0	L																					
0.0			0.0			0.0				0.0																						
0.9	L		23.2	I		2.2	L			1.5	L																					
0.0			0.0			0.0				0.0																						
0.0			0.0			6.2	M			16.3	I																					
2.2	L		2.4	L		4.0	M			12.3	M																					
0.9	L		1.0	L		0.0				1.5	L																					
4.8	M		0.5	L		1.8	L			0.5	L																					
0.0			0.0			0.0				0.5	L																					
0.0			0.0			0.0				3.0	M																					
0.0			4.3	M		35.1	I			13.3	M																					
0.0			0.0			0.0				0.0																						
0.0			0.0			0.4	L			0.0																						
1.8	L		3.9	M		0.4	L			1.0	L																					
84.2	D		58.5	D		42.7	I			22.2	I																					
0.0			0.5	L		0.0				0.0																						
5.3	M		1.9	L		4.0	M			26.1	I																					
0.0			0.0			0.0				0.0																						
0.0			0.0			0.0				0.0																						
100			100			100				100																						

Table 4.8. Pollen Occurrence in Honey Collected in Apiary 18 in the Piedmont Ecoregion during 2022

Apiary 18	Jan		J/F	Feb					F/M		Mar					M/A	Apr					
Week ▶	4	5	6	7		8		9	10		11		12		13	14	15	16	17		18	
Type of Data ▶				%	PC	%	PC		%	PC	%	PC	%	PC					%	PC	%	PC
Acanthus (Bear's breeches)				0.0		0.0			0.0				0.0						0.0		0.0	
Acer (Maple)				77.6	D	7.4	M		56.8	D			17.2	I					4.8	M	0.0	
Amaranthaceae (Goosefoot family)				1.9	L	0.0			0.0				0.6	L					1.0	L	0.0	
Asteraceae (Coneflower/Golden Rod/Sunflower family)				10.4	M	0.0			0.0				0.0						0.0		0.0	
Asteraceae (Dandelion/Cat's Ear/Endive family)				1.0	L	0.0			1.4	L			1.5	L					0.0		1.3	L
Asteraceae (Ragweed/Calendula/Solidago family)				3.2	L	0.0			0.0				0.0						1.4	L	0.0	
Betula (Birch)				0.0		0.0			0.0				0.0						0.5	L	0.0	
Brassicaceae (Mustard family)				0.0		0.0			0.0				0.0						0.0		0.0	
Carya (Hickory/Pecan)				0.0		0.0			0.0				0.0						0.0		0.0	
Centaurea (Knapweed)				0.0		0.0			0.0				0.0						0.0		0.0	
Cornus (Dogwood)				0.0		0.0			0.0				0.0						19.6	I	67.2	D
Diospyros (Persimmon)				0.0		0.0			0.0				0.0						0.0		0.0	
Ericaceae (Heath family)				0.0		0.0			0.0				0.0						0.0		0.0	
Fagus (Beech)				0.0		0.0			0.0				0.0						3.8	M	0.0	
Glycine max (Soybean)				0.0		0.0			0.0				0.0						0.0		0.0	
Hexasepalum teres (Poorjoe/Rough Buttonweed)				0.0		0.0			0.0				0.0						0.0		0.0	
Ilex (Holly)				0.0		0.0			0.0				0.0						0.0		0.4	L
Juglans (Walnut)				0.0		0.0			0.0				0.0						0.0		0.0	
Lagerstroemia (Crepe Myrtle)				0.0		0.0			0.0				0.3	L					0.0		0.0	
Lamiaceae (Mint family)				0.0		89.7	D		41.3	I			0.0						0.0		0.0	
Ligustrum (Privet)				0.0		0.0			0.0				0.0						0.0		0.0	
Liquidambar (Sweetgum)				0.0		0.0			0.0				0.0						6.7	M	3.5	M
Liriodendron (Tulip Poplar)				0.0		0.0			0.0				0.6	L					2.4	L	7.0	M
Lonicera (Honeysuckle)				1.0	L	0.3	L		0.0				0.0						0.0		0.0	
Magnolia				0.0		0.0			0.0				0.0						0.0		0.9	L
Malus (Apple/Crabapple)				0.0		0.0			0.0				0.0						1.9	L	0.0	
Nyssa (Tupelo/Black Gum)				0.0		0.0			0.0				0.0						0.0		0.0	
Oxydendrum (Sourwood)				0.0		0.0			0.0				0.0						0.0		0.0	
Parthenocissus (Virginia Creeper)				0.0		0.0			0.0				5.0	M					1.4	L	0.0	
Pinus (Pine)				0.0		0.0			0.0				0.0						0.0		0.0	
Plantago (Plantain)				2.9	L	0.3	L		0.0				6.1	M					0.5	L	0.0	
Poaceae (Grass family)				0.6	L	0.0			0.0				0.6	L					1.4	L	1.7	L
Populus (Aspen/Cottonwood/Poplar)				0.0		0.0			0.0				0.0						0.0		0.0	
Quercus (Oak)				0.0		0.0			0.0				0.0						1.9	L	1.7	L
Rhus/Toxicodendron (Sumac/Poison Ivy)				0.0		0.0			0.0				0.0						0.0		0.0	
Robinia (Locust)				0.0		0.0			0.0				0.0						0.0		0.0	
Rosaceae (Rose/Cherry/Plum/Peach/Blackberry family)				0.0		0.0			0.0				56.0	D					49.8	D	11.4	M
Salix (Willow)				0.0		0.0			0.0				0.0						1.0	L	0.0	
Tilia (Basswood)				0.0		0.0			0.0				0.0						0.0		0.0	
Trifolium/Melilotus (Red/White Clover)				0.0		0.3	L		0.0				10.5	M					1.9	L	4.8	M
Ulmus (Elm)				0.0		0.0			0.0				0.0						0.0		0.0	
Vernonia				0.0		0.0			0.0				0.0						0.0		0.0	
Vitis (Grape)				0.0		0.0			0.0				0.0						0.0		0.0	
Zea mays (Corn)				0.0		0.0			0.0				1.7	L					0.0		0.0	
Unknowns #3 and #9				1.3	L	1.9	L		0.5	L			0.0						0.0		0.0	
Totals ▶				100		100			100				100						100		100	

Key: *D* (predominant) >45% I (secondary) 16-45% M (important) 3-15% L (minor) <3%

May								M/J	Jun						J/J		Jul								J/A		Aug					A/S		Sep					S/O	Oct					O/N	Nov
19		20		21		22		23	24		25		26		27		28		29		30		31		32		33		34		35	36		37		38	39		40	41		42	43	44	45	46
%	PC	%	PC	%	PC	%	PC		%	PC	%	PC	%	PC	%	PC	%	PC	%	PC	%	PC	%	PC	%	PC	%	PC	%	PC		%	PC	%	PC		%	PC		%	PC					
0.9	L	0.0		0.0		0.0			0.0		0.0		0.0		0.0		0.0		0.0		0.0		0.0		0.0				0.0			0.0		0.0			0.0			0.0						
0.5	L	1.8	L	0.4	L	0.0			0.0		0.0		0.0		0.0		0.0		0.0		0.0		0.0		0.0				0.0			0.0		0.0			0.0			0.0						
0.0		0.0		0.0		0.0			0.0		0.0		0.0		0.0		0.0		0.0		0.0		0.0		0.0				32.6	I		34.3	I	0.0			0.0			1.5	L					
0.0		0.0		0.0		0.0			0.0		0.0		0.0		0.0		0.0		0.0		0.0		0.0		0.0				3.3	M		0.0		0.0			3.4	M		3.9	M					
0.0		0.9	L	3.7	M	0.0			3.9	M	4.3	M	6.6	M	1.9	L	2.9	L	3.5	M	4.3	M	8.6	M	2.9	L			0.0			3.9	M	0.9	L		6.5	M		2.5	L					
0.0		0.0		0.4	L	1.5	L		0.0		1.4	L	0.0		0.0		0.0		0.0		0.0		0.0		0.0				23.7	I		3.4	M	0.0			0.0			0.0						
0.0		0.0		0.0		0.0			0.0		0.0		0.0		0.0		0.0		0.0		0.0		0.0		0.0				0.0			0.0		0.0			0.0			0.0						
1.4	L	1.4	L	0.0		1.0	L		0.0		0.0		3.3	M	0.0		0.0		0.0		1.7	L	0.0		1.0	L			0.0			0.0		0.0			0.0			0.0						
0.0		0.0		1.2	L	0.0			0.0		0.0		0.0		0.0		0.0		0.0		0.0		0.0		0.0				0.0			0.0		0.0			0.0			0.0						
0.0		0.0		0.0		2.9	L		8.2	M	0.5	L	0.9	L	0.0		0.0		0.0		0.0		0.0		0.0				0.0			0.0		0.0			0.0			0.0						
15.1	M	3.6	M	4.9	M	2.4	L		5.8	M	0.0		0.9	L	0.0		1.3	L	0.0		0.0		0.5	L	0.0				0.0			0.0		0.0			0.0			0.0						
0.0		0.0		0.0		0.5	L		3.4	M	0.0		0.0		0.0		0.0		0.4	L	0.0		2.9	L	0.0				0.0			0.0		0.0			0.0			0.0						
0.0		0.0		0.0		0.0			0.0		0.0		0.0		0.0		0.0		0.0		0.0		0.0		0.0				0.0			0.0		0.0			4.3	M		21.2	I					
0.0		0.0		0.0		0.0			0.0		0.0		0.0		0.0		0.0		0.0		0.0		0.0		0.0				0.0			0.0		0.0			0.0			0.0						
0.0		0.0		0.0		0.0			0.0		0.0		0.0		0.0		0.0		0.0		0.0		0.0		0.0				0.0			0.5	L	0.0			0.4	L		0.0						
0.0		0.0		0.0		0.0			0.0		0.0		0.0		2.9	L	0.0		0.0		0.0		0.0		0.0				0.7	L		0.0		0.0			0.0			0.0						
2.8	L	9.5	M	3.7	M	0.0			1.0	L	1.4	L	1.9	L	0.0		2.9	L	2.2	L	0.9	L	2.4	L	0.0				0.0			0.5	L	1.3	L		0.0			3.0	M					
0.0		0.0		0.0		0.0			0.0		0.0		0.0		0.0		0.0		0.0		0.0		0.0		0.5	L			0.0			0.0		0.0			0.0			0.0						
0.0		1.4	L	2.5	L	1.5	L		1.0	L	6.8	M	2.8	L	5.8	M	2.9	L	28.9	I	34.8	I	7.2	M	6.7	M			7.0	M		1.4	L	0.0			77.2	D		42.4	I					
0.0		0.0		0.0		0.0			0.0		0.0		0.0		0.0		0.0		0.0		0.0		0.0		0.0				0.0			0.0		0.0			0.0			0.0						
0.0		0.9	L	17.3	I	15.5	M		13.0	M	0.0		10.4	M	0.0		6.7	M	12.3	M	10.0	M	10.5	M	10.5	M			0.0			2.4	L	0.0			0.4	L		2.0	L					
3.7	M	0.5	L	0.0		3.9	M		1.0	L	1.0	L	0.0		0.0		0.0		0.0		0.0		0.0		0.0				0.0			0.0		0.0			0.0			0.0						
4.1	M	0.0		3.7	M	0.0			1.0	L	0.5	L	1.9	L	0.0		0.0		0.0		0.4	L	0.0		0.0				0.0			0.0		0.0			0.0			0.0						
0.0		0.0		0.0		0.0			0.0		0.0		0.0		0.0		0.0		0.0		0.0		0.0		0.0				0.0			2.9	L	0.0			0.0			0.0						
13.8	M	7.2	M	2.5	L	0.0			0.5	L	0.5	L	0.9	L	0.0		0.0		0.0		0.9	L	1.9	L	0.0				0.0			0.0		0.0			0.0			0.0						
0.0		0.0		0.0		0.0			0.0		0.0		0.0		0.0		0.0		0.0		0.0		0.0		0.0				0.0			0.0		0.0			0.0			0.0						
0.0		0.0		0.0		0.0			0.0		0.0		0.0		0.0		0.4	L	0.0		0.0		0.5	L	1.9	L			0.0			0.0		0.0			0.0			0.0						
0.0		0.0		0.0		0.0			0.0		0.0		2.4	L	7.8	M	20.4	I	9.6	M	10.4	M	2.4	L	5.7	M			0.4	L		9.2	M	0.0			0.0			0.0						
0.0		0.5	L	0.0		0.0			0.0		0.0		0.0		0.0		0.0		0.0		3.5	M	2.4	L	3.3	M			0.0			2.9	L	2.2	L		0.0			3.4	M					
0.0		0.5	L	0.0		0.0			0.0		0.5	L	0.0		0.0		0.0		0.0		0.0		0.0		0.0				0.0			0.0		0.0			0.0			0.0						
0.0		0.0		0.0		0.0			9.2	M	14.0	M	22.3	I	22.8	I	26.7	I	6.1	M	6.1	M	14.4	M	16.7	I			1.5	L		3.4	M	16.1	I		3.9	M		14.3	M					
0.0		0.0		0.0		1.0	L		0.0		0.0		0.0		32.0	I	6.7	M	0.0		1.3	L	1.0	L	0.0				27.0	I		19.8	I	0.0			0.0			0.0						
0.0		0.0		0.0		0.0			4.3	M	3.4	M	0.9	L	0.0		0.0		0.0		0.0		0.0		0.0				0.0			0.0		0.0			0.0			0.0						
0.5	L	2.3	L	0.0		0.0			0.5	L	6.8	M	0.9	L	1.0	L	0.4	L	0.0		0.0		0.0		0.0				0.0			0.0		0.0			0.0			0.0						
0.9	L	1.4	L	4.5	M	3.9	M		0.5	L	1.0	L	0.0		0.0		0.0		3.1	M	0.4	L	0.0		0.0				0.0			0.0		0.4	L		0.0			0.0						
17.4	I	30.8	I	16.9	I	46.6	D		14.0	M	21.3	I	10.9	M	0.0		0.0		2.2	L	3.0	M	0.0		0.0				0.0			0.0		0.0			0.0			0.0						
15.6	I	18.6	I	0.0		7.8	M		7.7	M	2.4	L	4.7	M	0.0		1.3	L	4.8	M	3.5	M	14.8	M	16.7	I			0.0			2.9	L	2.2	L		0.0			0.0						
0.0		0.0		0.0		0.0			0.0		0.0		0.0		0.0		0.0		0.9	L	0.0		1.0	L	1.0	L			0.0			0.0		0.4	L		0.0			0.0						
0.0		0.0		0.0		0.0			0.0		0.0		0.0		0.0		0.0		0.0		0.4	L	0.0		0.0				0.0			0.0		0.0			0.0			0.0						
23.4	I	17.6	I	33.3	I	9.7	M		20.3	I	6.8	M	16.6	I	1.9	L	14.6	M	19.3	I	14.3	M	21.5	I	19.1	I			0.0			6.8	M	28.6	I		1.7	L		1.5	L					
0.0		0.0		0.0		0.0			0.0		0.0		0.5	L	0.0		0.0		0.0		0.0		0.0		0.0				0.0			0.0		0.0			0.0			0.0						
0.0		0.0		0.0		0.0			0.0		0.0		0.0		0.0		0.0		0.0		0.0		0.0		0.0				3.7	M		1.4	L	0.0			0.0			0.0						
0.0		0.0		0.0		0.0			2.9	L	3.9	M	7.1	M	0.0		0.0		1.3	L	1.3	L	0.0		1.9	L			0.0			0.0		16.5	I		0.4	L		3.0	M					
0.0		0.0		0.0		0.0			0.0		0.0		0.0		22.8	I	7.9	M	0.0		0.9	L	0.0		0.0				0.0			2.4	L	10.3	M		1.7	L		1.5	L					
0.0		1.4	L	4.9	M	1.9	L		1.9	L	23.7	I	3.8	M	1.0	L	5.0	M	5.3	M	1.7	L	8.1	M	12.0	M			0.0			1.9	L	21.0	I		0.0			0.0						
100		100		100		100			100		100		100		100		100		100		100		100		100				100			100		100			100			100						

Table 4.9. Pollen Occurrence in Honey Collected in Apiary 19 in the Piedmont Ecoregion during 2022

Apiary 19	Jan		J/F	Feb			F/M		Mar					M/A		Apr						
Week ▶	4	5	6	7	8	9	10		11	12		13		14		15		16		17	18	
Type of Data ▶							%	PC		%	PC	%	PC	%	PC	%	PC	%	PC		%	PC
Acer (Maple)							98.2	D		42.2	I	2.7	L			27.3	I	6.6	M		3.8	M
Asteraceae (Coneflower/Golden Rod/Sunflower family)							0.0			0.0		0.0				0.0		0.0			0.0	
Asteraceae (Dandelion/Cat's Ear/Endive family)							0.0			0.0		0.0				7.8	M	0.0			1.4	L
Asteraceae (Ragweed/Calendula/Solidago family)							0.5	L		0.0		0.0				0.0		0.0			0.0	
Betula (Birch)							0.0			0.0		0.0				0.0		0.0			0.0	
Brassicaceae (Mustard family)							0.0			3.4	M	0.0				1.5	L	0.0			0.0	
Carya (Hickory/Pecan)							0.0			0.0		0.0				0.0		0.0			0.0	
Caryophyllaceae (Carnation family)							0.0			0.0		0.0				0.0		0.0			0.0	
Centaurea (Knapweed)							0.0			0.0		0.0				0.0		0.0			0.0	
Cercis (Redbud)							0.0			0.0		0.0				0.0		0.9	L		0.0	
Cornus (Dogwood)							0.0			0.0		0.0				0.0		0.0			2.4	L
Diospyros (Persimmon)							0.0			0.0		0.0				0.0		0.0			0.0	
Fagus (Beech)							0.0			0.0		0.0				0.0		0.0			0.0	
Glycine max (Soybean)							0.0			0.0		0.0				0.0		0.0			0.0	
Ilex (Holly)							0.0			1.9	L	0.0				0.0		0.0			0.0	
Lagerstroemia (Crepe Myrtle)							0.5	L		0.0		1.3	L			0.0		0.0			0.0	
Ligustrum (Privet)							0.5	L		0.0		0.0				0.5	L	0.0			0.0	
Liriodendron (Tulip Poplar)							0.0			0.0		0.0				0.0		0.0			0.0	
Lonicera (Honeysuckle)							0.0			0.0		0.0				0.0		0.0			2.9	L
Magnolia							0.0			0.0		0.0				0.0		0.0			0.0	
Nyssa (Tupelo/Black Gum)							0.0			0.0		0.0				0.0		0.0			0.0	
Parthenocissus (Virginia Creeper)							0.0			0.0		0.0				0.0		0.0			0.0	
Pinus (Pine)							0.0			0.0		0.0				0.0		0.0			1.4	L
Pisum sativum (Pea)							0.0			0.0		0.0				0.0		0.0			0.0	
Plantago (Plantain)							0.5	L		0.0		0.0				0.0		0.0			0.0	
Poaceae (Grass family)							0.0			0.0		0.0				0.0		0.0			0.0	
Polygonum/Persicaria (Buckwheat/Knotweed)							0.0			0.0		0.0				0.0		0.0			0.0	
Quercus (Oak)							0.0			0.0		0.0				0.0		0.0			1.9	L
Rhus/Toxicodendron (Sumac/Poison Ivy)							0.0			0.0		0.0				0.0		0.0			0.0	
Rosaceae (Rose/Cherry/Plum/Peach/Blackberry family)							0.0			35.9	I	96.0	D			51.2	D	90.0	D		65.9	D
Salix (Willow)							0.0			0.0		0.0				0.0		0.0			0.5	L
Trifolium/Melilotus (Red/White Clover)							0.0			16.0	I	0.0				9.3	M	1.9	L		16.3	I
Ulmus (Elm)							0.0			0.5	L	0.0				2.4	L	0.0			0.0	
Vicia (Vetch)							0.0			0.0		0.0				0.0		0.0			0.0	
Vitis (Grape)							0.0			0.0		0.0				0.0		0.0			0.0	
Unknowns #3 and #9							0.0			0.0		0.0				0.0		0.0			0.0	
Various Unknown Palynomorphs							0.0			0.0		0.0				0.0		0.5	L		3.4	M
Totals ▶							100			100		100				100		100			100	

Key: *D* (predominant) >45% I (secondary) 16-45% M (important) 3-15% L (minor) <3%

May							M/J		Jun						J/J	Jul				J/A	Aug			A/S	Sep				S/O	Oct					O/N	Nov	
19		20		21	22		23		24		25		26		27	28	29	30	31	32	33	34	35	36	37	38	39		40	41	42		43	44	45	46	
%	PC	%	PC		%	PC	%	PC	%	PC	%	PC	%	PC													%	PC			%	PC				%	PC
1.0	L	0.0			0.0		0.0				0.0		0.0														0.0				0.0					0.0	
0.0		0.0			0.5	L	0.0				0.0		0.0														0.0				2.8	L				5.2	M
0.0		0.0			0.0		0.0				1.5	L	0.0														0.0				0.0					0.9	L
0.0		0.0			0.0		0.0				0.0		0.0														0.0				9.4	M				0.0	
0.0		3.4	M		0.0		0.9	L			0.0		0.0														0.0				0.0					0.0	
0.0		0.0			0.0		0.0				0.0		0.0														0.0				0.0					0.0	
0.0		0.0			0.0		0.0				0.0		0.5	L													0.0				0.0					0.0	
1.9	L	0.0			0.0		0.0				0.0		0.0														0.0				0.0					0.0	
0.0		0.0			5.7	M	5.5	M			5.8	M	0.5	L													0.5	L			0.5	L				3.3	M
0.0		0.0			0.0		0.0				0.0		0.0														0.0				0.0					0.0	
2.9	L	0.0			1.4	L	2.8	L			0.0		0.0														0.0				0.0					0.0	
3.4	M	0.0			1.4	L	1.8	L			0.5	L	2.4	L													0.0				0.0					0.0	
0.0		0.0			0.0		0.0				0.0		0.0														0.0				2.8	L				0.0	
0.0		0.0			0.0		0.0				0.0		1.0	L													0.0				0.0					0.0	
0.0		27.5	I		23.0	I	7.3	M			3.9	M	9.2	M													0.0				0.5	L				0.0	
0.0		0.0			1.4	L	8.3	M			6.8	M	3.9	M													9.4	M			38.2	I				9.0	M
0.0		0.0			0.0		6.9	M			1.0	L	2.9	L													1.0	L			0.0					1.9	L
1.5	L	3.9	M		3.8	M	0.0				0.0		5.3	M													0.0				0.0					0.0	
0.0		0.0			0.0		0.0				0.0		0.0														0.0				0.0					0.0	
0.0		0.0			0.0		1.4	L			0.0		1.4	L													0.0				0.0					0.0	
0.0		0.0			0.0		0.0				0.5	L	0.0														0.0				0.0					0.0	
0.0		0.0			0.0		0.0				2.4	L	0.0														0.0				1.9	L				0.0	
0.0		0.0			0.0		0.0				0.0		0.0														0.0				0.0					1.4	L
0.0		0.0			0.0		0.0				0.0		2.4	L													0.0				0.0					0.0	
0.0		0.0			0.0		0.0				1.9	L	0.0														0.0				0.0					0.0	
0.0		0.0			4.8	M	3.2	M			0.0		2.4	L													5.4	M			0.5	L				28.9	I
0.0		0.0			0.0		0.0				0.0		0.0														1.0	L			0.0					0.0	
5.3	M	6.9	M		1.0	L	0.0				0.0		15.5	M													3.0	M			0.0					0.0	
0.0		1.5	L		1.0	L	0.0				1.0	L	0.0														0.0				0.0					0.5	L
60.7	D	22.1	I		43.5	I	45.4	D			45.6	D	7.2	M													3.0	M			0.9	L				0.0	
1.5	L	12.7	M		0.0		0.0				0.0		0.5	L													64.4	D			14.2	M				33.6	I
18.0	I	13.2	M		9.6	M	6.4	M			16.0	I	40.1	I													0.0				0.5	L				1.4	L
0.0		0.0			0.0		0.0				0.0		0.0														4.0	M			20.3	I				7.6	M
0.0		6.9	M		0.0		10.1	M			1.0	L	0.0														0.0				0.0					0.0	
0.0		0.0			0.0		0.0				0.0		1.0	L													0.0				0.0					0.0	
0.0		0.0			0.0		0.0				11.2	M	0.5	L													2.5	L			7.5	M				1.4	L
3.9	M	2.0	L		2.9	L	0.0				1.0	L	3.4	M													5.9	M			0.0					4.7	M
100		100			100		100				100		100														100				100					100	

4.3 Southeastern Plains

The Southeastern Plains include the Sand Hills region and are a mosaic of plains between shallow stream valleys that developed atop Cretaceous and younger sands, silts, and clays that once formed coastal and near-shore deposits (Griffith et al., 2002). Annual precipitation in the Southeastern Plains ranges from 132–111.8 cm (52–44 in.) (Runkle et al., 2022) and peaks in March and July, although there is no true dry season. While rainfall is generally high, much of this region is very well-drained—especially the Sand Hills region just below the Fall Line. Closer to the southern boundary, soils contain significantly more clay and are loams and sandy loams. Much of this lower part of the ecoregion contains Carolina bays, which are elliptical, low boggy areas with sandy rims.

A stream near Camden, South Carolina, in the Southeastern Plains ecoregion. Photo credit: Rosalind Severt.

The Sand Hills region is dominated by fire-tolerant vegetation (SC DNR, 2005c), especially on ridgetops and upper hillslopes, including pine (*Pinus*) and oak (*Quercus*) with herbaceous plants and grasses. On lower hillslopes, oak-hickory (*Quercus-Carya*) forests similar to those in the Piedmont may develop. Seepage slopes within the Sand Hills are caused by kaolinite clay horizons or iron cement blocking downward percolation of water. In these moist areas, communities contain pines (*Pinus*), titi (*Cyrilla*), sand myrtle (*Leiophyllum*), mountain laurel (*Kalmia*), and inkberry (*Ilex*) in addition to grasses, sedges, and herbaceous plants. Around blackwater streams and in river bottoms, wet-tolerant trees are common, including swamp tupelo (*Nyssa*); bald cypress (*Taxodium*); water elm (*Ulmus*); pond and other pines (*Pinus*); water, willow, laurel, and cherrybark oaks (*Quercus*); and—in drier areas—sweetgum (*Liquidambar*), tulip poplar (*Liriodendron*), and red maple (*Acer*). The Atlantic white cedar (*Chamaecyperus*) grows where wildfires are especially common. While most of the Atlantic southern loam plains between the Sand Hills and the Middle Atlantic Coastal Plain are crop-dominated farmland, there are also grasslands, pine (*Pinus*) woodlands, and oak-hickory (*Quercus-Carya*) forests. Grass-sedge-herbaceous plant assemblages dominate the depression meadows of the Carolina bays, and bottomland forests similar to those in the Piedmont occur along riverways, dominated by tupelo (*Nyssa*) and bald cypress (*Taxodium*).

Hood (2006) notes 13 taxa of importance to honeybees in the Southeastern Plains. These are: red maple (*Acer*) from mid-January through February; tupelo (*Nyssa*) from mid-March through late April; poplar (*Populus*) from late March through early May; blackberry (Rosaceae) from early April through late June; American holly (*Ilex*) from mid-April through late May; sparkleberry (Ericaceae) from late April through May; vetch (*Vicia*) from mid-May through June; Chinese tallow (*Triadica*), sea sage (*Salvia*), and goldenrod (Asteraceae [coneflower/goldenrod/sunflower family]) from early September through mid-November; Mexican clover (*Richardia*) from late September though late November; and golden aster (Asteraceae [coneflower/goldenrod/sunflower family]) from late October through early December. Fifty-six genera important to beekeepers in Region 12, which includes the Southeastern Plains, are reported in the Honey Bee Forage Map (GSFC, 2007; Ayers & Harmon, 1992). Of these, five are considered important nectar sources: tupelo (*Nyssa*) from March through June, tulip poplar (*Liriodendron*) and holly (*Ilex*) from April through June, sourwood (*Oxydendrum*) from May through August, and goldenrod (Asteraceae [coneflower/goldenrod/sunflower family]) from July through November. Of the 59 genera noted by Smith (2021), stokes aster (Asteraceae [coneflower/goldenrod/sunflower family]), black cherry (Rosaceae), horsemint (*Monarda*), black locust (*Robinia*), maples (*Acer*), and sourwood (*Oxydendrum*) bloom up to a month earlier than in the Piedmont; persimmon (*Diospyros*) blooms up to two months earlier. Horsemint (*Monarda*), Joe-Pye weed, and sneezeweed (both Asteraceae [coneflower/goldenrod/sunflower family]), and sumac (*Rhus*/Toxicodendron) all bloom up to a month longer in the Southeastern Plains than in the Piedmont.

Hive weights for the studied apiaries in the Southeastern Plains were highly variable (like those in the Piedmont) and contained artifacts related to addition and removal of supers and, potentially, honey harvests. A weak increase in weight was seen in some hives in early March, but it is in early to mid-April that there were clear weight increases in all hives (Figure 4.4). Some hives had a significant weight increase in mid-August, while all had one in early October.

A total of 102 pollen taxa were found in honey samples from this region (Table 4.10). Twelve of the thirteen taxa of importance noted by Hood (2006) occurred; Chinese tallow (*Triadica*) did not. Bloom times, however, varied significantly from those predicted by that publication. Red maple (*Acer*) were forecast from mid-January through February; in our study, maple (*Acer*) pollen was observed from early February through May and later reappeared in July, October, and November, likely reflecting movement of honey within the hive—as weight gains were generally low during that period. Tupelo (*Nyssa*) occurrence was forecast from mid-March through late April; in our study, *Nyssa* appearance was both later and longer: mid-April through June. Holly (*Ilex*) was forecast from mid-April through late May; in our study, it appeared earlier (early February) but ended at the forecast time (later May). However, it appeared sporadically from June through October, again suggesting honey movement within the hive. Poplar (*Populus*) was forecast from late March through early May, but we only saw it in early to mid-June. It is possible that the poplar indicated in Hood's table referred to tulip poplar (*Liriodendron*), which appears from latest February through latest May/earliest June and then sporadically in late June, September, and November—again, likely indicating honey movement within the hive. Mexican clover (*Richardia*) was forecast from late September though late November. We saw it then, but also in late February through early March. This occurrence is not entirely surprising, as it has been observed blooming in March, in May, and from June through September (http://coastalplainplants.org/wiki/index.php/Richardia_scabra). The presence of Lamiaceae (mint family), representing both the *Salvia* reported by Hood and other members of the family, occurred on and off all year but almost continuously from February through June.

Several common taxa found in our samples were not identified by Hood (2006). Brassicaceae (mustard family) were present most of the spring and summer and off and on throughout the fall. *Cercis* (redbud) and *Cornus* (dogwood) first appeared in February and persisted through early May. *Diospyros* (persimmon) appeared in late April and persisted through June. Hypercicaceae (St. John's wort family) occurred on and off all year but almost continuously from March through early June. *Lagerstroemia* (crepe myrtle) occurred in almost every sample; this was surprising, as it is known to be a May- through early-fall-blooming shrub, and suggests at least some degree of either hive-robbing early in the year or migration of old honey out of the overwintering comb. Poaceae (grass family), Rosaceae (rose/cherry/plum/peach/blackberry family), and *Trifolium/Melilotus* (red/white clover) occurred in almost all samples but were continuous from late April through June and again from July through November. *Vitis* (grape) occurred in late-April through June and through most of the fall.

Our palynological data demonstrate that *Acer* (maple) significantly contributes to honey production in the Southeastern Plains, as it does in the two regions discussed previously. Five of the six participating apiaries produced monofloral *Acer* (maple) honey in at least one sample from early February through early March (Tables 4.11, 4.12, 4.13, 4.14,

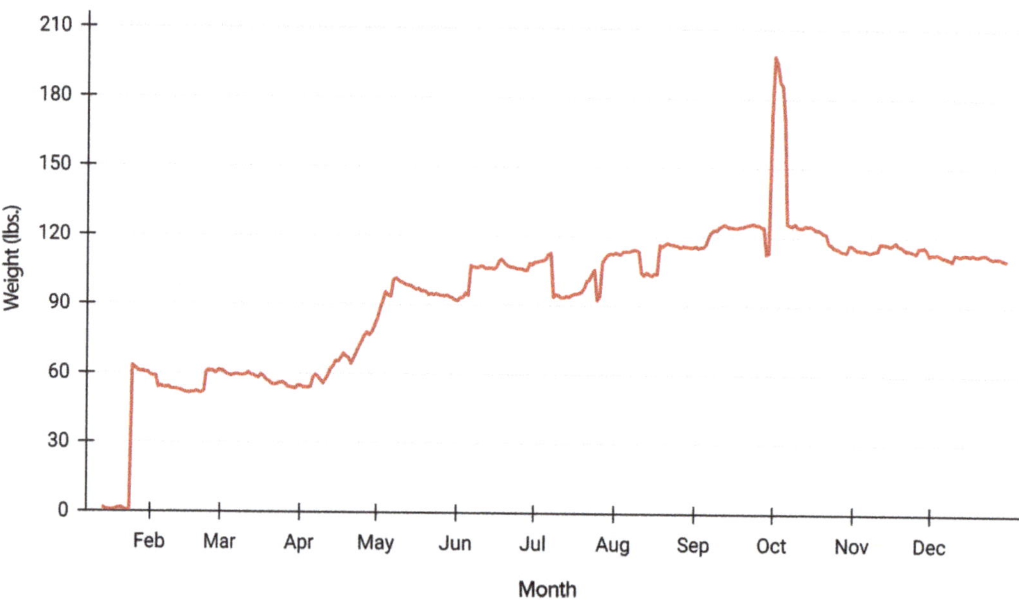

Figure 4.4. An example of BroodMinder hive weight data collected from Apiary 10 in the Southeastern Plains.
Note: Labels indicate the first day of each listed month.

and 4.15); only Apiary 7 did not (Table 4.16). Monofloral honeys are indicated with red color-coding in the apiary pollen occurrence tables.

In addition to *Acer* (maple) monofloral honey, other monofloral honeys were identified. Apiary 7 (Table 4.16) produced monofloral Rosaceae (rose/cherry/plum/peach/blackberry family) and *Robinia* (locust) honey in early March; *Ulmus* (elm) honey in late September; and Asteraceae (coneflower/goldenrod/sunflower family) honey in mid-October. Apiary 9 (Table 4.12) produced monofloral *Trifolium/Melilotus* (red/white clover) honey in mid-March. Apiary 10 (Table 4.13) produced monofloral *Nyssa* (tupelo/black gum) honey in mid-May, Rosaceae (rose/cherry/plum/peach/blackberry family) honey for seven samples spanning late March to late June, *Vitis* (grape) honey in early July, *Gelsemium* (yellow jessamine) honey in mid-July, *Glycine max* (soybean) honey in late August, and two samples of Amaranthaceae (goosefoot family) honey in early September. Apiary 13 (Table 4.15) produced monofloral *Cercis* (redbud) honey in early March. Beekeepers might want to note the diversity of taxa producing these monofloral honeys, their locations, and time of production—from early February off and on until mid-October.

There were 46 taxa that contributed to honey for at least three straight weeks in the Southeastern Plains (Table 4.10). This is the second-largest number of such taxa observed between the five ecoregions (the Piedmont had 47). Like the Piedmont, the Southeastern Plains contains a wide range of habitats, and it contained the most taxa present in the honey samples—with 102 total taxa present (see Chapter 5). Unlike in the Piedmont, where such taxa (those present for at least three straight weeks) made up 50.5% of the diversity, such taxa made up a slightly smaller proportion of the diversity in the Southern Plains—45.1%. These 46 taxa are: *Acer* (maple), Amaranthaceae (goosefoot family), Amaryllidaceae (amaryllis/onion family), Asteraceae (coneflower/goldenrod/sunflower family), Asteraceae (dandelion/cat's ear/endive family), Asteraceae (ragweed/calendula/solidago family), Brassicaceae (mustard family), *Camellia*, *Carya* (hickory/pecan), *Cercis* (redbud), *Cornus* (dogwood), *Diospyros* (persimmon), *Elaeagnus* (autumn/Russian olive), Fabaceae (pea family), *Fagopyrum* (buckwheat), *Gelsemium* (yellow jessamine), *Glycine max* (soybean), *Hamamelis* (witch hazel), Hypercicaceae (St. John's wort family), *Ilex* (holly), *Lagerstroemia* (crepe myrtle), Lamiaceae (mint family), *Ligustrum* (privet), *Liquidambar* (sweetgum), *Liriodendron* (tulip poplar), *Lonicera* (honeysuckle), *Magnolia*, *Nuphar* (water lily), *Nyssa* (tupelo/black gum), *Parthenocissus* (Virginia creeper), *Phlox*, *Phoardendron*, *Pinus* (pine), Poaceae (grass family), *Portulaca* (purslanes), *Quercus* (oak), *Rhus*/*Toxicodendron* (sumac/poison ivy), *Ribes* (currant), *Robinia* (locust), Rosaceae (rose/cherry/plum/peach/blackberry family), *Salix* (willow), *Trifolium/Melilotus* (red/white clover), *Ulmus* (elm), *Vitis* (grape), Unknown #1, and Unknowns #3 and #9. These taxa span the entire foraging season; the appearance of these pollen in the honey for such an extended period of time suggests that they are important forage for bees. As previously noted, *Pinus* (pine), Poaceae (grass family), *Quercus* (oak), and *Ulmus* (elm) do not contribute nectar, but the presence of their pollen grains in the honey samples indicates that the pollen was in the hive and available to the bees for their nutrition.

Table 4.10. Plant Taxa whose Pollen was Observed in Honey Collected from the Southeastern Plains Ecoregion during 2022

Southeastern Plains	Jan		J/F	Feb			F/M	Mar			M/A	Apr			
Week ▶	4	5	6	7	8	9	10	11	12	13	14	15	16	17	18
Acer (Maple)															
Aesculus (Buckeye)															
Alnus (Alder)															
Alternanthera (Joyweed)															
Amaranthaceae (Goosefoot family)															
Amaryllidaceae (Amaryllis/Onion family)															
Apiaceae (Carrot/Parsely/Queen Anne's Lace family)															
Arecaceae (Palm family)															
Asteraceae (Coneflower/Golden Rod/Sunflower family)															
Asteraceae (Dandelion/Cat's Ear/Endive family)															
Asteraceae (Ragweed/Calendula/Solidago family)															
Baptisia (Wild Indigo)															
Berberis (Barberry)															
Berchemia (Rattan Vine)															
Betula (Birch)															
Bidens (Tickseed)															
Boraginaceae (Forget-me-not family)															
Brassicaceae (Mustard family)															
Buddleja (Butterfly Bush)															
Camellia															
Carya (Hickory/Pecan)															
Castanea (Chinkapin/Chestnut)															
Casuarina (invasive Australian Pine)															
Ceanothus															
Celtis (Hackberry)															
Cephalanthus (Buttonbush)															
Cercis (Redbud)															
Cornus (Dogwood)															
Corylus/Carpinus (Hazel/Hornbeam)															
Cucumis (Cucumber/Melon)															
Cyperaceae (Sedge family)															
Dalea (Prairie Clover)															
Diospyros (Persimmon)															
Elaeagnus (Autumn/Russian Olive)															
Ericaceae (Heath family)															
Erodium															
Fabaceae (Pea family)															
Fagopyrum (Buckwheat)															
Fagus (Beech)															
Gelsemium (Yellow Jessamine)															
Glycine max (Soybean)															
Hamamelis (Witch Hazel)															
Hexasepalum teres (Poorjoe/Rough Buttonweed)															
Hypercicaceae (St. John's Wort family)															
Hyssopus (Hyssop)															
Ilex (Holly)															
Impatiens (Touch-me-not)															
Lagerstroemia (Crepe Myrtle)															
Lamiaceae (Mint family)															
Ligustrum (Privet)															
Liliaceae (Lily family)															
Liquidambar (Sweetgum)															
Liriodendron (Tulip Poplar)															
Lonicera (Honeysuckle)															
Ludwigia															
Magnolia															

May				M/J	Jun			J/J	Jul				J/A	Aug			A/S	Sep			S/O	Oct				O/N	Nov
19	20	21	22	23	24	25	26	27	28	29	30	31	32	33	34	35	36	37	38	39	40	41	42	43	44	45	46

(continued)

Table 4.10. (*continued*)

Southeastern Plains	Jan		J/F	Feb			F/M	Mar			M/A	Apr			
Week ▶	4	5	6	7	8	9	10	11	12	13	14	15	16	17	18
Malvaviscus (Wax Mallow)			□		□		□	□					□		
Mimosa (Sensitive Plant)							□	□					□		□
Morus (Mulberry)			□		■		□	■		□	□	□	□	□	□
Myrtaceae (Myrtle family)			□		□		□	□		□	□	■	□	□	□
Nuphar (Water Lily)						□	□			□		■		□	■
Nyssa (Tupelo/Black Gum)			□		□	□	□	■		□	□	□	■	■	■
Onagraceae (Primrose family)						□	□			□		□		□	□
Oxydendrum (Sourwood)			□		□	■	□	□		□	□	□	□	□	□
Parthenocissus (Virginia Creeper)			□		□	□	■	■		□	□	■	■	■	■
Phlox			□		□		■	■					■		□
Phoardendron			□		□		□	■		■	□	□	□	□	□
Pinus (Pine)			□		□	■	□	□		■	□	■	□	■	■
Plantago (Plantain)			□		□	□	□	■		□	□	□	■	□	■
Poaceae (Grass family)			■		□	■	■	■		□	■	■	■	□	■
Polygonum/Persicaria (Buckwheat/Knotweed)			□		□	■	□	□		□		■	□	□	■
Populus (Aspen/Cottonwood/Poplar)							□	□					□		□
Portulaca (Purslanes)			□		□	□	□	□		□	□	□	□	□	□
Primulaceae (Primrose family)			□		□		□	□		□	□	□	□	□	■
Quercus (Oak)			□		□	□	□	□		■	■	□	■	■	■
Rhus/Toxicodendron (Sumac/Poison Ivy)			□		□	■	■	■		□	□	■	■	■	■
Ribes (Currant)			■		□		□	■		■	□	□	■	■	■
Richardia			□		□	■	□	■		□	□	□	□	□	□
Robinia (Locust)			■		□	□	■	■		■	□	■	■	■	■
Rosaceae (Rose/Cherry/Plum/Peach/Blackberry family)			■		■	■	■	■		■	■	■	■	■	■
Rubus (Blackberry/Dewberry)							□	□					□		□
Rumex			□		□		■	□					□		
Salix (Willow)			■		■	■	■	■		□	□	■	■	■	■
Saururus (Lizard's Tail)			□		□		□	□					□		□
Solanum (Tomato)							□	□					□		□
Spathiphyllum			□		□		□	□		□	□	□	□	□	□
Tilia (Basswood)			□		□	□	□	■					□	□	□
Tillandsia							■	□					□		□
Trifolium/Melilotus (Red/White Clover)			■		□	■	■	■		■	■	■	■	■	■
Typha			□		□		□	□		□	□	□	□	□	□
Ulmus (Elm)			■		■	■	□	■		□	□	■	□	□	■
Urtica (Nettles)							□	□					□		□
Verbena			□		□	□	□	□					□	□	□
Vernonia						□	□			□		□		□	□
Vicia (Vetch)						□	□	□						□	□
Viola (Violet)			■		□	■	■	□		□	□	■	□	□	■
Vitis (Grape)			□		□	□	□	□		□	□	■	□	□	■
Wisteria			□		□		□	□		■	■	□	■	□	□
Zanthoxylum (Prickly Ash)			□		□		□	□					□		
Zea mays (Corn)			□		□	□	□	□		□	□	■	□	□	□
Unknown #1			□		■		■	□		□	□	□	□	■	□
Unknowns #3 and #9			□		□	■	■	■		□	□	□	■	■	■
Various Unknown Palynomorphs						□	□			□		□		□	■
Broken Palynomorphs			□		□	■	■	■		□	■	■	■	■	■

Key: *no data collected* (blank) *taxon absent on this week* □ *taxon present on this week* ■

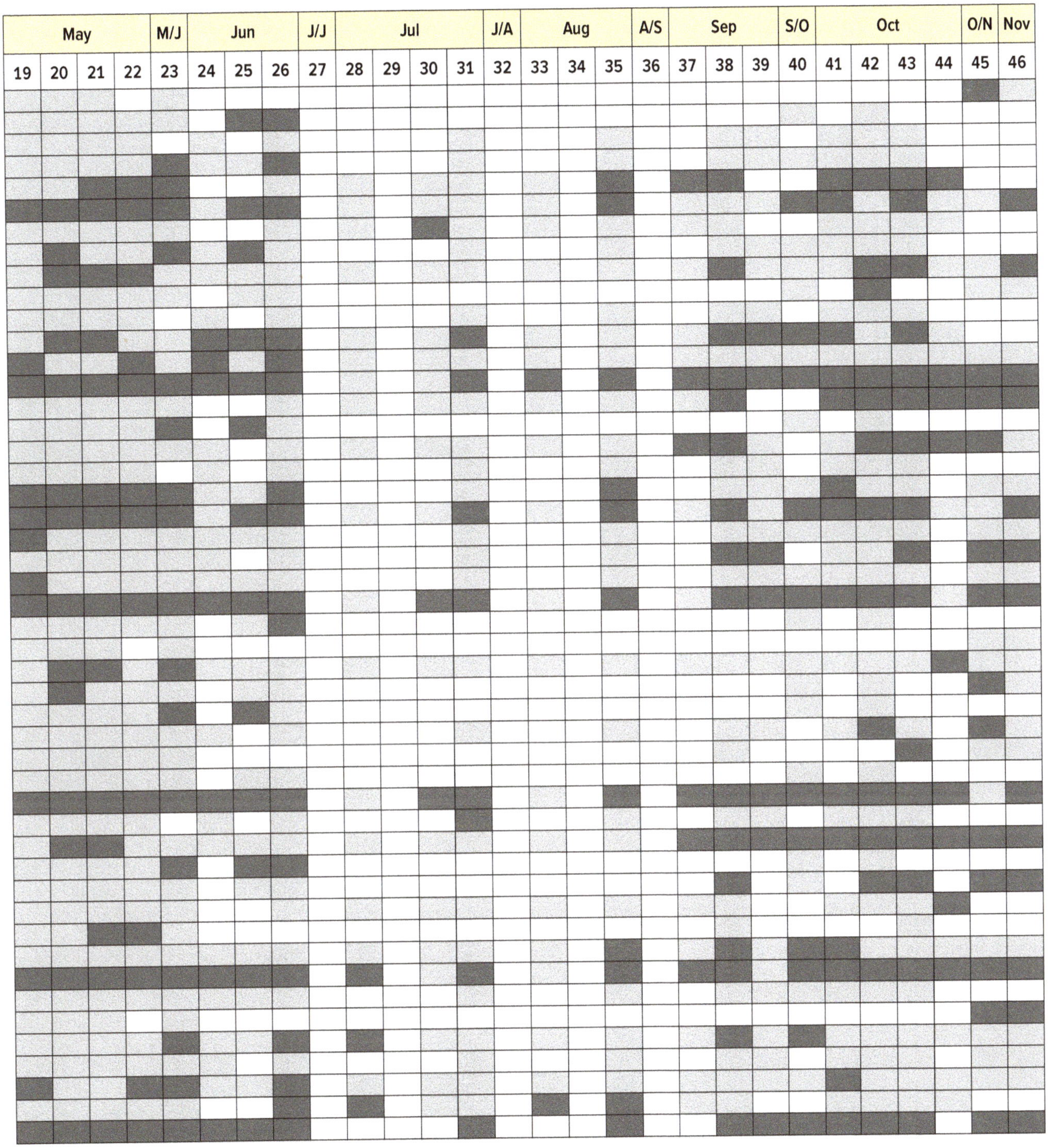

May				M/J	Jun			J/J	Jul				J/A	Aug			A/S	Sep			S/O	Oct				O/N	Nov
19	20	21	22	23	24	25	26	27	28	29	30	31	32	33	34	35	36	37	38	39	40	41	42	43	44	45	46

Table 4.11. Pollen Occurrence in Honey Collected in Apiary 6 in the Southeastern Plains Ecoregion during 2022

Apiary 6	Jan		J/F	Feb				F/M		Mar				M/A	Apr					
Week ▶	4	5	6	7	8	9		10		11		12	13	14	15	16	17		18	
Type of Data ▶						%	PC	%	PC	%	PC						%	PC	%	PC
Acer (Maple)						48.0	D	78.1	D	63.7	D						0.6	L	0.0	
Aesculus (Buckeye)						0.0		0.0		0.0							0.0		0.0	
Alnus (Alder)						0.9	L	0.0		0.0							0.0		0.0	
Alternanthera (Joyweed)						0.0		0.0		0.0							0.6	L	0.0	
Amaryllidaceae (Amaryllis/Onion family)						0.0		0.0		0.0							0.0		0.0	
Apiaceae (Carrot/Parsely/Queen Anne's Lace family)						0.0		0.0		0.0							0.0		0.0	
Asteraceae (Coneflower/Golden Rod/Sunflower family)						3.3	M	0.0		4.6	M						0.6	L	1.1	L
Asteraceae (Dandelion/Cat's Ear/Endive family)						0.0		0.0		0.4	L						0.0		0.0	
Asteraceae (Ragweed/Calendula/Solidago family)						0.0		0.0		0.0							0.3	L	0.0	
Betula (Birch)						0.0		0.0		0.4	L						0.0		0.0	
Brassicaceae (Mustard family)						0.7	L	11.9	M	1.5	L						27.4	I	1.2	L
Carya (Hickory/Pecan)						0.2	L	0.0		0.0							0.3	L	1.2	L
Casuarina (invasive Australian Pine)						0.0		0.0		0.0							0.3	L	0.0	
Celtis (Hackberry)						1.5	L	0.0		0.0							0.0		0.0	
Cercis (Redbud)						0.2	L	0.0		8.1	M						7.2	M	0.6	L
Cornus (Dogwood)						0.0		0.0		0.4	L						1.6	L	2.9	L
Cyperaceae (Sedge family)						0.0		0.0		0.0							0.0		0.0	
Diospyros (Persimmon)						0.2	L	0.0		0.0							0.0		0.0	
Elaeagnus (Autumn/Russian Olive)						0.2	L	0.0		0.0							0.0		0.0	
Erodium						0.0		0.0		0.0							0.0		0.0	
Fagus						0.0		0.0		0.0							3.7	M	0.0	
Gelsemium (Yellow Jessamine)						0.0		0.0		0.0							0.0		0.0	
Hexasepalum teres (Poorjoe/Rough Buttonweed)						0.4	L	0.0		0.0							0.0		0.0	
Hypercicaceae (St. John's Wort family)						0.0		0.0		0.0							0.0		0.0	
Ilex (Holly)						0.7	L	0.0		0.8	L						10.0	M	1.0	L
Lagerstroemia (Crepe Myrtle)						4.4	M	0.0		1.2	L						0.9	L	0.3	L
Lamiaceae (Mint family)						14.5	M	5.2	M	0.8	L						0.3	L	0.1	L
Ligustrum (Privet)						0.0		0.0		0.0							0.0		0.0	
Liquidambar (Sweetgum)						0.0		0.0		0.0							0.0		0.0	
Liriodendron (Tulip Poplar)						0.0		0.0		0.4	L						26.8	I	11.1	M
Ludwigia						0.0		0.0		0.0							0.0		0.0	
Magnolia						0.4	L	0.0		0.0							0.0		0.0	
Nyssa (Tupelo/Black Gum)						0.0		0.0		0.0							0.9	L	72.3	D
Oxydendrum (Sourwood)						0.7	L	0.0		0.0							0.0		0.0	
Parthenocissus (Virginia Creeper)						0.0		0.0		5.0	M						0.0		0.0	
Pinus (Pine)						0.4	L	0.0		0.0							0.0		0.0	
Poaceae (Grass family)						1.1	L	0.0		2.3	L						0.0		0.0	
Quercus (Oak)						0.0		0.0		0.0							1.9	L	0.5	L
Rhus/Toxicodendron (Sumac/Poison Ivy)						3.5	M	0.0		0.0							0.0		0.0	
Richardia						0.7	L	0.0		0.4	L						0.0		0.0	
Robinia (Locust)						0.0		0.0		0.0							1.2	L	0.2	L
Rosaceae (Rose/Cherry/Plum/Peach/Blackberry family)						1.5	L	0.0		1.9	L						6.9	M	0.0	
Salix (Willow)						8.1	M	3.3	M	2.7	L						5.6	M	0.0	
Tilia (Basswood)						0.0		0.0		0.0							0.0		0.0	
Trifolium/Melilotus (Red/White Clover)						4.8	M	0.0		0.0							0.3	L	0.0	
Ulmus (Elm)						1.5	L	0.0		1.2	L						0.0		0.0	
Verbena						0.0		0.0		0.0							0.0		0.0	
Vicia						0.0		0.0		0.0							0.0		0.0	
Viola (Violet)						0.7	L	1.4	L	0.0							0.0		0.0	
Vitis (Grape)						0.0		0.0		0.0							0.0		0.0	
Broken Palynomorphs						1.3	L	0.0		4.2	M						2.5	L	7.5	M
Totals ▶						100		100		100							100		100.0	

Key: *D* (predominant) >45%   I (secondary) 16-45%   M (important) 3-15%   L (minor) <3%

May								M/J		Jun			J/J	Jul				J/A	Aug			A/S	Sep				S/O	Oct					O/N	Nov
19		20		21		22		23		24	25	26	27	28	29	30	31	32	33	34	35	36	37	38		39	40	41	42	43		44	45	46
%	PC	%	PC	%	PC	%	PC	%	PC															%	PC					%	PC			
0.0		0.0		0.0		0.0		0.3	L															0.0						0.0				
0.0		0.0		0.0		0.0		0.3	L															0.0						0.0				
0.0		0.0		0.0		0.0		0.0																0.0						0.0				
0.0		0.0		0.0		0.0		0.0																0.0						0.0				
0.0		0.0		0.0		0.0		0.0																0.3	L					0.0				
0.0		0.0		0.0		0.0		0.0																0.0						1.1	L			
1.2	L	1.0	L	1.3	L	0.3	L	0.0																0.0						10.7	M			
0.0		0.0		0.4	L	1.7	L	5.5	M															0.0						0.0				
0.0		0.0		0.0		0.0		0.0																0.0						0.0				
0.0		0.0		0.0		0.0		0.0																0.0						0.0				
10.0	M	4.2	M	40.1	I	6.4	M	48.2	D															1.5	L					0.8	L			
0.0		0.3	L	0.0		0.0		0.3	L															0.0						0.0				
0.0		0.0		0.0		0.0		0.0																0.0						0.0				
0.0		0.0		0.0		0.0		1.6	L															0.0						6.9	M			
0.0		0.0		0.0		0.0		0.0																0.0						0.0				
0.0		0.0		0.0		0.0		0.0																0.0						0.0				
0.0		0.0		0.0		1.3	L	0.0																0.0						0.0				
0.0		0.0		3.0	M	6.4	M	1.6	L															0.0						0.8	L			
0.0		0.0		0.0		0.0		0.0																0.0						0.0				
0.0		0.7	L	0.0		0.0		0.0																0.0						0.0				
0.0		0.0		0.0		0.0		0.0																0.0						0.0				
0.0		0.0		0.0		0.7	L	0.0																0.0						0.0				
0.0		0.0		0.0		0.0		0.0																0.0						0.0				
0.0		0.0		0.0		6.1	M	0.0																0.0						1.9	L			
3.8	M	1.6	L	0.0		2.0	L	0.5	L															0.0						0.0				
0.6	L	0.7	L	0.8	L	0.0		0.0																2.1	L					1.5	L			
1.5	L	0.3	L	0.0		0.0		0.0																0.0						0.0				
1.5	L	1.3	L	0.0		1.0	L	0.0																0.0						0.0				
0.0		0.0		0.0		0.0		0.0																0.3	L					0.4	L			
0.0		7.2	M	3.4	M	0.7	L	0.5	L															0.0						0.0				
0.0		0.0		0.0		0.7	L	0.0																0.0						0.0				
1.8	L	8.2	M	2.1	L	0.0		0.5	L															0.0						0.0				
23.6	I	19.3	I	2.5	L	12.1	M	0.5	L															0.0						1.5	L			
0.0		0.0		0.0		0.0		0.0																0.0						0.0				
0.0		0.0		0.0		0.3	L	0.0																0.0						0.0				
0.0		0.0		0.0		0.0		0.0																0.0						0.0				
0.0		0.0		1.3	L	1.3	L	0.3	L															15.1	M					39.3	I			
0.9	L	12.4	M	0.4	L	0.0		0.0																0.0						0.0				
0.0		0.0		0.4	L	0.0		0.0																27.6	I					5.3	M			
0.0		0.0		0.0		0.0		0.0																11.3	M					3.1	M			
0.0		0.0		0.0		0.0		0.0																0.0						0.0				
33.6	I	17.0	I	2.5	L	24.2	I	0.8	L															0.3	L					3.1	M			
0.0		0.0		0.0		0.0		0.0																0.0						0.0				
0.0		0.0		0.0		0.0		0.0																0.0						0.4	L			
10.9	M	11.4	M	2.1	L	6.4	M	3.6	M															0.3	L					1.5	L			
0.0		0.3	L	0.0		0.0		0.0																3.6	M					3.4	M			
0.0		0.0		0.0		0.0		0.0																0.3	L					1.5	L			
0.0		0.0		1.7	L	1.0	L	0.0																0.0						0.0				
0.0		0.0		0.0		0.0		0.0																0.3	L					0.0				
1.8	L	0.7	L	26.6	I	12.5	M	34.5	I															0.9	L					0.0				
8.8	M	13.4	M	11.4	M	14.8	M	0.8	L															36.2	I					16.8	I			
100		100		100		100		100																100						100				

Table 4.12. Pollen Occurrence in Honey Collected in Apiary 9 in the Southeastern Plains Ecoregion during 2022

Apiary 9	Jan		J/F	Feb				F/M		Mar					M/A	Apr				
Week ▶	4	5	6	7	8	9		10		11		12		13	14	15	16		17	18
Type of Data ▶						%	PC	%	PC	%	PC	%	PC				%	PC		
Acer (Maple)						73.1	D	28.0	I	4.0	M	1.3	L				0.0			
Aesculus (Buckeye)						0.7	L	0.0		0.0		0.0					0.0			
Amaryllidaceae (Amaryllis/Onion family)						0.0		0.0		0.0		0.0					0.0			
Apiaceae (Carrot/Parsely/Queen Anne's Lace family)						0.3	L	0.0		0.0		0.0					0.0			
Artemisia (Wormwood/Absinthe)						0.0		0.0		0.0		0.0					0.0			
Asteraceae (Coneflower/Golden Rod/Sunflower family)						0.0		7.1	M	1.0	L	0.4	L				0.4	L		
Asteraceae (Dandelion/Cat's Ear/Endive family)						0.0		0.0		0.0		0.0					0.0			
Berberis (Barberry)						0.0		0.0		0.0		0.0					0.0			
Bidens (Tickseed)						3.0	M	2.9	L	0.5	L	0.0					0.0			
Brassicaceae (Mustard family)						0.0		0.0		0.5	L	0.0					2.0	L		
Buddleja (Butterfly Bush)						0.3	L	0.0		0.0		0.0					0.0			
Camellia						5.6	M	2.9	L	0.0		0.0					0.0			
Carex						0.0		0.0		2.0	L	0.0					0.0			
Carya (Hickory/Pecan)						0.3	L	0.0		0.0		0.0					0.0			
Castanea (Chinkapin/Chestnut)						3.3	M	7.1	M	1.0	L	0.0					0.0			
Cercis (Redbud)						0.0		3.7	M	1.0	L	0.0					7.7	M		
Cornus (Dogwood)						0.0		1.8	L	34.7	I	0.0					4.9	M		
Dalea (Prairie Clover)						0.0		0.3	L	0.0		0.0					0.0			
Diospyros (Persimmon)						0.0		0.0		0.0		0.0					0.0			
Elaeagnus (Autumn/Russian Olive)						0.0		0.5	L	0.0		0.0					0.0			
Fabaceae (Pea family)						0.0		0.0		19.3	I	0.0					0.0			
Fraxinus (Ash)						0.7	L	0.0		0.0		0.0					0.0			
Gelsemium (Yellow Jessamine)						0.0		0.0		0.0		0.0					0.0			
Hamamelis (Witch Hazel)						0.0		0.0		0.0		0.0					0.0			
Hypercicaceae (St. John's Wort family)						1.0	L	1.8	L	0.0		0.0					0.0			
Ilex (Holly)						0.0		3.4	M	0.0		0.9	L				0.8	L		
Lagerstroemia (Crepe Myrtle)						0.0		0.0		0.0		0.0					0.0			
Lamiaceae (Mint family)						0.0		1.6	L	0.0		0.0					2.8	L		
Ligustrum (Privet)						0.0		0.0		0.0		0.0					0.0			
Linaria						0.0		0.0		0.0		2.2	L				0.0			
Liquidambar (Sweetgum)						0.0		0.0		8.9	M	0.4	L				10.1	M		
Liriodendron (Tulip Poplar)						0.0		0.3	L	0.0		0.4	L				5.3	M		
Lonicera (Honeysuckle)						1.0	L	0.0		0.0		0.0					0.0			
Magnolia						0.0		0.0		0.0		0.0					10.5	M		
Nyssa (Tupelo/Black Gum)						0.0		0.0		0.0		0.0					1.6	L		
Parthenocissus (Virginia Creeper)						0.3	L	4.2	M	3.0	M	0.0					0.4	L		
Phlox						0.0		0.0		0.0		0.0					0.0			
Pinus (Pine)						0.0		0.0		0.0		0.0					0.0			
Poaceae (Grass family)						0.3	L	4.0	M	4.0	M	0.0					0.0			
Quercus (Oak)						0.3	L	0.8	L	4.5	M	3.4	M				6.5	M		
Rhus/Toxicodendron (Sumac/Poison Ivy)						0.0		0.0		0.0		0.0					0.0			
Ribes (Currant)						0.0		0.0		0.0		0.0					0.4	L		
Robinia (Locust)						0.0		0.0		0.0		0.0					0.0			
Rosaceae (Rose/Cherry/Plum/Peach/Blackberry family)						0.0		5.3	M	15.8	I	0.0					25.1	I		
Salix (Willow)						1.0	L	2.9	L	0.0		3.4	M				0.0			
Trifolium/Melilotus (Red/White Clover)						0.0		0.3	L	0.0		87.5	D				5.7	M		
Ulmus (Elm)						6.9	M	5.5	M	0.0		0.0					0.0			
Vitis (Grape)						0.0		0.0		0.0		0.0					2.4	L		
Unknowns #3 and #9						0.0		0.0		0.0		0.0					0.0			
Various Unknown Palynomorphs						2.0	L	15.6	I	0.0		0.0					13.4	M		
Totals ▶						100		100		100		100					100			

Key: *D* (predominant) >45%   I (secondary) 16-45%   M (important) 3-15%   L (minor) <3%

May							M/J		Jun			J/J	Jul				J/A	Aug			A/S	Sep			S/O	Oct				O/N	Nov
19		20	21		22		23		24	25	26	27	28	29	30	31	32	33	34	35	36	37	38	39	40	41	42	43	44	45	46
%	PC		%	PC	%	PC	%	PC																							
0.0			0.0		0.0		0.0																								
0.0			0.0		0.3	L	0.3	L																							
0.0			0.0		1.3	L	0.0																								
0.0			0.0		0.3	L	0.0																								
0.0			0.0		0.3	L	0.0																								
0.0			0.0		0.3	L	0.0																								
0.0			0.0		0.8	L	0.0																								
0.0			0.0		0.0		0.6	L																							
0.0			0.0		0.0		0.0																								
0.0			0.3	L	3.0	M	0.3	L																							
0.0			0.0		0.0		0.0																								
0.0			0.0		0.0		0.0																								
0.0			0.0		0.0		0.0																								
0.0			0.0		0.5	L	0.0																								
0.0			0.0		0.0		0.0																								
15.8	I		0.3	L	0.0		0.0																								
0.0			0.0		0.0		0.0																								
0.0			0.0		0.0		0.0																								
0.0			0.0		4.8	M	1.3	L																							
0.0			0.0		0.0		0.0																								
0.0			0.0		3.0	M	2.2	L																							
0.0			0.0		0.0		0.0																								
0.0			0.0		3.3	M	0.6	L																							
0.0			0.0		0.5	L	0.0																								
0.0			0.0		1.5	L	2.5	L																							
23.5	I		13.3	M	3.3	M	15.0	M																							
0.0			0.0		4.0	M	0.0																								
6.4	M		3.5	M	0.0		0.0																								
0.4	L		0.0		1.3	L	2.5	L																							
0.0			0.0		0.0		0.0																								
0.0			0.0		0.3	L	0.0																								
0.0			8.5	M	5.6	M	0.6	L																							
0.0			0.0		0.0		0.0																								
0.0			2.5	L	0.0		0.0																								
2.1	L		5.4	M	2.3	L	1.3	L																							
0.0			0.0		0.5	L	0.3	L																							
0.0			0.0		0.0		0.3	L																							
0.0			0.0		0.3	L	0.6	L																							
1.3	L		0.0		9.3	M	0.3	L																							
0.4	L		0.3	L	0.3	L	0.0																								
1.7	L		0.0		2.0	L	1.0	L																							
0.0			0.0		1.5	L	0.0																								
0.0			0.0		0.0		0.6	L																							
35.0	I		39.9	I	18.2	I	40.4	I																							
0.0			0.0		0.0		0.0																								
8.5	M		25.3	I	2.0	L	12.1	M																							
0.4	L		0.0		0.0		0.0																								
1.7	L		0.6	L	11.1	M	12.7	M																							
0.0			0.0		0.0		0.6	L																							
2.6	L		0.0		18.4	I	3.5	M																							
100			100		100		100																								

Table 4.13. Pollen Occurrence in Honey Collected in Apiary 10 in the Southeastern Plains Ecoregion during 2022

Apiary 10	Jan		J/F	Feb				F/M		Mar				M/A	Apr						
Week ▶	4	5	6	7	8	9		10		11	12	13		14	15		16	17		18	
Type of Data ▶						%	PC	%	PC			%	PC		%	PC		%	PC	%	PC
Acer (Maple)						72.4	D	80.9	D			0.8	L		0.0			0.0		0.0	
Alnus (Alder)						3.1	M	0.0				0.0			0.0			0.0		0.0	
Alternanthera (Joyweed)						0.0		0.0				0.0			0.0			0.0		0.0	
Amaranthaceae (Goosefoot family)						0.0		0.0				0.0			0.0			0.0		0.0	
Asteraceae (Coneflower/Golden Rod/Sunflower family)						0.0		0.0				0.0			0.0			0.0		0.0	
Asteraceae (Dandelion/Cat's Ear/Endive family)						0.0		0.0				0.0			0.4	L		0.0		2.4	L
Asteraceae (Ragweed/Calendula/Solidago family)						2.2	L	0.0				0.0			6.8	M		0.0		5.7	M
Brassicaceae (Mustard family)						2.2	L	10.4	M			0.4	L		41.3	I		0.0		3.3	M
Carya (Hickory/Pecan)						0.0		0.0				0.0			0.0			0.0		0.0	
Cucumis (Cucumber/Melon)						0.0		0.0				0.0			0.0			0.0		0.0	
Diospyros (Persimmon)						0.0		0.0				0.0			0.0			0.0		0.0	
Fagopyrum (Buckwheat)						0.0		0.0				0.0			0.0			0.0		0.0	
Gelsemium (Yellow Jessamine)						0.0		0.0				0.0			0.0			0.0		0.8	L
Glycine max (Soybean)						0.0		0.4	L			0.0			0.0			0.0		0.0	
Hexasepalum teres (Poorjoe/Rough Buttonweed)						0.0		0.0				0.0			0.0			0.0		0.0	
Hyssopus (Hyssop)						0.0		0.0				0.0			0.0			0.0		0.0	
Ilex (Holly)						0.0		0.0				0.0			0.0			0.0		0.4	L
Impatiens (Touch-me-not)						0.0		0.0				0.0			0.0			0.0		0.0	
Lagerstroemia (Crepe Myrtle)						1.8	L	0.0				0.0			3.4	M		0.0		6.1	M
Lamiaceae (Mint family)						0.0		0.0				0.0			0.0			0.0		1.6	L
Ligustrum (Privet)						0.0		0.0				0.0			0.4	L		0.0		0.4	L
Liliaceae (Lily family)						0.0		0.0				0.0			0.0			0.0		0.0	
Liquidambar (Sweetgum)						0.4	L	0.0				0.0			0.4	L		0.0		0.0	
Liriodendron (Tulip Poplar)						0.0		0.0				0.0			0.4	L		2.0	L	1.2	L
Lonicera (Honeysuckle)						0.0		0.0				0.0			0.0			0.0		0.0	
Magnolia						0.0		0.0				0.0			0.4	L		0.0		0.0	
Nuphar (Water Lily)						0.0		0.0				0.0			1.3	L		0.0		1.6	L
Nyssa (Tupelo/Black Gum)						0.0		0.0				0.0			0.0			0.4	L	4.5	M
Onagraceae (Primrose family)						0.0		0.0				0.0			0.0			0.0		0.0	
Parthenocissus (Virginia Creeper)						0.0		1.2	L			0.0			0.0			0.0		0.0	
Pinus (Pine)						0.0		0.0				0.4	L		0.4	L		0.8	L	0.0	
Plantago (Plantain)						0.0		0.0				0.0			0.0			0.0		1.2	L
Poaceae (Grass family)						7.1	M	0.8	L			0.0			2.6	L		0.0		2.4	L
Polygonum/Persicaria (Buckwheat/Knotweed)						1.3	L	0.0				0.0			2.1	L		0.0		0.4	L
Portulaca (Purslanes)						0.0		0.0				0.0			0.0			0.0		0.0	
Rhus/Toxicodendron (Sumac/Poison Ivy)						0.9	L	0.0				0.0			0.4	L		0.0		0.0	
Rosaceae (Rose/Cherry/Plum/Peach/Blackberry family)						2.2	L	0.0				98.4	D		31.5	I		75.1	D	51.0	D
Salix (Willow)						0.0		4.8	M			0.0			2.6	L		0.0		0.0	
Trifolium/Melilotus (Red/White Clover)						0.4		0.4	L			0.0			3.8	M		16.7	I	4.9	M
Ulmus (Elm)						2.2	L	0.0				0.0			0.0			0.0		0.0	
Vernonia						0.0		0.0				0.0			0.0			0.0		0.0	
Viola (Violet)						0.0		0.0				0.0			1.3	L		0.0		0.4	L
Vitis (Grape)						0.0		0.0				0.0			0.4	L		0.0		2.0	L
Zea mays (Corn)						0.0		0.0				0.0			0.0			0.0		0.0	
Unknowns #3 and #9						3.6	M	1.2	L			0.0			0.0			4.9	M	9.0	M
Various Unknown Palynomorphs						0.0		0.0				0.0			0.0			0.0		0.4	L
Totals ▶						100		100				100			100			100		100	

Key: *D* (predominant) >45% I (secondary) 16-45% M (important) 3-15% L (minor) <3%

May				M/J	Jun			J/J	Jul				J/A	Aug			A/S	Sep			S/O	Oct				O/N	Nov
19	20	21	22	23	24	25	26	27	28	29	30	31	32	33	34	35	36	37	38	39	40	41	42	43	44	45	46
% PC	% PC	% PC	% PC	% PC			% PC		% PC		% PC	% PC		% PC		% PC		% PC	% PC			% PC	% PC	% PC	% PC		
0.0	0.0	0.7 L	0.0	0.0			0.0		0.0		0.0	1.4 L		0.0		0.0		0.0	0.0			0.0	0.0	0.0	0.0		
0.0	0.0	0.0	0.0	0.0			0.0		0.0		0.0	0.0		0.0		0.0		0.0	0.0			0.0	0.0	0.0	0.0		
0.0	0.0	0.0	0.0	2.7 L			0.0		0.0		0.0	0.0		0.9 L		0.0		0.0	0.0			0.0	0.0	0.0	0.0		
0.0	0.0	0.0	0.0	0.0			0.0		0.0		0.0	5.5 M		22.5 I		32.6 I		66.2 D	51.9 D			23.3 I	44.7 I	11.4 M	16.5 I		
0.0	0.0	0.0	0.0	0.0			0.0		0.0		0.0	0.0		0.0		0.0		0.0	11.0 M			29.7 I	4.4 M	29.3 I	40.3 I		
0.0	0.0	0.0	0.0	0.0			0.4 L		0.0		0.0	0.0		0.0		0.0		0.0	0.0			0.0	0.0	0.0	0.0		
0.4 L	0.0	0.9 L	0.0	0.0			0.0		0.0		11.6 M	20.5 I		6.8 M		1.7 L		1.3 L	0.0			0.0	0.0	0.0	0.0		
1.7 L	19.1 I	8.1 M	4.3 M	0.0			0.4 L		0.0		0.0	0.0		0.5 L		0.0		0.0	0.5 L			0.0	0.0	0.4 L	0.0		
0.8 L	0.0	0.0	0.0	0.0			0.0		0.0		0.0	0.0		0.0		0.0		0.0	0.0			0.0	0.0	0.0	0.0		
0.0	0.0	0.0	0.0	0.0			0.0		0.0		0.0	0.0		0.0		0.0		0.0	0.0			0.0	0.0	0.0	2.9 L		
0.8 L	4.8 M	3.1 M	0.0	0.0			0.0		0.0		0.0	0.0		0.0		0.0		0.0	0.0			0.0	0.0	0.0	0.0		
0.0	0.0	0.0	1.0 L	0.0			0.0		0.0		0.0	0.0		0.0		2.1 L		0.0	0.0			0.0	10.7 M	6.6 M	0.5 L		
0.0	0.0	0.0	0.0	0.0			1.8 L		0.0		51.9 D	11.0 M		0.5 L		0.0		0.4 L	0.0			0.0	0.0	0.0	0.0		
0.0	0.0	0.0	0.0	0.0			0.0		0.0		0.0	1.4 L		35.1 I		51.3 D		6.6 M	9.0 M			5.5 M	5.3 M	10.0 M	3.4 M		
0.0	0.0	0.0	0.0	0.0			0.0		0.0		0.0	0.0		0.0		0.0		0.4 L	0.0			1.8 L	2.4 L	0.0	2.9 L		
0.0	0.0	0.0	0.0	0.0			0.0		0.0		0.0	0.0		0.0		0.0		9.6 M	2.4 L			0.0	0.0	0.0	0.0		
0.8 L	0.0	1.5 L	2.4 L	3.1 M			0.0		0.0		0.0	1.4 L		0.0		0.0		0.0	0.0			0.0	0.0	0.0	0.0		
0.0	0.0	0.0	0.0	0.0			0.0		0.0		2.6 L	4.1 M		0.0		0.0		0.0	0.0			0.0	0.0	0.0	0.0		
0.4 L	0.0	0.7 L	0.0	0.0			0.0		0.9 L		0.0	11.0 M		12.6 M		0.8 L		0.9 L	0.5 L			0.5 L	1.0 L	0.9 L	0.5 L		
0.0	0.0	0.0	0.0	0.0			0.0		0.0		0.0	0.0		0.0		0.0		0.0	0.0			0.0	0.0	0.0	0.0		
2.5 L	0.0	3.7 M	3.9 M	1.8 L			2.7 L		0.4 L		0.0	0.0		0.0		0.0		0.0	0.0			0.5 L	0.0	0.0	0.0		
0.0	0.0	0.0	0.0	0.0			0.0		0.0		1.3 L	0.0		0.0		0.0		0.0	0.0			0.0	0.0	0.0	0.0		
0.0	0.0	0.0	0.0	0.0			0.0		0.0		0.0	0.0		0.0		0.0		0.0	0.0			0.0	0.0	0.0	0.0		
1.7 L	0.0	0.0	0.0	0.0			0.0		0.0		0.0	0.0		0.0		0.0		0.0	0.0			0.0	0.0	0.0	0.0		
0.0	0.0	0.0	0.0	0.0			0.0		0.0		15.0 M	8.2 M		4.1 M		2.1 L		0.0	0.0			0.0	0.0	0.0	0.0		
0.0	0.0	0.4 L	0.0	0.4 L			0.0		0.0		0.0	0.0		0.0		0.0		0.0	0.0			0.0	0.0	0.0	0.0		
0.0	0.0	0.9 L	1.0 L	0.9 L			0.0		0.0		0.0	0.0		0.0		0.8 L		0.9 L	2.4 L			11.9 M	9.7 M	2.2 L	3.9 M		
10.1 M	74.8 D	11.9 M	2.9 L	5.4 M			0.4 L		0.0		0.0	0.0		0.0		0.0		0.0	0.0			0.0	0.0	0.0	0.0		
0.0	0.0	0.0	0.0	0.0			0.0		0.0		6.8 M	0.0		0.0		0.0		0.0	0.0			0.0	0.0	0.0	0.0		
0.0	0.0	0.0	0.0	0.0			0.0		0.0		0.0	0.0		0.0		0.0		0.0	0.0			0.0	0.0	0.0	0.0		
0.0	0.0	0.2 L	0.0	0.0			0.0		0.0		0.0	0.0		0.0		0.0		0.0	0.0			0.0	0.0	0.0	0.0		
0.0	0.0	0.0	0.0	0.0			0.0		0.0		0.0	0.0		0.0		0.0		0.0	0.0			0.0	0.0	0.0	0.0		
0.0	0.0	0.9 L	0.5 L	0.0			0.0		0.0		0.0	15.1 M		12.2 M		6.8 M		9.6 M	6.7 M			6.4 M	3.4 M	2.2 L	5.3 M		
0.0	0.0	0.0	0.0	0.0			0.0		0.0		0.0	0.0		0.0		0.0		0.0	3.3 M			6.8 M	4.9 M	2.2 L	3.4 M		
0.0	0.0	0.0	0.0	0.0			0.0		0.0		0.0	0.0		0.0		0.0		2.6 L	5.2 M			0.0	0.0	2.6 L	1.5 L		
2.5 L	0.0	6.4 M	4.8 M	0.9 L			0.4 L		0.0		0.0	0.0		0.0		0.0		0.0	0.0			0.0	0.0	0.0	0.0		
67.5 D	1.3 L	49.0 D	60.9 D	44.6 D			66.5 D		0.0		8.2 M	11.0 M		0.0		0.0		0.0	0.0			0.0	0.0	0.0	0.0		
0.0	0.0	3.9 M	0.0	0.4 L			0.0		0.0		0.0	0.0		0.0		0.0		0.0	0.0			0.0	0.0	0.0	0.5 L		
0.8 L	0.0	4.8 M	3.4 M	1.3 L			20.1 I		0.0		2.6 L	2.7 L		0.0		0.4 L		0.4 L	2.9 L			0.0	0.0	0.0	3.4 M		
0.0	0.0	2.4 L	0.0	0.0			0.0		0.0		0.0	0.0		0.0		0.0		0.4 L	2.4 L			13.7 M	9.2 M	32.3 I	10.2 M		
0.0	0.0	0.0	0.0	0.0			0.0		0.0		0.0	0.0		0.0		0.0		0.0	0.0			0.0	0.0	0.0	1.9 L		
0.0	0.0	0.0	0.0	0.0			0.0		0.0		0.0	0.0		0.0		0.0		0.0	0.0			0.0	0.0	0.0	0.0		
0.0	0.0	0.4 L	0.0	0.0			0.0		94.4 D		0.0	6.8 M		0.0		0.0		0.4 L	1.9 L			0.0	4.4 M	0.0	2.9 L		
0.0	0.0	0.0	0.0	0.0			4.0 M		0.4 L		0.0	0.0		0.0		0.0		0.0	0.0			0.0	0.0	0.0	0.0		
9.7 M	0.0	0.0	15.0 M	38.4 I			0.9 L		0.0		0.0	0.0		0.0		0.0		0.0	0.0			0.0	0.0	0.0	0.0		
0.0	0.0	0.0	0.0	0.0			2.2 L		3.9 M		0.0	0.0		5.0 M		1.3 L		0.0	0.0			0.0	0.0	0.0	0.0		
100	100	100	100	100			100		100		100	100		100		100		100	100			100	100	100	100		

Table 4.14. Pollen Occurrence in Honey Collected in Apiary 11 in the Southeastern Plains Ecoregion during 2022

Apiary 11	Jan		J/F		Feb				F/M		Mar				M/A	Apr				
Week ▶	4	5	6		7	8		9	10		11		12	13	14	15	16		17	18
Type of Data ▶			%	PC		%	PC		%	PC	%	PC					%	PC		
Acer (Maple)			96.0	D		88.4	D		36.3	I	48.3	D					0.0			
Aesculus (Buckeye)			0.0			0.0			0.0		0.0						0.0			
Alnus (Alder)			0.0			0.3	L		0.0		0.0						0.0			
Amaranthaceae (Goosefoot family)			0.0			0.0			0.0		0.0						0.0			
Amaryllidaceae (Amaryllis/Onion family)			0.0			0.0			0.0		0.0						0.0			
Arecaceae (Palm family)			0.0			0.0			0.0		0.0						0.0			
Asteraceae (Coneflower/Golden Rod/Sunflower family)			0.0			0.0			0.9	L	2.7	L					0.8	L		
Berberis (Barberry)			0.0			0.0			0.0		0.0						0.0			
Brassicaceae (Mustard family)			0.3	L		0.0			0.0		7.2	M					7.4	M		
Camellia			0.0			0.0			0.0		1.9	L					1.1	L		
Castanea (Chinkapin/Chestnut)			0.0			0.0			0.0		0.0						0.0			
Casuarina (invasive Australian Pine)			0.0			0.0			0.0		0.0						0.0			
Cercis (Redbud)			0.0			0.0			28.3	I	3.5	M					4.6	M		
Cornus (Dogwood)			0.0			0.0			0.0		0.0						9.3	M		
Corylus/Carpinus (Hazel/Hornbeam)			0.0			0.0			0.0		0.0						0.0			
Cyperaceae (Sedge family)			0.0			0.0			0.0		0.0						0.0			
Diospyros (Persimmon)			0.0			0.0			0.0		0.0						0.0			
Fabaceae (Pea family)			0.0			0.0			2.4	L	0.5	L					0.0			
Fagus			0.0			0.0			0.0		0.0						0.8	L		
Gelsemium (Yellow Jessamine)			0.0			0.0			0.0		0.0						0.0			
Hypercicaceae (St. John's Wort family)			0.7	L		1.2	L		2.4	L	1.9	L					0.0			
Ilex (Holly)			0.0			0.0			0.0		0.3	L					1.4	L		
Lagerstroemia (Crepe Myrtle)			0.0			0.0			0.0		0.3	L					0.0			
Lamiaceae (Mint family)			2.7	M		6.5	M		9.8	M	2.7	L					0.0			
Ligustrum (Privet)			0.0			0.0			0.0		0.0						0.0			
Liliaceae (Lily family)			0.0			0.0			0.0		0.0						0.0			
Liquidambar (Sweetgum)			0.0			0.0			1.8	L	0.0						0.0			
Liriodendron (Tulip Poplar)			0.0			0.0			0.0		2.9	L					9.0	M		
Magnolia			0.0			0.0			0.0		0.0						0.0			
Malvaviscus (Wax Mallow)			0.0			0.0			0.0		0.0						0.0			
Nyssa (Tupelo/Black Gum)			0.0			0.0			0.0		0.3	L					0.3	L		
Parthenocissus (Virginia Creeper)			0.0			0.0			0.0		0.5	L					0.0			
Phlox			0.0			0.0			0.3	L	1.9	L					0.3	L		
Plantago (Plantain)			0.0			0.0			0.0		0.5	L					0.0			
Poaceae (Grass family)			0.0			0.0			0.0		1.9	L					0.0			
Polygonum/Persicaria (Buckwheat/Knotweed)			0.0			0.0			0.0		0.0						0.0			
Portulaca (Purslanes)			0.0			0.0			0.0		0.0						0.0			
Quercus (Oak)			0.0			0.0			0.0		0.0						2.2	L		
Rhus/Toxicodendron (Sumac/Poison Ivy)			0.0			0.0			0.0		0.8	L					4.1	M		
Ribes (Currant)			0.0			0.0			0.0		0.5	L					0.5	L		
Richardia			0.0			0.0			0.0		0.0						0.0			
Robinia (Locust)			0.0			0.0			0.0		6.7	M					6.6	M		
Rosaceae (Rose/Cherry/Plum/Peach/Blackberry family)			0.0			0.0			0.0		0.8	L					10.1	M		
Rumex			0.0			0.0			0.9	L	0.0						0.0			
Salix (Willow)			0.0			0.0			0.3	L	0.0						41.0	I		
Saururus (Lizard's Tail)			0.0			0.0			0.0		0.0						0.0			
Spathiphyllum			0.0			0.0			0.0		0.0						0.0			
Tilia (Basswood)			0.0			0.0			0.0		0.3	L					0.0			
Trifolium/Melilotus (Red/White Clover)			0.0			0.0			0.3	L	13.1	M					0.0			
Ulmus (Elm)			0.0			0.3	L		0.0		0.0						0.0			
Verbena			0.0			0.0			0.0		0.0						0.0			
Viola (Violet)			0.3	L		0.0			0.0		0.0						0.0			
Vitis (Grape)			0.0			0.0			0.0		0.0						0.0			
Zanthoxylum (Prickly Ash)			0.0			0.0			0.0		0.0						0.0			
Zea mays (Corn)			0.0			0.0			0.0		0.0						0.0			
Unknown #1			0.0			3.3	M		14.9	M	0.0						0.0			
Unknowns #3 and #9			0.0			0.0			1.5	L	0.8	L					0.5	L		
Broken Palynomorphs			0.0			0.0			0.0		0.0						0.0			
Totals ▶			100			100			100		100						100			

May							M/J		Jun			J/J	Jul				J/A	Aug			A/S	Sep			S/O	Oct				O/N		Nov	
19		20		21		22	23		24	25	26	27	28	29	30	31	32	33	34	35	36	37	38	39	40	41	42	43	44	45		46	
%	PC	%	PC	%	PC		%	PC																						%	PC	%	PC
0.0		0.0		0.0			0.0																							0.0		0.5	L
0.0		0.0		0.0			0.0																							0.3	L	0.0	
0.0		0.0		0.0			0.0																							0.0		0.0	
0.0		0.0		0.0			0.0																							11.6	M	5.0	M
0.2	L	0.0		0.0			0.6	L																						0.3	L	0.9	L
0.0		0.0		0.0			0.0																							0.7	L	0.0	
0.8	L	0.6	L	0.0			0.0																							22.8	I	18.7	I
0.0		0.0		0.0			0.0																							0.0		0.5	L
30.0	I	26.3	I	33.8	I		6.2	M																						0.0		3.7	M
0.2	L	0.0		0.0			0.0																							0.0		0.0	
0.8	L	0.0		0.0			0.0																							0.0		0.0	
0.4	L	0.0		0.0			0.0																							0.0		0.0	
0.4	L	0.0		0.0			0.0																							0.0		0.0	
0.4	L	0.0		0.0			0.0																							0.0		0.0	
0.0		0.0		0.0			0.0																							0.0		0.5	L
0.8	L	0.0		0.0			0.0																							0.0		0.0	
0.0		1.2	L	0.0			0.0																							0.0		0.0	
0.2	L	0.0		0.0			0.9	L																						0.0		0.0	
0.0		0.0		0.0			0.0																							0.0		0.0	
0.0		0.0		0.0			0.0																							0.0		0.9	L
0.8	L	1.2	L	1.3	L		0.3	L																						0.0		0.5	L
1.8	L	0.6	L	1.3	L		0.3	L																						0.0		0.0	
0.4	L	0.0		0.0			0.6	L																						3.1	M	21.0	I
1.0	L	0.0		0.9	L		0.0																							0.0		0.0	
6.5	M	14.1	M	4.8	M		0.0																							0.0		3.2	M
0.2	L	0.0		0.0			0.0																							0.0		0.0	
0.0		0.0		0.0			0.0																							0.0		0.0	
16.7	I	10.1	M	3.5	M		2.2	L																						0.0		0.5	L
6.2	M	0.9	L	0.4	L		7.1	M																						0.0		0.5	L
0.0		0.0		0.0			0.0																							2.0	L	0.0	
8.9	M	10.1	M	2.2	L		11.5	M																						0.0		2.3	L
0.0		1.8	L	0.4	L		0.0																							0.0		3.7	M
0.0		0.0		0.0			0.0																							0.0		0.0	
0.0		0.0		0.0			0.0																							0.0		0.0	
0.2	L	1.8	L	0.4	L		0.3	L																						22.4	I	10.5	M
0.0		0.0		0.0			0.0																							9.9	M	1.4	L
0.0		0.0		0.0			0.0																							0.3	L	0.0	
1.8	L	3.7	M	0.9	L		0.0																							0.0		0.0	
1.8	L	0.6	L	0.0			0.0																							0.0		0.9	L
0.2	L	0.0		0.0			0.0																							0.0		0.0	
0.0		0.0		0.0			0.0																							5.1	M	0.5	L
0.0		0.0		0.0			0.0																							0.0		0.0	
14.8	M	12.8	M	21.9	I		43.0	I																						1.7	L	2.7	L
0.0		0.0		0.0			0.0																							0.0		0.0	
0.0		0.0		0.0			0.0																							0.0		0.0	
0.0		0.0		0.0			0.0																							0.3	L	0.0	
0.0		0.0		0.0			0.0																							0.3	L	0.0	
0.0		0.0		0.0			0.0																							0.0		0.0	
0.4	L	6.1	M	13.2	M		10.5	M																						0.0		0.9	L
0.0		0.0		0.0			0.0																							0.3	L	2.3	L
0.0		0.0		0.0			0.0																							1.4	L	0.5	L
0.0		0.0		0.0			0.0																							0.0		0.0	
2.4	L	5.2	M	3.1	M		14.9	M																						5.4	M	10.5	M
0.0		0.0		0.0			0.0																							0.3	L	0.5	L
0.0		0.0		0.0			0.6	L																						0.0		0.0	
0.0		0.0		0.0			0.0																							0.0		0.0	
0.0		0.0		0.0			0.0																							0.0		0.0	
1.8	L	2.8	L	11.8	M		0.9	L																						11.6	M	7.3	M
100		100		100			100																							100		100	

Table 4.15. Pollen Occurrence in Honey Collected in Apiary 13 in the Southeastern Plains Ecoregion during 2022

Apiary 13	Jan		J/F	Feb			F/M		Mar				M/A	Apr					
Week ▶	4	5	6	7	8	9	10		11		12	13	14	15	16		17	18	
Type of Data ▶							%	PC	%	PC					%	PC		%	PC
Acer (Maple)							46.9	D	0.0						24.0	I		1.6	L
Alnus (Alder)							0.0		0.0						0.0			0.0	
Amaranthaceae (Goosefoot family)							0.1	L	0.0						0.6	L		0.0	
Amaryllidaceae (Amaryllis/Onion family)							0.0		0.0						0.0			0.0	
Asteraceae (Coneflower/Golden Rod/Sunflower family)							0.4	L	0.0						0.0			3.7	M
Asteraceae (Dandelion/Cat's Ear/Endive family)							0.3	L	0.0						0.3	L		2.9	L
Berchemia (Rattan Vine)							0.0		0.0						0.0			0.0	
Bidens (Tickseed)							0.3	L	0.0						0.0			0.0	
Brassicaceae (Mustard family)							18.9	I	0.0						1.8	L		2.1	L
Camellia							1.3	L	0.0						0.0			0.8	L
Corylus/Carpinus (Hazel/Hornbeam)							0.0		0.0						0.0			0.8	L
Carya (Hickory/Pecan)							0.0		0.0						0.0			0.5	L
Castanea (Chinkapin/Chestnut)							0.0		0.0						0.0			0.0	
Cercis (Redbud)							0.0		46.8	D					20.5	I		0.0	
Cornus (Dogwood)							0.1	L	26.2	I					3.6	M		0.0	
Diospyros (Persimmon)							0.0		0.0						0.0			4.8	M
Elaeagnus (Autumn/Russian Olive)							0.1	L	0.0						0.0			0.0	
Ericaceae (Heath family)							0.0		0.0						0.0			0.0	
Erodium							0.0		0.0						0.0			0.3	L
Fabaceae (Pea family)							0.0		0.0						0.0			0.0	
Fagopyrum (Buckwheat)							0.0		0.0						0.0			0.0	
Gelsemium (Yellow Jessamine)							0.0		0.0						0.0			0.0	
Hamamelis (Witch Hazel)							0.0		0.0						0.0			0.5	L
Hypercicaceae (St. John's Wort family)							0.0		0.0						3.0	M		0.0	
Ilex (Holly)							2.2	L	0.0						2.7	L		10.8	M
Lagerstroemia (Crepe Myrtle)							3.6	M	1.2	L					5.0	M		4.8	M
Lamiaceae (Mint family)							0.3	L	0.0						0.0			0.8	L
Ligustrum (Privet)							0.0		0.0						0.0			1.3	L
Liliaceae (Lily family)							0.0		0.0						0.0			0.3	L
Liquidambar (Sweetgum)							0.3	L	0.6	L					0.0			0.0	
Liriodendron (Tulip Poplar)							2.6	L	0.9	L					1.8	L		0.0	
Lonicera (Honeysuckle)							0.1	L	0.0						0.0			0.0	
Magnolia							0.1	L	0.0						0.0			4.0	M
Mimosa (Sensitive Plant)							0.0		0.0						0.0			0.0	
Myrtaceae (Myrtle family)							0.0		0.0						0.0			0.0	
Nyssa (Tupelo/Black Gum)							0.0		0.0						1.2	L		1.1	L
Oxydendrum (Sourwood)							0.0		0.0						0.0			0.0	
Parthenocissus (Virginia Creeper)							2.1	L	0.0						0.0			2.9	L
Phlox							0.4	L	0.0						0.0			0.0	
Pinus (Pine)							0.0		0.0						0.0			0.0	
Plantago (Plantain)							0.0		0.0						0.0			0.0	
Poaceae (Grass family)							0.0		0.0						0.3	L		9.8	M
Populus (Aspen/Cottonwood/Poplar)							0.0		0.0						0.0			0.0	
Quercus (Oak)							0.0		0.0						0.0			0.0	
Rhus/Toxicodendron (Sumac/Poison Ivy)							1.4	L	0.0						0.0			2.6	L
Ribes (Currant)							0.0		0.0						0.0			1.1	L
Robinia (Locust)							1.0	L	15.4	M					1.5	L		0.0	
Rosaceae (Rose/Cherry/Plum/Peach/Blackberry family)							6.6	M	0.6	L					1.8	L		21.2	I
Rubus (Blackberry/Dewberry)							0.0		0.0						0.0			0.0	
Salix (Willow)							0.1	L	8.0	M					15.1	M		0.3	L
Saururus (Lizard's Tail)							0.0		0.0						0.0			0.0	
Solanum							0.0		0.0						0.0			0.0	
Tillandsia							0.8	L	0.0						0.0			0.0	
Trifolium/Melilotus (Red/White Clover)							0.3	L	0.0						5.0	M		1.1	L
Ulmus (Elm)							0.0		0.3	L					0.0			0.3	L

May								M/J		Jun					J/J	Jul				J/A	Aug			A/S	Sep			S/O		Oct					O/N	Nov
19		20		21		22		23		24	25		26		27	28	29	30	31	32	33	34	35	36	37	38	39	40		41	42		43	44	45	46
%	PC	%	PC	%	PC	%	PC	%	PC		%	PC	%	PC														%	PC		%	PC				
0.0		0.4	L	0.7	L	4.4	M	1.3	L		0.4	L	0.0															0.0			0.4	L				
0.0		0.0		0.0		0.0		0.0			0.4	L	0.0															0.0			0.0					
0.0		0.0		0.0		0.0		0.0			0.0		0.0															0.0			0.0					
0.0		0.0		0.0		0.0		0.0			0.0		1.9	L														0.0			0.0					
0.5	L	1.6	L	1.5	L	0.7	L	0.0			0.4	L	0.7	L														19.4	I		3.9	M				
0.5	L	0.8	L	0.0		0.3	L	0.0			0.7	L	0.0															1.4	L		1.3	L				
0.0		0.0		0.4	L	0.0		0.0			0.0		0.0															0.0			0.0					
0.0		0.0		0.0		0.0		0.0			0.0		0.0															0.0			0.0					
0.0		0.8	L	2.2	L	1.7	L	1.3	L		1.8	L	1.5	L														0.0			0.0					
0.0		0.0		0.0		0.0		0.0			0.0		0.0															0.0			0.0					
0.0		0.0		0.0		0.3	L	0.0			0.0		0.0															0.0			0.0					
0.5	L	0.0		0.4	L	0.0		0.0			0.0		0.0															0.0			0.0					
6.3	M	7.7	M	0.0		0.0		0.0			0.0		0.0															0.0			0.0					
0.5	L	0.4	L	0.0		0.0		0.0			0.0		0.0															0.0			0.0					
0.0		0.0		0.0		0.0		0.9	L		0.4	L	0.4	L														0.0			0.0					
0.0		0.0		8.5	M	4.7	M	2.2	L		3.2	M	7.9	M														4.2	M		3.5	M				
0.0		0.0		0.0		0.0		0.0			0.0		0.0															0.0			0.0					
0.0		0.0		18.5	I	2.7	L	0.0			0.0		0.0															0.0			0.0					
0.0		0.0		0.0		0.0		0.0			0.0		0.0															0.0			0.0					
2.7	L	8.9	M	0.0		0.0		1.3	L		1.4	L	0.0															0.0			0.0					
0.0		0.0		0.0		0.0		0.9	L		0.0		0.0															0.0			0.0					
0.0		0.0		1.5	L	20.6	I	32.6	I		15.1	M	10.1	M														0.0			1.3	L				
0.5	L	0.8	L	0.0		0.0		0.0			0.0		0.0															0.0			0.0					
5.0	M	0.0		0.0		8.4	M	0.4	L		0.0		0.0															0.0			1.3	L				
4.1	M	17.9	I	1.9	L	0.3	L	0.4	L		0.0		0.7	L														0.0			2.2	L				
3.6	M	2.0	L	0.7	L	0.0		1.8	L		4.3	M	5.2	M														9.7	M		32.6	I				
0.0		2.4	L	0.0		1.4	L	0.0			1.1	L	3.3	M														0.0			0.4	L				
11.3	M	24.8	I	0.7	L	0.0		3.5	M		9.0	M	0.4	L														0.0			1.7	L				
0.0		0.0		0.0		0.0		0.0			0.0		0.0															0.0			0.0					
0.0		0.0		0.0		0.3	L	0.0			0.0		0.4	L														0.0			0.0					
3.2	M	0.8	L	0.7	L	0.0		0.4	L		0.0		0.0															0.0			0.0					
0.0		0.0		0.0		0.0		0.0			0.0		0.0															0.0			0.0					
0.0		0.0		1.9	L	0.7	L	0.0			1.8	L	0.7	L														4.2	M		1.3	L				
0.0		0.0		0.0		0.0		0.0			1.1	L	0.4	L														0.0			0.0					
0.0		0.0		0.0		0.0		0.4	L		0.0		1.1	L														0.0			0.0					
2.7	L	0.8	L	13.3	M	3.4	M	0.0			1.1	L	0.4	L														1.4	L		0.0					
0.0		0.0		0.0		0.0		0.4	L		1.8	L	0.0															0.0			0.0					
0.0		0.0		0.4	L	0.3	L	0.0			0.0		0.0															0.0			0.0					
0.0		0.0		0.0		0.0		0.0			0.0		0.0															0.0			0.9	L				
0.0		0.0		0.0		0.0		0.0			0.7	L	0.0															1.4	L		0.0					
0.5	L	0.0		0.0		0.0		0.0			0.0		0.0															0.0			0.0					
3.2	M	1.6	L	0.0		0.0		0.0			0.7	L	1.9	L														1.4	L		0.4	L				
0.0		0.0		0.0		0.0		0.9	L		0.4	L	0.0															0.0			0.0					
0.5	L	0.0		2.2	L	2.0	L	0.4	L		0.0		0.7	L														0.0			0.0					
0.9	L	0.0		1.5	L	1.0	L	0.4	L		0.7	L	0.4	L														6.9	M		10.0	M				
0.0		0.0		0.0		0.0		0.0			0.0		0.0															0.0			0.0					
0.0		0.0		0.0		0.0		0.0			0.0		0.0															0.0			0.0					
25.7	I	19.9	I	13.3	M	14.5	M	11.5	M		11.1	M	22.8	I														11.1	M		8.7	M				
0.0		0.0		0.0		0.0		0.0			0.0		0.4	L														0.0			0.0					
0.0		0.0		0.0		0.0		0.0			0.0		0.0															0.0			0.0					
0.0		0.4	L	0.0		0.0		0.0			0.0		0.0															0.0			0.0					
0.0		0.0		0.0		0.0		6.2	M		2.2	L	0.0															0.0			0.0					
0.0		0.0		0.0		0.0		0.0			0.0		0.0															0.0			0.0					
18.5	I	7.7	M	2.2	L	1.4	L	1.8	L		3.6	M	9.0	M														20.8	I		14.8	M				
0.0		0.0		0.0		0.0		0.0			0.0		0.0															8.3	M		2.2	L				

(continued)

Table 4.15. (*continued*)

Apiary 13	Jan		J/F	Feb			F/M		Mar				M/A	Apr				
Week ▶	4	5	6	7	8	9	10		11		12	13	14	15	16		17	18	
Type of Data ▶							%	PC	%	PC					%	PC		%	PC
Urtica (Nettles)							0.0		0.0						0.0			0.0	
Verbena							0.0		0.0						0.0			0.0	
Viola (Violet)							0.0		0.0						0.0			0.0	
Vitis (Grape)							0.0		0.0						0.0			0.0	
Wisteria							0.0		0.0						1.8	L		0.0	
Zea mays (Corn)							0.0		0.0						0.0			0.0	
Unknowns #3 and #9							0.0		0.0						0.0			0.0	
Broken Palynomorphs							9.6	M	0.0						10.1	M		19.8	I
Totals ▶							100		100						100			100	

Key: *D* (predominant) >45% I (secondary) 16-45% M (important) 3-15% L (minor) <3%

Table 4.16. Pollen Occurrence in Honey Collected in Apiary 7 in the Southeastern Plains Ecoregion during 2022

Apiary 7	Jan		J/F		Feb				F/M		Mar					M/A		Apr							
Week ▶	4	5	6		7	8		9	10		11		12	13		14		15		16		17		18	
Type of Data ▶			%	PC		%	PC		%	PC	%	PC		%	PC	%	PC	%	PC	%	PC	%	PC	%	PC
Acer (Maple)			13.5	M		21.7	I		8.4	M	8.9	M		4.3	M	0.0		2.3	L	3.7	M	0.0		0.0	
Aesculus (Buckeye)			0.0			0.5	L		0.0		0.0			0.0		0.0		0.0		0.0		0.0		0.0	
Alnus (Alder)			0.0			0.0			0.0		0.0			0.5	L	0.0		0.0		0.0		0.0		0.0	
Alternanthera (Joyweed)			0.0			0.0			0.0		0.0			0.0		0.0		0.0		0.0		0.0		0.0	
Amaranthaceae (Goosefoot family)			0.0			0.0			0.4	L	0.0			0.0		0.0		0.0		0.0		0.0		0.0	
Amaryllidaceae (Amaryllis/Onion family)			0.0			0.0			0.0		0.0			0.0		0.0		1.7	L	0.0		0.0		0.0	
Apiaceae (Carrot/Parsely/Queen Anne's Lace family)			0.0			0.0			0.0		0.0			0.0		0.0		0.0		0.0		0.0		0.0	
Arecaceae (Palm family)			0.0			0.0			0.0		0.0			0.0		0.0		0.0		0.0		0.0		0.0	
Asteraceae (Coneflower/Golden Rod/Sunflower family)			0.0			0.0			0.0		0.0			0.0		0.0		3.1	M	0.0		0.0		3.0	M
Asteraceae (Dandelion/Cat's Ear/Endive family)			0.0			0.0			0.0		0.0			0.0		0.0		0.0		0.0		1.3	L	0.8	L
Baptisia (Wild Indigo)			0.0			0.0			0.0		0.0			1.9	L	0.0		0.0		0.0		0.0		0.0	
Boraginaceae (Forget-me-not family)			0.0			0.0			0.0		0.0			0.0		0.0		0.0		0.0		0.0		0.0	
Brassicaceae (Mustard family)			12.6	M		7.2	M		0.0		0.0			2.9	L	0.0		1.1	L	0.4	L	4.7	M	3.4	M
Buddleja (Butterfly Bush)			0.0			0.0			0.8	L	0.8	L		0.0		0.0		0.0		0.0		0.0		0.0	
Camellia			1.0	L		12.6	M		3.2	M	7.7	M		1.4	L	0.0		0.0		0.0		0.0		0.0	
Corylus/Carpinus (Hazel/Hornbeam)			0.0			0.0			0.0		0.0			0.0		0.0		0.0		0.0		0.0		0.0	
Carya (Hickory/Pecan)			0.0			0.0			0.0		0.0			0.5	L	0.0		0.8	L	0.0		0.4	L	1.3	L
Castanea (Chinkapin/Chestnut)			0.0			0.0			0.0		0.0			0.0		0.0		0.0		0.0		0.0		0.0	
Ceanothus			0.0			0.0			0.0		0.0			0.0		0.0		0.0		0.0		0.0		0.0	
Celtis (Hackberry)			0.0			0.0			0.0		0.0			0.0		0.0		0.0		0.0		0.0		0.0	
Cephalanthus (Buttonbush)			0.0			0.0			0.0		0.0			0.0		0.0		0.0		0.0		0.0		0.0	
Cercis (Redbud)			0.0			0.0			20.8	I	33.5	I		24.8	I	0.0		9.9	M	19.4	I	6.9	M	0.8	L
Cornus (Dogwood)			0.0			0.0			0.0		0.0			0.0		2.8	L	0.6	L	1.1	L	1.7	L	0.0	
Cyperaceae (Sedge family)			0.0			0.0			0.0		0.0			0.0		0.0		0.6	L	0.7	L	0.0		0.0	
Dalea (Prairie Clover)			0.0			0.0			0.0		0.0			0.0		0.0		0.0		0.0		0.0		0.0	
Diospyros (Persimmon)			0.0			0.0			0.0		0.0			0.0		0.0		0.0		0.0		0.0		0.0	
Elaeagnus (Autumn/Russian Olive)			0.0			0.0			0.0		0.4	L		0.5	L	0.0		0.0		0.0		0.0		0.0	
Ericaceae (Heath family)			0.0			0.0			0.0		0.0			0.0		0.0		0.0		0.0		0.0		0.0	
Fabaceae (Pea family)			0.0			0.5	L		2.0	L	0.0			0.0		0.0		0.0		0.0		13.4	M	0.0	
Gelsemium (Yellow Jessamine)			0.0			0.0			0.0		0.0			0.0		0.0		0.0		0.0		0.0		0.0	
Glycine max (Soybean)			0.0			0.0			0.0		0.0			0.0		0.0		0.0		0.0		0.4	L	0.0	
Hexasepalum teres (Poorjoe/Rough Buttonweed)			0.0			0.0			0.0		0.0			0.0		6.9	M	0.0		0.0		0.0		0.0	

May				M/J	Jun			J/J	Jul				J/A	Aug			A/S	Sep			S/O	Oct				O/N	Nov
19	20	21	22	23	24	25	26	27	28	29	30	31	32	33	34	35	36	37	38	39	40	41	42	43	44	45	46
% PC	% PC	% PC	% PC	% PC		% PC	% PC														% PC		% PC				
0.0	0.0	0.0	0.0	0.4 L		0.4 L	0.7 L														0.0		0.0				
0.0	0.0	0.0	0.0	0.0		0.0	0.0														0.0		1.7 L				
0.0	0.0	0.0	0.0	0.0		0.0	0.0														1.4 L		0.0				
0.0	0.0	20.4 I	15.5 M	18.9 I		8.6 M	13.1 M														4.2 M		1.3 L				
0.0	0.0	0.0	0.0	0.0		0.0	0.0														0.0		0.0				
0.0	0.0	0.0	0.0	0.0		0.0	0.0														1.4 L		0.0				
0.0	0.0	0.0	0.0	0.4 L		0.0	0.0														0.0		0.0				
9.9 M	0.0	7.0 M	15.2 M	11.0 M		28.0 I	15.7 I														2.8 L		10.0 M				
100	100	100	1	100.0		100	100														100		100				

May				M/J	Jun			J/J	Jul				J/A	Aug			A/S	Sep			S/O	Oct				O/N	Nov
19	20	21	22	23	24	25	26	27	28	29	30	31	32	33	34	35	36	37	38	39	40	41	42	43	44	45	46
% PC	% PC	% PC	% PC		% PC		% PC					% PC				% PC			% PC	% PC		% PC	% PC	% PC			
0.0	1.9 L	0.0	0.0		0.0		0.0					0.0				0.0			0.0	0.0		0.6 L	0.0	0.0			
0.3 L	0.0	0.0	0.0		0.0		0.0					0.0				0.0			0.0	0.0		0.0	0.0	0.0			
0.0	0.0	0.0	0.0		0.0		0.0					0.0				0.0			0.0	0.0		0.0	0.0	0.0			
0.0	0.0	0.0	0.0		0.0		0.8 L					0.0				0.0			0.0	0.0		0.0	0.0	0.0			
0.0	0.0	0.0	0.0		0.0		0.0					1.6 L				0.9 L			0.3 L	0.0		0.8 L	0.0	0.3 L			
0.0	0.0	0.0	0.0		0.0		0.0					0.0				0.4 L			0.8 L	0.0		0.0	0.0	0.0			
0.0	0.0	0.0	0.0		0.0		0.0					1.6 L				0.0			0.0	1.2 L		0.0	0.0	0.0			
0.0	0.0	0.0	0.0		0.0		0.0					0.0				0.0			0.0	0.0		0.0	0.0	2.4 L			
0.3 L	0.0	0.3 L	0.7 L		1.3 L		1.7 L					6.0 M				6.4 M			34.6 I	0.8 L		1.7 L	64.9 D	32.3 I			
0.0	0.0	0.0	0.0		2.6 L		0.0					0.3 L				0.0			0.3 L	0.4 L		0.3 L	0.4 L	0.0			
0.0	0.0	0.0	0.0		0.0		0.0					0.0				0.0			0.0	0.0		0.0	0.0	0.0			
0.0	0.0	0.0	0.0		0.0		0.0					0.0				0.0			0.0	0.0		0.0	0.0	0.5 L			
0.0	0.0	3.7 M	11.6 M		0.0		0.0					1.9 M				1.7 L			1.7 L	1.6 L		5.0 M	4.1 M	0.8 L			
0.0	0.0	0.0	0.0		0.0		0.0					0.0				0.0			0.3 L	0.0		0.0	0.4 L	0.0			
0.0	0.0	0.0	0.0		0.0		0.0					0.0				0.0			0.0	0.0		0.0	0.0	0.0			
0.0	0.0	0.0	0.2 L		0.0		0.0					0.0				0.0			0.0	0.0		0.0	0.0	0.0			
0.3 L	0.0	1.7 L	0.2 L		0.7 L		0.0					0.0				0.0			0.0	0.0		0.0	0.0	0.0			
8.9 M	2.3 L	0.0	0.0		0.0		0.0					0.0				0.0			0.0	0.0		0.0	0.0	0.0			
0.0	0.0	0.0	0.0		0.0		0.0					3.8 M				0.0			0.0	0.0		0.0	0.0	0.0			
0.0	0.0	0.0	0.0		0.0		0.0					0.0				0.0			0.3 L	0.0		0.0	0.4 L	0.0			
0.6 L	0.9 L	0.0	0.0		0.0		0.0					0.0				0.0			0.0	0.0		0.0	0.0	0.0			
5.7 M	1.4 L	0.0	0.0		0.0		0.0					0.0				0.0			0.0	0.0		0.0	0.0	0.0			
1.0 L	0.0	0.0	0.0		0.0		0.0					0.0				0.4 L			0.0	0.0		0.0	0.0	0.0			
0.0	0.0	0.0	0.0		0.0		0.0					0.0				0.0			0.0	0.0		0.0	0.0	0.0			
0.3 L	0.0	0.0	0.0		0.0		0.0					0.0				0.0			0.0	0.0		0.0	0.0	0.0			
0.0	0.0	3.1 M	2.1 L		2.0 L		0.8 L					0.3 L				0.0			0.3 L	0.0		0.0	0.0	0.0			
0.0	0.0	0.0	0.5 L		0.0		0.0					0.3 L				0.0			0.0	0.0		0.0	0.0	0.0			
0.0	0.0	0.0	3.8 M		0.0		0.0					0.0				0.0			0.0	0.0		0.0	0.0	0.0			
0.3 L	8.9 M	0.0	0.0		0.0		0.0					0.0				0.0			0.0	0.0		0.0	0.0	0.0			
0.0	0.0	0.0	0.0		1.3 L		0.8 L					2.5 L				0.0			0.0	0.0		0.3 L	0.0	0.0			
0.0	0.0	0.0	0.0		0.0		0.0					0.0				0.0			0.0	0.0		0.0	0.0	0.0			
0.0	0.0	0.0	0.0		0.0		0.0					0.0				0.0			0.0	0.0		0.0	0.0	0.0			

(continued)

Table 4.16. (*continued*)

Apiary 7	Jan		J/F			Feb			F/M		Mar					M/A		Apr							
Week ▸	**4**	**5**	**6**		**7**	**8**		**9**	**10**		**11**		**12**	**13**		**14**		**15**		**16**		**17**		**18**	
Type of Data ▸			%	PC		%	PC		%	PC	%	PC		%	PC	%	PC	%	PC	%	PC	%	PC	%	PC
Hypercicaceae (St. John's Wort family)			1.4	L		0.0			0.0		0.0			1.0	L	0.0		2.0	L	0.0		0.9	L	1.7	L
Ilex (Holly)			1.0	L		0.0			1.2	L	3.2	M		3.8	M	6.9	M	0.6	L	2.2	L	4.7	M	11.0	M
Lagerstroemia (Crepe Myrtle)			0.0			0.0			0.0		0.0			0.0		1.4	L	4.5	M	1.9	L	0.0		3.0	M
Lamiaceae (Mint family)			19.4	I		4.8	M		5.6	M	8.1	M		15.7	I	21.1	I	0.0		0.0		0.8	L	3.4	M
Ligustrum (Privet)			0.0			0.0			0.0		0.0			0.0		0.0		0.0		0.0		0.9	L	2.1	L
Liliaceae (Lily family)			0.0			0.0			1.2	L	0.4	L		0.0		0.0		0.0		0.0		0.0		0.0	
Liquidambar (Sweetgum)			0.0			0.5	L		0.4	L	0.0			0.0		0.0		0.6	L	0.4	L	0.0		0.0	
Liriodendron (Tulip Poplar)			0.0			0.0			0.0		0.0			0.0		0.0		1.7	L	0.4	L	1.3	L	1.7	L
Lonicera (Honeysuckle)			0.0			0.0			0.0		0.0			0.0		0.0		0.0		0.0		0.0		0.0	
Magnolia			0.0			0.5	L		0.0		0.4	L		0.0		0.0		0.0		0.0		0.0		0.0	
Morus (Mulberry)			0.0			9.2	M		0.0		0.4	L		0.0		0.0		0.0		0.0		0.0		0.0	
Myrtaceae (Myrtle family)			0.0			0.0			0.0		0.0			0.0		0.0		0.3	L	0.0		0.0		0.0	
Nyssa (Tupelo/Black Gum)			0.0			0.0			0.0		0.0			0.0		0.0		0.0		0.0		0.0		5.1	M
Oxydendrum (Sourwood)			0.0			0.0			0.0		0.0			0.0		0.0		0.0		0.0		0.0		0.0	
Parthenocissus (Virginia Creeper)			0.0			0.0			0.0		0.0			0.0		0.0		0.6	L	0.7	L	6.0	M	0.0	
Phoardendron			0.0			0.0			0.0		0.8	L		0.5	L	0.0		0.0		0.0		0.0		0.0	
Pinus (Pine)			0.0			0.0			0.0		0.0			0.0		0.0		0.3	L	0.0		0.4	L	3.8	M
Plantago (Plantain)			0.0			0.0			0.0		0.0			0.0		0.0		0.0		1.1	L	0.0		0.0	
Poaceae (Grass family)			0.5	L		0.0			0.0		0.0			0.0		0.5	L	0.0		0.4	L	0.0		1.3	L
Portulaca (Purslanes)			0.0			0.0			0.0		0.0			0.0		0.0		0.0		0.0		0.0		0.0	
Primulaceae (Primrose family)			0.0			0.0			0.0		0.0			0.0		0.0		0.0		0.0		0.0		0.4	L
Quercus (Oak)			0.0			0.0			0.0		0.0			0.5	L	9.2	M	0.0		0.0		1.7	L	2.1	L
Rhus/Toxicodendron (Sumac/Poison Ivy)			0.0			0.0			0.0		0.0			0.0		0.0		0.0		0.0		8.2	M	0.4	L
Ribes (Currant)			1.4	L		0.0			0.0		0.0			2.9	L	0.0		0.0		0.0		0.4	L	0.0	
Richardia			0.0			0.0			0.0		0.0			0.0		0.0		0.0		0.0		0.0		0.0	
Robinia (Locust)			2.9	L		0.0			6.8	M	1.2	L		1.4	L	0.0		48.6	D	19.0	I	0.4	L	0.0	
Rosaceae (Rose/Cherry/Plum/Peach/Blackberry family)			13.1	M		29.5	I		49.2	D	21.8	I		23.8	I	33.0	I	1.4	L	14.6	M	15.1	M	25.3	I
Salix (Willow)			14.0	M		13.0	M		0.0		0.0			0.0		0.0		2.3	L	0.4	L	0.0		2.1	L
Spathiphyllum			0.0			0.0			0.0		0.0			0.0		0.0		0.0		0.0		0.0		0.0	
Trifolium/Melilotus (Red/White Clover)			2.4	L		0.0			0.0		0.0			10.0	M	14.7	M	1.7	L	10.8	M	15.9	I	11.4	M
Typha			0.0			0.0			0.0		0.0			0.0		0.0		0.0		0.0		0.0		0.0	
Ulmus (Elm)			16.9	M		0.0			0.0		0.0			0.0		0.0		0.6	L	0.0		0.0		0.0	
Viola (Violet)			0.0			0.0			0.0		0.0			0.0		0.0		0.0		0.0		0.0		0.0	
Vitis (Grape)			0.0			0.0			0.0		0.0			0.0		0.0		0.0		0.0		0.0		0.0	
Wisteria			0.0			0.0			0.0		0.0			3.8	M	1.4	L	0.0		1.5	L	0.0		0.0	
Zea mays (Corn)			0.0			0.0			0.0		0.0			0.0		0.0		0.3	L	0.0		0.0		0.0	
Unknown #1			0.0			0.0			0.0		0.0			0.0		0.0		0.0		0.0		0.9	L	0.0	
Unknowns #3 and #9			0.0			0.0			0.0		0.0			0.0		0.0		0.0		0.0		0.0		0.0	
Broken Palynomorphs			0.0			0.0			0.0		12.5	M		0.0		2.3	L	14.7	M	21.3	I	13.4	M	16.0	I
Totals ▸			100			100			100		100			100		100		100		100		100		100	

Key: *D* (predominant) >45% I (secondary) 16-45% M (important) 3-15% L (minor) <3%

May								M/J	Jun					J/J	Jul					J/A	Aug					A/S	Sep					S/O	Oct							O/N	Nov
19		20		21		22		23	24		25	26		27	28	29	30	31		32	33	34	35		36	37	38		39		40	41		42		43		44	45	46
%	PC	%	PC	%	PC	%	PC		%	PC		%	PC					%	PC				%	PC			%	PC	%	PC		%	PC	%	PC	%	PC			
3.5	M	2.3	L	0.0		0.0			3.3	M		1.7	L					2.5	L				3.9	M			0.3	L	0.0			3.6	M	0.0		0.3	L			
9.6	M	3.3	M	5.4	M	1.7	L		0.0			0.0						0.3	L				0.9	L			0.6	L	0.0			3.3	M	0.0		0.0				
0.0		0.0		0.0		0.2	L		16.4	I		12.5	M					21.1	I				24.9	I			8.6	M	4.0	M		3.6	M	0.0		34.0	I			
0.0		0.0		1.7	L	0.2	L		0.7	L		0.0						0.0					1.3	L			0.8	L	0.0			0.6	L	0.0		0.0				
7.6	M	3.7	M	5.8	M	1.9	L		2.0	L		29.2	I					0.6	L				0.4	L			0.0		0.0			3.9	M	0.4	L	0.5	L			
0.0		0.0		0.0		0.0			0.0			0.0						0.0					0.0				0.0		0.0			0.0		0.0		0.0				
0.0		0.0		0.0		0.0			0.0			0.0						0.3	L				0.0				0.0		0.4	L		0.3	L	0.0		0.0				
0.0		36.0	I	0.0		45.6	D		0.0			1.7	L					0.0					0.0				0.3	L	0.0			0.0		0.0		0.0				
0.0		0.0		0.0		0.0			0.0			0.0						0.0					0.0				0.3	L	0.0			0.0		0.0		0.0				
0.6	L	0.0		0.3	L	0.5	L		0.7	L		2.5	L					0.3	L				0.9	L			0.0		0.0			0.6	L	0.0		0.0				
0.0		0.0		0.0		0.0			0.0			0.0						0.0					0.0				0.0		0.0			0.0		0.0		0.0				
0.0		0.0		0.0		0.0			0.0			0.0						0.0					0.0				0.0		0.0			0.0		0.0		0.0				
14.0	M	4.2	M	54.8	D	14.9	M		0.0			0.0						0.0					2.6	L			0.0		0.0			0.6	L	0.0		1.9	L			
0.0		2.8	L	0.0		0.0			0.0			0.0						0.0					0.0				0.0		0.0			0.0		0.0		0.0				
0.0		0.0		0.0		0.0			0.0			0.0						0.0					0.0				0.3	L	0.0			0.0		0.4	L	0.3	L			
0.0		0.0		0.0		0.0			0.0			0.0						0.0					0.0				0.0		0.0			0.0		0.0		0.0				
0.0		0.9	L	1.4	L	0.0			2.6	L		1.7	L					0.3	L				0.0				0.3	L	0.4	L		0.3	L	0.0		0.3	L			
0.0		0.0		0.0		1.2	L		2.6	L		1.7	L					0.0					0.0				0.0		0.0			0.0		0.0		0.0				
0.0		2.3	L	1.0	L	0.5	L		0.7	L		0.8	L					8.8	M				2.6	L			9.7	M	4.0	M		3.3	M	1.1	L	5.7	M			
0.0		0.0		0.0		0.0			0.0			0.0						0.0					0.0				0.0		0.0			0.0		16.0	I	0.3	L			
0.0		0.0		0.0		0.0			0.0			0.0						0.0					0.0				0.0		0.0			0.0		0.0		0.0				
1.9	L	0.0		0.3	L	0.2	L		0.0			0.0						0.0					0.9	L			0.0		0.0			0.3	L	0.0		0.0				
0.0		0.0		1.4	L	0.0			0.0			1.7	L					0.3	L				2.1	L			0.6	L	0.0			2.2	L	0.4	L	0.0				
0.0		0.0		0.0		0.0			0.0			0.0						0.0					0.0				0.0		0.0			0.0		0.0		0.0				
0.0		0.0		0.0		0.0			0.0			0.0						0.0					0.0				0.0		0.8	L		0.0		0.0		2.2	L			
0.6	L	0.0		0.0		0.0			0.0			0.0						0.0					0.0				0.0		0.0			0.0		0.0		0.0				
26.1	I	7.0	M	5.1	M	4.3	M		11.8	M		10.0	M					6.9	M				10.7	M			1.1	L	0.8	L		19.0	I	0.7	L	0.8	L			
0.0		0.5	L	0.0		0.0			0.0			0.0						0.0					0.0				0.0		0.0			0.0		0.0		0.0				
0.0		0.0		0.0		0.0			0.0			0.0						0.0					0.0				0.0		0.0			0.0		0.7	L	0.0				
2.2	L	0.0		4.1	M	1.2	L		5.3	M		4.2	M					0.9	L				1.7	L			0.6	L	0.8	L		4.4	M	0.4	L	0.0				
0.0		0.0		0.0		0.0			0.0			0.0						1.3	L				0.0				0.0		0.0			0.0		0.0		0.0				
0.0		0.0		0.0		0.0			0.0			0.0						0.0					0.0				6.6	M	64.8	D		0.0		2.6	L	0.8	L			
0.0		0.0		0.0		0.0			0.0			0.0						0.0					0.4	L			0.3	L	0.0			0.3	L	0.0		0.0				
0.0		0.0		0.0		3.3	M		12.5	M		11.7	M					1.3	L				1.7	L			0.8	L	0.0			1.1	L	0.0		0.3	L			
0.0		0.0		0.0		0.0			0.0			0.0						0.0					0.0				0.0		0.0			0.0		0.0		0.0				
0.0		0.0		0.0		0.0			0.0			0.0						0.0					0.0				0.3	L	0.0			0.0		0.0		0.0				
0.0		0.0		0.0		0.0			0.0			0.0						0.0					0.0				0.0		0.0			0.0		0.0		0.0				
0.0		0.0		0.0		0.0			0.0			0.0						0.0					0.0				0.0		0.0			0.3	L	0.0		0.0				
15.9	I	21.5	I	9.9	M	5.2	M		33.6	I		16.7	I					36.8	I				35.2	I			30.2	I	20.2	I		44.1	I	7.1	M	16.4	I			
100		100		100		100			100			100						100					100				100		100			100		100		100				

4.4 Middle Atlantic Coastal Plain

Sample hives in the Middle Atlantic Coastal Plain ecoregion. Photo credit: Karen Hilbourn.

The Middle Atlantic Coastal Plain ecoregion is comprised of flat, low-elevation plains containing numerous wetland environments including swamps, marshes, and estuaries that occur between the Southeastern Plains and the Southern Coastal Plain (Griffith et al., 2002). Most of the sediments that make up these complex marine and terrestrial wetland and fluvial environments were deposited during the Pleistocene and Holocene epochs; they are among the youngest sediments in the state. Annual precipitation in this region generally ranges from 112–142 cm (44–56 in.), and the greatest precipitation occurs along the northeastern coast while the least occurs along the border with the Southeastern Plains (Runkle et al., 2022). Soils developed atop the underlying unconsolidated sediments and are significantly finer than those in the Southeastern Plains, leading to relatively poorly-drained fine-loamy and coarse-loamy soils on the flatlands, with clayey and organic soils developing in wetlands and swamps (such as in Carolina bays and the freshwater evergreen shrub wetlands called pocosins) and newly-forming sandy, clayey, and organic soils along the river valleys (Griffith et al., 2002).

The upper portion of the ecoregion is characterized by remnant pine flatwoods, pine savannahs, freshwater marshes, ponds, and pocosins; it has been heavily impacted by artificial drainage for forestry and agriculture, and much of the region is now devoted to loblolly pine (*Pinus*) plantations and agricultural tracts. This region is dissected by floodplains and low terraces containing and bordering large slow-flowing rivers and deepwater swamps; oxbow lakes occur adjacent to many of the rivers. Like the adjacent Southeastern Plains, this part of the ecoregion contains bottomland hardwood forests composed of wetland oaks *(Quercus)*, hickories (*Carya*), green ash (*Fraxinus*), and red maple (*Acer*) trees (SC DNR, 2005d). Cypress (*Taxodium*) and sweetgum (*Nyssa*) trees dominate the swamps. The northeastern coast is characterized by barrier islands and coastal marshes (SC DNR, 2005d). Forests found in these settings include live oak and laurel oak (*Quercus*), red cedar (*Juniperus*), loblolly pine (*Pinus*), yaupon holly (*Ilex*), wax myrtle (*Morella*), and dwarf palmetto and cabbage palm (*Sabal*) (SC DNR, 2005d). The region is biologically diverse (Griffth et al., 2002) and hosts a wide range of herbaceous plants, especially in wetlands and coastal areas.

Hood (2006) notes 21 taxa of importance in the Coastal Plain, which includes both the Middle Atlantic Coastal Plain ecoregion and the Southern Coastal Plain ecoregion. These are: red maple (*Acer*) from mid-January through February; yellow jessamine (*Gelsemium*) and wild plum (Rosaceae [rose/cherry/plum/peach/blackberry family]) from mid-February through March; sassafras (*Sassafras*) and willow (*Salix*) from mid-March through April; tupelo (*Nyssa*) and poplar (*Populus*) from late March through late April; blackberry (*Rubus*) from late March through late June; American holly (*Ilex*) from mid-April through mid-May; gallberry (*Ilex*) from late April through late May; snow vine (*Porana*) and samson snakeroot (Fabaceae) from mid-April through late May; pepper bush (*Clethra*) from late May through mid-July; Mexican clover (*Richardia*) from early July through late September; goldenrod (Asteraceae [coneflower/goldenrod/sunflower family]) from mid-September through late October; cabbage palmetto (*Sabal*) from early September through early October; heartsease (*Viola*) from early September through late October; flat top goldenrod (Asteraceae [coneflower/goldenrod/sunflower family]) from early September through mid-October; Spanish needle (*Yucca*) from early October through early December; and golden aster (Asteraceae [coneflower/goldenrod/sunflower family]) from early October through mid-December. Like the Southeastern Plains, the Middle Atlantic Coastal Plain falls into Region 12 of the Honey Bee Forage Map (Ayers & Harmon, 1992). Of the 56 taxa they reported, four are considered important nectar sources in this region: tupelo (*Nyssa*) from March through June, tulip poplar (*Liriodendron*) and holly (*Ilex*) from April through June, and goldenrod (Asteraceae [coneflower/goldenrod/sunflower family]) from July through November. Sourwood (*Oxydenrum*) is unlikely to occur, as this area is outside of its native range. Of the 59 genera noted by Smith (2021), none are noted with specific occurrences in this ecoregion; however, it is assumed that they are present and that, like in the Southeastern Plains, many have extended bloom times.

Hive weight gains in the Middle Atlantic Coastal Plain were small in the early spring (Figure 4.5), with fairly consistent small increases noted in early February and early to mid-April. An early- to mid-May weight increase ranged from small in most hives to substantial in hive 4. All hives showed substantial weight gains in mid-late June except hive 5, in which weight gains were very small until the late July-early August increase seen in all hives. Weight increases were seen again in late September-early October and again in early November.

Eight of the 21 important taxa noted in Hood (2006) for the Coastal Plain section of South Carolina were observed in the Middle Atlantic Coastal Plain ecoregion (Table 4.17). These included: *Acer* (maple), Asteraceae (coneflower/goldenrod/sunflower family), *Gelsemium* (yellow jessamine), *Ilex* (holly), *Nyssa* (tupelo/black gum), *Populus* (aspen/cottonwood/poplar), Rosaceae (rose/cherry/plum/peach/blackberry family), and *Salix* (willow). All taxa were observed at roughly the same period of the year as noted by Hood but were generally found earlier and later than previously noted. The exception was *Populus* (aspen/cottonwood/poplar), which was observed in early October instead of April.

For this region, too, all four of the participating apiaries produced monofloral *Acer* (maple) honey for at least one sample from mid-February through early May (Tables 4.18, 4.19, 4.20, and 4.21). Monofloral honeys are indicated with red color-coding in the apiary pollen occurrence tables.

And in this region, like the others, *Acer* (maple) honey was not the only monofloral honey produced. Of the 54 taxa observed in the Middle Atlantic Coastal Plain, 13 taxa produced monofloral honey samples. Apiary 2 (Table 4.18) produced monofloral Rosaceae (rose/cherry/plum/peach/blackberry family) honey in mid-May and *Ilex* (holly) honey in mid-July. Apiary 4 (Table 4.19) produced monofloral Unknowns #3 and #9 honey in early April, *Trifolium/Melilotus* (red/white clover) honey in late April and early May (two samples), and *Ulmus* (elm) honey in early October. Apiary 5 (Table 4.20) produced monofloral Rosaceae (rose/cherry/plum/peach/blackberry family) honey throughout March (three samples), *Cornus* (dogwood) honey in early April, Unknowns #3 and #9 honey in early and mid-May (two samples), *Nyssa* (tupelo/black gum) honey in late May and late October (one sample each), *Glycine max* (soybean) honey in late September, Asteraceae (ragweed/calendula/solidago family) honey in early October (two samples), and *Ulmus*

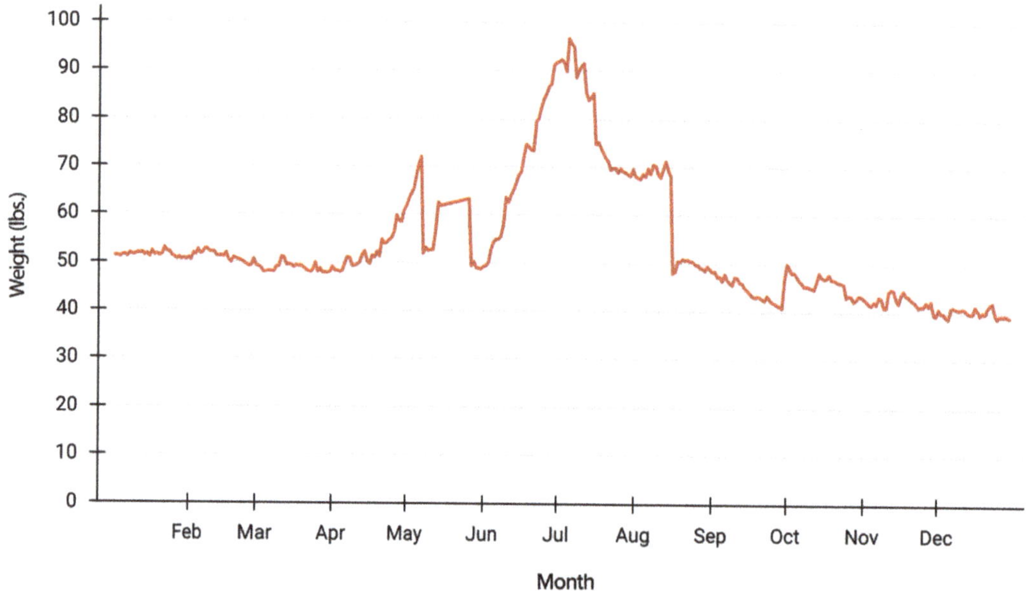

Figure 4.5. An example of BroodMinder hive weight data collected from Apiary 4 in the Middle Atlantic Coastal Plain.

Note: Labels indicate the first day of each listed month.

(elm) honey in late September. Apiary 8 (Table 4.21) produced monofloral Rosaceae (rose/cherry/plum/peach/blackberry family) honey in early April, early May, and early October (one sample each); *Magnolia* honey in late April through early June (four samples); and *Ilex* (holly) honey in late May and mid-August (one sample each). While Poaceae (grass family) pollen was dominant in mid-September, most grasses produce little nectar, and this ought not be classified as a truly monofloral honey. Apiarists should note that the pollen content of the honey reflects the plants that contribute to the nectar the bees have foraged to generate that honey. Likewise, the proportion of pollen observed for each plant taxon provides information about the contribution that plant made to the honey. Bees in the Middle Atlantic Coastal Plain were making monofloral honey throughout the year (from mid-February into early October) from a wide variety of plants.

The Middle Atlantic Coastal Plain also had many plants/taxa contributing to the honey for extended periods of time. These plants are blooming for a sufficiently long time to be available forage for bees. There were 30 taxa in the Middle Atlantic Coastal Plain that contributed to honey for at least three straight weeks (Table 4.17). While this is not as many taxa as observed in the Piedmont or Southeastern Plains, it is still more than half (55.6%) of the 54 total taxa observed in this ecoregion (see Chapter 5). These 30 taxa were: *Acer* (maple), Amaryllidaceae (amaryllis/onion family), Asteraceae (coneflower/goldenrod/sunflower family), Asteraceae (dandelion/cat's ear/endive family), Asteraceae (ragweed/calendula/solidago family), Brassicaceae (mustard family), *Camellia*, Caryophyllaceae (carnation family), *Cornus* (dogwood), *Diospyros* (persimmon), *Gelsemium* (yellow jessamine), *Glycine max* (soybean), *Hexasepalum teres* (poor Joe/rough buttonweed), *Ilex* (holly), *Lagerstroemia* (crepe myrtle), *Ligustrum* (privet), *Liriodendron* (tulip poplar), *Magnolia*, *Nyssa* (tupelo/black gum), *Parthenocissus* (Virginia creeper), *Pinus* (pine), Poaceae (grass family), *Quercus* (oak), *Rhus*/*Toxicodendron* (sumac/poison ivy), Rosaceae (rose/cherry/plum/peach/blackberry family), *Salix* (willow), *Trifolium*/*Melilotus* (red/white clover), *Ulmus* (elm), *Vitis* (grape), and Unknowns #3 and #9.

Table 4.17. Plant Taxa whose Pollen was Observed in Honey Collected from the Middle Atlantic Coastal Plain Ecoregion during 2022

Middle Atlantic Coastal Plain	Jan		J/F	Feb			F/M	Mar			M/A	Apr			
Week ▸	4	5	6	7	8	9	10	11	12	13	14	15	16	17	18
Acer (Maple)															
Alnus (Alder)															
Alternanthera (Joyweed)															
Amaranthaceae (Goosefoot family)															
Amaryllidaceae (Amaryllis/Onion family)															
Asteraceae (Coneflower/Golden Rod/Sunflower family)															
Asteraceae (Dandelion/Cat's Ear/Endive family)															
Asteraceae (Ragweed/Calendula/Solidago family)															
Betula (Birch)															
Brassicaceae (Mustard family)															
Camellia															
Canna															
Carya (Hickory/Pecan)															
Caryophyllaceae (Carnation family)															
Cornus (Dogwood)															
Diospyros (Persimmon)															
Ericaceae (Heath family)															
Fagus (Beech)															
Gelsemium (Yellow Jessamine)															
Glycine max (Soybean)															
Gossypium (Cotton)															
Hexasepalum teres (Poorjoe/Rough Buttonweed)															
Hypercicaceae (St. John's Wort family)															
Ilex (Holly)															
Lagerstroemia (Crepe Myrtle)															
Ligustrum (Privet)															
Liliaceae (Lily family)															
Liquidambar (Sweetgum)															
Liriodendron (Tulip Poplar)															
Lonicera (Honeysuckle)															
Magnolia															
Morus (Mulberry)															
Narcissus (Daffodil)															
Nyssa (Tupelo/Black Gum)															
Parthenocissus (Virginia creeper)															
Pinus (Pine)															
Plantago (Plantain)															
Poaceae (Grass family)															
Polygonum/Persicaria (Buckwheat/Knotweed)															
Populus (Aspen/Cottonwood/Poplar)															
Quercus (Oak)															
Rhus/Toxicodendron (Sumac/Poison Ivy)															
Rosaceae (Rose/Cherry/Plum/Peach/Blackberry family)															
Salix (Willow)															
Sambucus (Elderberry)															
Solanum (Tomato)															
Symphoricarpos (Snowberry)															
Tradescantia (Spiderwort)															
Trifolium/Melilotus (Red/White Clover)															
Ulmus (Elm)															
Vitis (Grape)															
Zea mays (Corn)															
Unknowns #3 and #9															
Unknown #13															
Various Unknown Palynomorphs															
Broken Palynomorphs															

Key: *no data collected* ☐ *taxon absent on this week* ☐ *taxon present on this week* ■

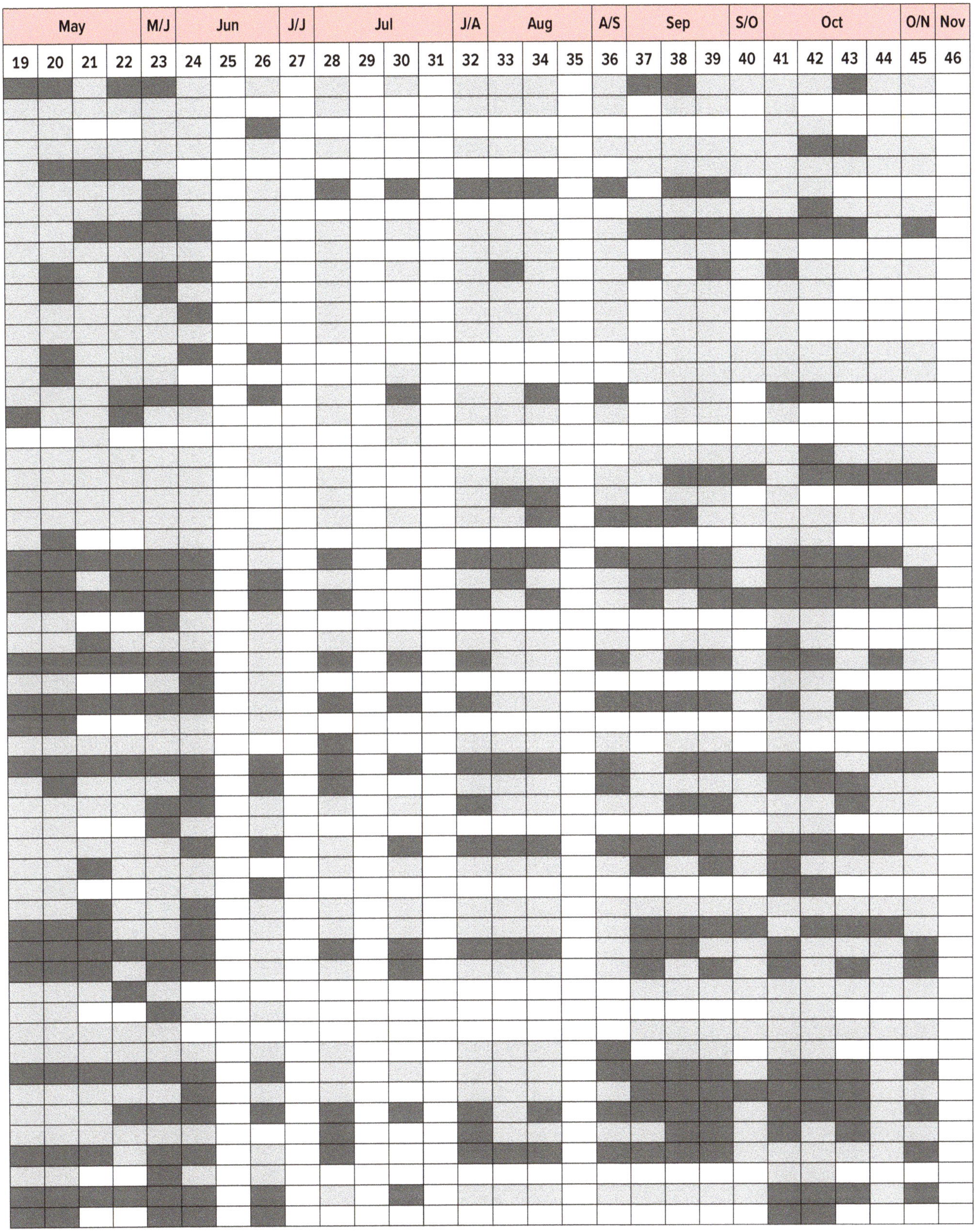

May				M/J	Jun			J/J	Jul				J/A	Aug			A/S	Sep			S/O	Oct				O/N	Nov
19	20	21	22	23	24	25	26	27	28	29	30	31	32	33	34	35	36	37	38	39	40	41	42	43	44	45	46

Table 4.18. Pollen Occurrence in Honey Collected in Apiary 2 in the Middle Atlantic Coastal Plain Ecoregion during 2022

Apiary 2	Jan		J/F	Feb			F/M	Mar				M/A		Apr					
Week ▶	4	5	6	7	8	9	10	11	12	13		14		15		16	17	18	
Type of Data ▶										%	PC	%	PC	%	PC			%	PC
Acer (Maple)										71.3	D			6.5	M			0.0	
Asteraceae (Coneflower/Golden Rod/Sunflower family)										0.0				0.0				0.0	
Asteraceae (Ragweed/Calendula/Solidago family)										10.2	M			2.8	L			0.0	
Brassicaceae (Mustard family)										0.0				0.9	L			0.0	
Cornus (Dogwood)										0.0				3.3	M			0.0	
Diospyros (Persimmon)										0.0				0.0				0.0	
Ericaceae (Heath family)										0.5	L			0.0				0.0	
Fagus (Beech)										0.0				0.0				1.4	L
Hexasepalum teres (Poorjoe/Rough Buttonweed)										0.0				1.4	L			0.5	L
Ilex (Holly)										0.5	L			0.0				0.9	L
Liquidambar (Sweetgum)										0.0				2.8	L			0.5	L
Liriodendron (Tulip Poplar)										0.0				0.5	L			6.4	M
Magnolia										0.0				0.0				0.0	
Nyssa (Tupelo/Black Gum)										0.0				0.0				24.5	I
Pinus (Pine)										0.5	L			0.0				0.5	L
Poaceae (Grass family)										5.6	M			0.0				0.0	
Polygonum/Persicaria (Buckwheat/Knotweed)										9.3	M			1.9	L			0.0	
Quercus (Oak)										0.0				34.1	I			38.2	I
Rhus/Toxicodendron (Sumac/Poison Ivy)										0.0				0.0				2.3	L
Rosaceae (Rose/Cherry/Plum/Peach/Blackberry family)										0.0				0.5	L			2.3	L
Salix (Willow)										0.0				44.9	I			22.3	I
Trifolium/Melilotus (Red/White Clover)										2.3	L			0.5	L			0.0	
Vitis (Grape)										0.0				0.0				0.0	
Various Unknown Palynomorphs										0.0				0.0				0.5	L
Totals ▶										100				100				100	

Key: *D* (predominant) >45%   I (secondary) 16-45%   M (important) 3-15%   L (minor) <3%

May					M/J	Jun			J/J	Jul					J/A	Aug			A/S	Sep			S/O	Oct				O/N	Nov
19	20	21		22	23	24	25	26	27	28	29	30		31	32	33	34	35	36	37	38	39	40	41	42	43	44	45	46
		%	PC									%	PC																
		0.0										0.0																	
		0.0										5.1	M																
		1.9	L									0.0																	
		0.0										0.0																	
		0.0										0.0																	
		0.0										0.5	L																
		0.0										0.0																	
		0.0										0.0																	
		0.0										0.0																	
		0.0										59.9	D																
		2.9	L									0.0																	
		0.0										0.5	L																
		0.0										0.5	L																
		0.5	L									21.7	I																
		0.0										0.0																	
		0.0										4.6	M																
		0.5	L									0.0																	
		0.5	L									0.0																	
		0.0										0.0																	
		81.3	D									0.5	L																
		4.8	M									0.5	L																
		1.0	L									0.0																	
		0.0										0.9	L																
		6.7	M									5.5	M																
		100										100																	

Table 4.19. Pollen Occurrence in Honey Collected in Apiary 4 in the Middle Atlantic Coastal Plain Ecoregion during 2022

Apiary 4	Jan		J/F		Feb				F/M	Mar					M/A		Apr							
Week ▶	4	5	6		7	8		9	10	11	12	13		14		15		16		17		18	
Type of Data ▶			%	PC		%	PC					%	PC	%	PC	%	PC	%	PC	%	PC	%	PC
Acer (Maple)			0.0			84.3	D					92.6	D	78.9	D	27.9	I	0.0		0.9	L	0.0	
Alternanthera (Joyweed)			0.0			0.0						0.0		0.0		0.0		0.0		0.0		0.0	
Amaranthaceae (Goosefoot family)			0.5	L		0.0						0.0		0.0		0.4	L	0.0		0.0		0.0	
Asteraceae (Coneflower/Golden Rod/Sunflower family)			33.0	I		0.0						0.0		0.0		0.4	L	0.0		0.0		0.0	
Asteraceae (Dandelion/Cat's Ear/Endive family)			0.0			0.7	L					0.5	L	0.0		1.2	L	0.0		0.0		0.0	
Asteraceae (Ragweed/Calendula/Solidago family)			0.0			0.0						0.0		0.0		0.0		0.0		0.0		0.0	
Brassicaceae (Mustard family)			0.0			0.2	L					0.0		0.0		0.0		0.0		0.0		0.0	
Camellia			28.2	I		0.0						0.5	L	0.4	L	2.4	L	0.0		0.0		0.0	
Caryophyllaceae (Carnation family)			0.0			0.0						0.0		0.8	L	0.0		0.0		0.0		0.0	
Diospyros (Persimmon)			0.0			0.0						0.0		0.0		0.0		0.0		0.0		0.0	
Gelsemium (Yellow Jessamine)			31.6	I		0.0						0.0		0.0		1.2	L	0.0		0.0		0.0	
Hypercicaceae (St. John's Wort family)			0.0			0.0						0.0		0.0		0.0		0.0		0.0		0.0	
Ilex (Holly)			0.0			8.0	M					0.0		0.0		6.5	M	0.0		0.0		35.1	I
Lagerstroemia (Crepe Myrtle)			0.0			0.0						0.0		0.0		0.8	L	0.0		0.0		1.0	L
Ligustrum (Privet)			0.0			0.0						0.0		0.0		1.2	L	0.0		0.0		1.0	L
Liliaceae (Lily family)			0.0			0.0						0.0		0.0		0.0		0.0		0.0		0.0	
Liquidambar (Sweetgum)			0.0			0.0						0.0		0.0		0.8	L	0.0		0.0		0.0	
Liriodendron (Tulip Poplar)			0.0			0.0						0.0		0.0		4.5	M	0.4	L	0.4	L	0.0	
Lonicera (Honeysuckle)			0.0			0.0						0.0		0.0		0.0		0.0		0.0		0.0	
Magnolia			0.0			0.0						0.0		0.0		0.8	L	0.0		0.0		0.0	
Morus (Mulberry)			0.0			0.2	L					0.0		0.0		0.0		0.0		0.0		0.0	
Nyssa (Tupelo/Black Gum)			0.0			0.0						0.0		0.0		0.0		0.0		0.0		0.5	L
Parthenocissus (Virginia creeper)			0.0			0.0						0.0		0.0		0.0		31.8	I	7.9	M	2.9	L
Pinus (Pine)			0.0			0.0						0.0		0.0		1.2	L	0.0		0.0		0.0	
Plantago (Plantain)			0.0			0.0						0.0		0.0		0.4	L	0.0		0.0		0.0	
Poaceae (Grass family)			0.0			0.0						0.0		0.0		0.0		0.0		0.0		0.0	
Populus (Aspen/Cottonwood/Poplar)			0.0			0.0						0.0		0.0		0.0		0.0		0.0		0.0	
Quercus (Oak)			0.0			0.0						0.0		0.0		0.0		0.0		0.0		0.0	
Rhus/Toxicodendron (Sumac/Poison Ivy)			0.0			0.0						0.0		0.0		0.0		0.0		0.0		0.0	
Rosaceae (Rose/Cherry/Plum/Peach/Blackberry family)			0.0			0.0						0.0		0.4	L	13.4	M	2.5	L	0.0		0.0	
Salix (Willow)			0.0			0.0						2.8	L	1.2	L	2.0	L	0.0		10.9	M	0.0	
Solanum (Tomato)			0.0			0.0						0.0		0.0		0.0		0.0		0.0		0.0	
Tradescantia (Spiderwort)			0.0			0.0						0.0		2.1	L	0.8	L	0.0		0.4	L	0.0	
Trifolium/Melilotus (Red/White Clover)			0.0			0.0						0.0		0.0		12.1	M	3.2	M	25.8	I	49.5	D
Ulmus (Elm)			0.0			3.1	M					0.0		1.7	L	0.0		0.0		0.0		0.0	
Vitis (Grape)			0.0			0.0						0.0		0.0		0.0		0.0		0.0		3.4	M
Unknowns #3 and #9			0.0			0.0						0.0		0.0		0.0		59.9	D	41.5	I	4.8	M
Unknown #13			0.0			0.0						0.0		0.0		0.0		0.0		0.0		0.0	
Various Unknown Palynomorphs			3.3	M		1.2	L					0.0		0.8	L	3.2	M	0.0		0.9	L	0.0	
Broken Palynomorphs			3.3	M		2.2	L					3.7	M	13.6	M	18.6	I	2.2	L	11.4	M	1.9	L
Totals ▶			100			100						100		100		100		100		100		100	

Key: *D* (predominant) >45% I (secondary) 16-45% M (important) 3-15% L (minor) <3%

May						M/J		Jun					J/J	Jul				J/A	Aug			A/S	Sep			S/O	Oct						O/N	Nov
19		20		21	22	23		24		25	26		27	28	29	30	31	32	33	34	35	36	37	38	39	40	41		42		43	44	45	46
%	PC	%	PC			%	PC	%	PC		%	PC															%	PC	%	PC				
76.9	D	0.0				29.8	I	0.0			0.0																0.0		0.0					
0.0		0.0				0.0		0.0			0.5	L															0.0		0.0					
0.0		0.0				0.0		0.0			0.0																0.0		0.4	L				
0.0		0.0				0.9	L	0.0			0.0																0.0		0.0					
0.0		0.0				1.3	L	0.0			0.0																0.0		0.4	L				
0.0		0.0				1.3	L	0.9	L		0.0																2.2	L	1.3	L				
0.0		17.5	I			2.1	L	1.7	L		0.0																0.9	L	0.0					
0.0		0.5	L			0.4	L	0.0			0.0																0.0		0.0					
0.0		0.0				0.0		1.3	L		0.5	L															0.0		0.0					
0.0		0.0				5.1	M	19.7	I		74.1	D															12.0	M	36.9	I				
0.0		0.0				0.0		0.0			0.0																0.0		3.0	M				
0.0		0.5	L			0.0		0.0			0.0																0.0		0.0					
0.2	L	0.9	L			4.3	M	1.7	L		0.0																0.0		1.7	L				
0.0		0.0				1.3	L	13.1	M		2.3	L															0.9	L	7.2	M				
0.0		0.0				0.9	L	0.4	L		5.0	M															6.2	M	0.8	L				
0.0		0.0				0.4	L	0.0			0.0																0.0		0.0					
0.0		0.0				0.0		0.0			0.0																0.4	L	0.0					
0.0		0.0				0.0		0.0			0.0																0.0		0.4	L				
0.0		0.0				0.0		0.9	L		0.0																0.0		0.0					
0.0		0.0				0.0		0.0			0.0																0.0		0.0					
0.2	L	0.5	L			0.0		0.0			0.0																0.0		0.0					
0.0		0.0				2.1	L	7.0	M		0.9	L															0.4	L	6.4	M				
0.0		1.4	L			0.0		3.1	M		2.7	L															1.8	L	5.1	M				
0.0		0.0				1.3	L	3.1	M		0.0																0.0		0.0					
0.0		0.0				0.4	L	0.0			0.0																0.0		0.0					
0.0		0.0				0.0		2.6	L		0.5	L															0.4	L	1.7	L				
0.0		0.0				0.0		0.0			0.5	L															1.8	L	0.8	L				
0.0		0.0				0.0		3.1	M		0.0																0.0		0.0					
0.0		0.0				0.0		0.0			0.0																0.0		2.1	L				
18.3	I	0.0				3.4	M	0.4	L		0.0																0.0		0.0					
0.2	L	3.8	M			3.4	M	0.4	L		0.0																0.9	L	0.0					
0.0		0.0				0.4	L	0.0			0.0																0.0		0.0					
0.0		0.0				0.0		0.0			0.0																0.0		0.0					
1.2	L	64.6	D			15.7	I	22.7	I		1.4	L															11.6	M	12.3	M				
0.0		0.0				0.0		0.4			0.0																47.1	D	0.4	L				
0.0		0.0				2.1	L	0.0			1.4	L															0.4	L	0.4	L				
0.0		2.8	L			0.4	L	0.4	L		0.0																0.0		0.0					
0.0		0.0				3.0	M	0.0			0.0																0.0		0.0					
0.2	L	2.4	L			5.1	M	4.4	M		1.8	L															2.7	L	4.7	M				
2.6	L	5.2	M			14.9	M	12.7	M		8.6	M															10.2	M	14.0	M				
100		100				100		100			100																100		100					

Table 4.20. Pollen Occurrence in Honey Collected in Apiary 5 in the Middle Atlantic Coastal Plain Ecoregion during 2022

Apiary 5	Jan		J/F	Feb				F/M		Mar					M/A	Apr						
Week ▶	4	5	6	7	8		9	10		11	12		13		14	15		16		17		18
Type of Data ▶					%	PC		%	PC		%	PC	%	PC		%	PC	%	PC	%	PC	
Acer (Maple)					99.4	D		0.5	L		0.0		1.3	L		0.0		3.4	M	20.5	I	
Amaranthaceae (Goosefoot family)					0.0			0.0			0.0		0.9	L		0.0		0.0		0.0		
Amaryllidaceae (Amaryllis/Onion family)					0.0			0.0			0.0		0.0			0.0		0.0		0.0		
Asteraceae (Dandelion/Cat's Ear/Endive family)					0.6	L		0.9	L		0.4	L	0.9	L		0.0		0.0		0.0		
Asteraceae (Ragweed/Calendula/Solidago family)					0.0			0.0			0.0		0.0			0.0		0.0		0.0		
Brassicaceae (Mustard family)					0.0			0.0			0.0		0.0			3.8	M	5.2	M	0.0		
Caryophyllaceae (Carnation family)					0.0			0.0			0.0		5.4	M		2.5	L	0.0		0.0		
Cornus (Dogwood)					0.0			0.0			0.0		3.6	M		64.2	D	3.4	M	5.1	M	
Gelsemium (Yellow Jessamine)					0.0			17.1	I		4.0		4.0	M		0.0		0.0		3.3	M	
Glycine max (Soybean)					0.0			0.0			0.0		0.0			0.0		0.0		0.0		
Hexasepalum teres (Poorjoe/Rough Buttonweed)					0.0			0.0			0.0		0.0			0.0		0.0		0.0		
Ilex (Holly)					0.0			0.0			0.0		0.0			0.0		12.3	M	2.8	L	
Lagerstroemia (Crepe Myrtle)					0.0			0.0			0.0		0.0			0.0		0.0		0.0		
Ligustrum (Privet)					0.0			0.0			0.0		0.0			0.0		0.0		7.9	M	
Liriodendron (Tulip Poplar)					0.0			0.0			0.0		0.0			0.0		7.1	M	0.0		
Magnolia					0.0			0.0			0.0		0.0			0.0		1.5	L	0.9	L	
Nyssa (Tupelo/Black Gum)					0.0			0.0			0.0		0.0			0.0		0.0		4.2	M	
Parthenocissus (Virginia Creeper)					0.0			0.0			0.0		0.0			0.0		0.0		0.0		
Pinus (Pine)					0.0			0.0			0.0		0.0			0.0		0.0		0.9	L	
Poaceae (Grass family)					0.0			0.0			0.0		0.0			0.0		0.0		0.0		
Polygonum/Persicaria (Buckwheat/Knotweed)					0.0			0.0			0.0		0.0			0.0		0.0		0.0		
Quercus (Oak)					0.0			0.0			0.0		0.0			0.4	L	4.1	M	2.8	L	
Rhus/Toxicodendron (Sumac/Poison Ivy)					0.0			0.0			0.0		0.0			0.0		0.0		0.0		
Rosaceae (Rose/Cherry/Plum/Peach/Blackberry family)					0.0			75.6	D		94.7	D	82.5	D		23.8	I	31.3	I	7.9	M	
Salix (Willow)					0.0			0.0			0.0		0.0			0.0		1.9	L	29.8	I	
Sambucus (Elderberry)					0.0			0.0			0.0		0.0			0.0		0.0		0.0		
Symphoricarpos (Snowberry)					0.0			5.1	M		0.0		1.3	L		2.5	L	0.0		0.0		
Trifolium/Melilotus (Red/White Clover)					0.0			0.0			0.0		0.0			0.0		0.0		5.1	M	
Ulmus (Elm)					0.0			0.0			0.0		0.0			0.0		0.0		0.9	L	
Vitis (Grape)					0.0			0.0			0.0		0.0			0.0		0.0		0.9	L	
Zea mays (Corn)					0.0			0.0			0.0		0.0			0.0		0.0		0.0		
Unknowns #3 and #9					0.0			0.0			0.0		0.0			0.0		0.0		7.0	M	
Various Unknown Palynomorphs					0.0			0.9	L		0.9	L	0.0			2.9	L	29.9	I	0.0		
Totals ▶					100			100			100		100			100		100		100		

Key: *D* (predominant) >45%   I (secondary) 16-45%   M (important) 3-15%   L (minor) <3%

May								M/J		Jun			J/J	Jul				J/A	Aug			A/S	Sep						S/O		Oct								O/N		Nov
19		20		21		22		23		24	25	26	27	28	29	30	31	32	33	34	35	36	37		38		39		40		41		42		43		44		45		46
%	PC	%	PC	%	PC	%	PC	%	PC														%	PC	%	PC	%	PC	%	PC	%	PC	%	PC	%	PC	%	PC	%	PC	
1.4	L	2.2	L	0.0		5.0	M	0.0															1.0	L	2.3	L	0.0		0.0		0.0		0.0		1.4	L	0.0		0.0		
0.0		0.0		0.0		0.0		0.0															0.0		0.0		0.0		0.0		0.0		0.0		1.4	L	0.0		0.0		
0.0		0.9	L	31.1	I	0.8	L	0.0															0.0		0.0		0.0		0.0		0.0		0.0		0.0		0.0		0.0		
0.0		0.0		0.0		0.0		0.0															0.0		0.0		0.0		0.0		0.0		0.0		0.0		0.0		0.0		
0.0		0.0		0.0		2.9	L	2.9	L														5.5	M	0.9	L	1.4	L	18.8	I	100.0	D	79.1	D	17.7	I	0.0		4.2	M	
0.0		1.3	L	0.0		5.8	M	0.4	L														2.0	L	0.0		0.9	L	0.0		0.0		0.0		0.0		0.0		0.0		
0.0		0.4	L	0.0		0.0		0.0															0.0		0.0		0.0		0.0		0.0		0.0		0.0		0.0		0.0		
0.0		1.3	L	0.0		0.0		0.0															0.0		0.0		0.0		0.0		0.0		0.0		0.0		0.0		0.0		
0.0		0.0		0.0		0.0		0.0															0.0		0.0		0.0		0.0		0.0		0.0		0.0		0.0		0.0		
0.0		0.0		0.0		0.0		0.0															0.0		0.0		49.5	D	13.5	M	0.0		15.5	M	22.0	I	5.6	M	24.5	I	
0.0		0.0		0.0		0.0		0.0															0.5	L	0.0		0.0		0.0		0.0		0.0		0.0		0.0		0.0		
0.5	L	4.4	M	3.8	M	10.8	M	16.0	I														1.5	L	17.4	I	1.9	L	0.0		0.0		0.0		2.4	L	5.1	M	0.0		
0.0		4.9	M	0.0		18.3	I	11.5	M														9.5	M	0.5	L	10.4	M	0.0		0.0		0.0		26.8	I	0.0		9.0	M	
1.4	L	8.4	M	1.0	L	3.3	M	0.0															2.0	L	0.0		0.0		0.5	L	0.0		0.9	L	2.9	L	3.3	M	0.9	L	
1.8	L	0.0		1.0	L	0.0		8.6	M														0.0		0.0		0.0		0.0		0.0		0.0		0.0		9.8	M	0.0		
0.5	L	1.3	L	2.9	L	0.0		0.0															3.0	M	0.0		0.9	L	0.0		0.0		0.0		0.5	L	1.4	L	0.0		
3.6	M	6.2	M	11.0	M	6.7	M	52.0	D														0.0		34.9	I	1.4	L	0.5	L	0.0		0.0		0.0		47.9	D	0.5	L	
0.0		0.0		0.0		0.0		0.0															0.0		0.0		0.0		0.0		0.0		0.0		1.4	L	0.0		0.0		
0.0		0.0		0.0		0.0		0.0															0.0		0.0		0.0		0.0		0.0		0.0		0.5	L	0.0		0.0		
0.0		0.0		0.0		0.0		0.0															4.5	M	0.0		15.6	I	0.0		0.0		2.7	L	3.3	M	3.3	M	0.0		
0.0		0.0		0.0		0.0		0.0															1.0	L	0.0		0.0		0.0		0.0		0.0		0.0		0.0		0.0		
0.0		0.0		0.0		0.0		0.0															0.0		0.0		0.0		0.0		0.0		0.0		0.0		0.0		0.0		
0.0		0.0		2.4	L	0.0		0.0															10.4	M	2.8	L	4.2	M	2.4	L	0.0		0.5	L	5.7	M	23.7	I	0.0		
0.0		0.0		0.0		18.3	I	0.0															3.5	M	35.8	I	0.0		0.0		0.0		0.0		0.0		0.0		3.8	M	
0.0		29.2	I	0.0		0.0		0.0															3.0	M	0.0		6.6	M	0.0		0.0		0.0		1.9	L	0.0		20.8	I	
0.0		0.0		0.0		1.3	L	0.0															0.0		0.0		0.0		0.0		0.0		0.0		0.0		0.0		0.0		
0.0		0.0		0.0		0.0		0.0															0.0		0.0		0.0		0.0		0.0		0.0		0.0		0.0		0.0		
0.0		1.3	L	0.0		26.7	I	2.9	L														11.4	M	3.7	M	0.9	L	0.0		0.0		0.0		0.5	L	0.0		2.8	L	
0.0		0.0		0.0		0.0		0.0															40.3	I	1.8	L	2.4	L	64.3	D	0.0		0.9	L	2.9	L	0.0		0.0		
0.0		0.0		0.0		0.0		0.0															0.5	L	0.0		0.5	L	0.0		0.0		0.5	L	2.9	L	0.0		2.8	L	
0.0		0.0		0.0		0.0		0.0															0.0		0.0		2.8	L	0.0		0.0		0.0		1.0	L	0.0		0.0		
90.9	D	38.1	I	46.9	D	0.0		0.0															0.5	L	0.0		0.5	L	0.0		0.0		0.0		0.0		0.0		26.9	I	
0.0		0.0		0.0		0.0		5.7	M														0.0		0.0		0.0		0.0		0.0		0.0		4.8	M	0.0		3.8	M	
100		100		100		100		100															100		100		100		100		100		100		100		100		100		

Table 4.21. Pollen Occurrence in Honey Collected in Apiary 8 in the Middle Atlantic Coastal Plain Ecoregion during 2022

Apiary 8	Jan		J/F	Feb				F/M		Mar			M/A		Apr					
Week ▶	4	5	6	7	8		9	10		11	12	13	14		15	16	17		18	
Type of Data ▶					%	PC		%	PC				%	PC			%	PC	%	PC
Acer (Maple)					85.6	D		99.6	D				9.5	M			10.6	M	0.0	
Alnus (Alder)					1.4	L		0.0					0.0				0.0		0.0	
Amaranthaceae (Goosefoot family)					4.1	M		0.0					0.0				0.0		0.0	
Asteraceae (Coneflower/Golden Rod/Sunflower family)					0.0			0.0					0.0				0.0		0.0	
Asteraceae (Ragweed/Calendula/Solidago family)					1.8	L		0.0					0.0				0.0		0.0	
Betula (Birch)					1.4	L		0.0					0.0				0.0		0.0	
Brassicaceae (Mustard family)					2.7	L		0.0					0.0				2.6	L	3.3	M
Camellia					1.4	L		0.4	L				0.0				0.0		0.0	
Canna					0.0			0.0					0.0				0.0		0.0	
Carya (Hickory/Pecan)					0.9	L		0.0					0.0				0.0		0.0	
Diospyros (Persimmon)					0.0			0.0					0.0				1.3	L	0.0	
Ericaceae (Heath family)					0.0			0.0					0.0				0.0		0.0	
Glycine max (Soybean)					0.0			0.0					0.0				0.0		0.0	
Gossypium (Cotton)					0.0			0.0					0.0				0.0		0.0	
Hexasepalum teres (Poorjoe/Rough Buttonweed)					0.5	L		0.0					0.0				0.0		0.0	
Ilex (Holly)					0.0			0.0					0.0				0.0		0.0	
Lagerstroemia (Crepe Myrtle)					0.0			0.0					0.0				0.0		0.0	
Ligustrum (Privet)					0.0			0.0					0.0				0.0		0.0	
Liquidambar (Sweetgum)					0.0			0.0					1.8	L			0.0		0.0	
Liriodendron (Tulip Poplar)					0.0			0.0					0.0				4.0	M	9.3	M
Magnolia					0.0			0.0					0.0				32.6	I	74.8	D
Narcissus (Daffodil)					0.0			0.0					0.0				0.0		0.0	
Nyssa (Tupelo/Black Gum)					0.0			0.0					0.0				0.0		0.0	
Parthenocissus (Virginia creeper)					0.0			0.0					0.0				0.0		0.0	
Pinus (Pine)					0.0			0.0					0.9	L			2.6	L	0.0	
Poaceae (Grass family)					0.0			0.0					0.0				0.0		0.0	
Polygonum/Persicaria (Buckwheat/Knotweed)					0.0			0.0					0.0				0.0		0.0	
Quercus (Oak)					0.0			0.0					16.7	I			20.3	I	1.4	L
Rhus/Toxicodendron (Sumac/Poison Ivy)					0.0			0.0					0.0				0.0		2.3	L
Rosaceae (Rose/Cherry/Plum/Peach/Blackberry family)					0.0			0.0					70.3	D			20.3	I	0.0	
Salix (Willow)					0.0			0.0					0.0				0.9	L	0.0	
Tradescantia (Spiderwort)					0.0			0.0					0.9	L			0.0		0.0	
Trifolium/Melilotus (Red/White Clover)					0.5	L		0.0					0.0				3.5	M	0.9	L
Ulmus (Elm)					0.0			0.0					0.0				1.3	L	0.0	
Vitis (Grape)					0.0			0.0					0.0				0.0		0.0	
Zea mays (Corn)					0.0			0.0					0.0				0.0		0.0	
Unknowns #3 and #9					0.0			0.0					0.0				0.0		0.0	
Various Unknown Palynomorphs					0.0			0.0					0.0				0.0		7.9	M
Totals ▶					100			100					100				100		100	

Key: *D* (predominant) >45% I (secondary) 16-45% M (important) 3-15% L (minor) <3%

May								M/J		Jun				J/J	Jul					J/A		Aug					A/S		Sep					S/O	Oct					O/N	Nov
19		20		21		22		23		24		25	26	27	28		29	30	31	32		33		34		35	36		37	38		39		40	41		42	43	44	45	46
%	PC	%	PC	%	PC	%	PC	%	PC	%	PC				%	PC				%	PC	%	PC	%	PC		%	PC		%	PC	%	PC		%	PC					
0.0		0.0		0.0		0.0		0.0		0.0					0.0					0.0		0.0		0.0			0.0			0.0		0.0			0.0						
0.0		0.0		0.0		0.0		0.0		0.0					0.0					0.0		0.0		0.0			0.0			0.0		0.0			0.0						
0.0		0.0		0.0		0.0		0.0		0.0					0.0					0.0		0.0		0.0			0.0			0.0		0.0			0.0						
0.0		0.0		0.0		0.0		0.0		0.0					29.3	I				12.2	M	17.6	I	3.9	M		22.0	I		5.2	M	7.4	M		0.0						
0.0		0.0		0.0		0.0		0.0		0.0					0.0					0.0		0.0		0.0			0.0			0.0		1.9	L		7.5	M					
0.0		0.0		0.0		0.0		0.0		0.0					0.0					0.0		0.0		0.0			0.0			0.0		0.0			0.0						
0.0		3.1	M	0.0		0.3	L	0.5	L	0.0					0.0					0.0		1.9	L	0.0			0.0			0.0		0.0			0.0						
0.0		0.0		0.0		0.0		0.0		0.0					0.0					0.0		0.0		0.0			0.0			0.0		0.0			0.0						
0.0		0.0		0.0		0.0		0.0		0.9	L				0.0					0.0		0.0		0.0			0.0			0.0		0.0			0.0						
0.0		0.0		0.0		0.0		0.0		0.0					0.0					0.0		0.0		0.0			0.0			0.0		0.0			0.0						
0.0		0.0		0.0		7.0	M	0.0		0.0					0.0					0.0		0.0		1.5	L		2.4	L		0.0		0.0			0.4	L					
0.8	L	0.0		0.0		0.3	L	0.0		0.0					0.0					0.0		0.0		0.0			0.0			0.0		0.0			0.0						
0.0		0.0		0.0		0.0		0.0		0.0					0.0					0.0		0.0		0.0			0.0			1.9	L	0.5	L		0.0						
0.0		0.0		0.0		0.0		0.0		0.0					0.0					0.0		1.4	L	1.5	L		0.0			0.0		0.0			0.0						
0.0		0.0		0.0		0.0		0.0		0.0					0.0					0.0		0.0		7.8	M		1.0	L		0.9	L	0.0			0.0						
0.0		2.7	L	0.0		55.4	D	5.5	M	0.0					3.4	M				20.7	I	40.7	I	51.2	D		15.8	I		3.3	M	3.2	M		4.4	M					
0.8	L	0.0		0.0		0.0		0.0		0.0					0.0					0.0		1.4	L	0.0			0.0			0.0		0.0			0.0						
0.8	L	10.7	M	5.4	M	1.6	L	7.3	M	4.3	M				0.5	L				0.5	L	0.0		0.5	L		0.0			0.0		2.8	L		0.0						
0.0		0.0		0.0		0.0		0.0		0.0					0.0					0.0		0.0		0.0			0.0			0.0		0.0			0.0						
6.2	M	6.7	M	11.7	M	0.9	L	8.2	M	13.7	M				5.3	M				2.3	L	0.0		0.0			0.5	L		2.8	L	22.7	I		0.4	L					
30.6	I	46.7	D	42.6	I	2.8	L	60.3	D	48.8	D				28.8	I				13.1	M	0.0		0.0			1.4	L		6.6	M	0.0			0.9	L					
0.0		0.0		0.0		0.0		0.0		0.0					1.4	L				0.0		0.0		0.0			0.0			0.0		0.0			0.0						
0.4	L	0.9	L	0.4	L	9.2	M	3.7	M	1.9	L				0.5	L				6.3	M	0.5	L	10.2	M		3.3	M		1.9	L	0.9	L		2.7	L					
0.0		0.0		0.0		0.0		0.0		0.0					1.4	L				0.0		0.0		0.0			0.5	L		0.0		0.0			0.0						
0.0		0.0		0.0		0.0		0.5	L	0.0					0.0					0.5	L	0.0		0.0			0.0			0.5	L	0.5	L		0.0						
0.0		0.0		0.0		0.0		0.0		0.0					0.0					16.2	I	31.9	I	17.1	I		14.8	M		66.5	D	24.1	I		24.3	I					
0.0		0.0		0.0		0.0		0.0		0.0					0.0					0.0		0.0		0.0			0.0			0.0		0.5	L		2.7	L					
0.0		0.0		0.0		0.0		0.0		0.0					0.0					0.0		0.0		0.0			0.0			0.0		0.0			0.0						
8.7	M	1.8	L	3.6	M	0.0		0.0		0.9	L				0.0					0.0		0.0		0.0			0.0			0.0		0.0			0.0						
50.4	D	27.1	I	36.3	I	1.9	L	14.2	M	28.0	I				5.3	M				5.0	M	1.9	L	3.9	M		0.0			0.0		0.0			49.6	D					
0.0		0.0		0.0		0.0		0.0		0.0					0.0					0.0		0.0		0.0			0.0			0.0		0.0			0.0						
0.0		0.0		0.0		0.0		0.0		0.0					0.0					0.0		0.0		0.0			0.5	L		0.0		0.0			0.0						
0.0		0.4	L	0.0		0.0		0.0		0.9	L				0.0					0.0		0.0		0.0			0.5	L		1.4	L	0.0			0.0						
0.0		0.0		0.0		0.0		0.0		0.0					0.0					0.0		0.0		0.0			0.0			0.0		0.0			0.0						
0.0		0.0		0.0		17.1	I	0.0		0.5	L				7.7	M				4.1	M	0.0		1.0	L		17.7	I		7.5	M	35.2	I		6.6	M					
0.0		0.0		0.0		0.0		0.0		0.0					1.4	L				0.5	L	0.0		0.0			0.0			0.9	L	0.0			0.4	L					
1.2	L	0.0		0.0		0.0		0.0		0.0					14.9	M				18.9	I	2.8	L	1.5	L		19.6	I		0.5	L	0.5	L		0.0						
0.0		0.0		0.0		3.5	M	0.0		0.0					0.0					0.0		0.0		0.0			0.0			0.0		0.0			0.0						
10		100		100		100		100		100					100					100		100		100			100			100		100			100						

4.5 Southern Coastal Plain

Sample hives in Mt. Pleasant, South Carolina, in the Southern Coastal Plain ecoregion. Photo credit: Tim Liptak.

This is the northernmost part of the much larger Southern Coastal Plain ecoregion, which extends from the central South Carolina coast to Georgia, the central Florida peninsula and panhandle, southern Alabama, and southern Mississippi. The region is largely flat-lying but also contains barrier islands, coastal lagoons, marshes, and swampy lowlands; in some places it is dissected by broad river valleys (Griffth et al., 2002). Most of the unconsolidated sediment was deposited relatively recently. The region receives somewhat more precipitation on average than the Middle Atlantic Coastal Plain (Runckle et al., 2022), and soils in the Southern Coastal Plain tend to be wetter than in other ecoregions in the state. Like the Middle Atlantic Coastal Plain, much of the Southern Coastal Plain has been significantly impacted by development. Slash and loblolly pines (*Pinus*) and grasses dominate the plains, while terrestrial wetlands are dominated by river swamp forests containing bald cypress (*Taxodium*), water tupelo (*Nyssa*), and oak (*Quercus*) (SC DNR, 2005e; Griffith et al., 2002). Barrier islands and coastal marshes contain very different plants from the rest of the Southern Coastal Plain (Griffith et al., 2002). Maritime forests on the barrier islands, in hummocks on salt marsh islands, and in sea-spray influenced parts of the mainland are dominated by live oaks (*Quercus*), southern magnolia (*Magnolia*), and various pine species (*Pinus*) and have an understory of shrubs and small trees including red cedar (*Juniperus*), cabbage palm (*Sabal*), American and yaupon hollies (*Ilex*), red bay (*Persea*), and wax myrtle (*Morella*).

In interior areas away from direct salt spray influence, the forests are more diverse (SC DNR, 2005e) and contain multiple oak species (*Quercus*), sugarberry (*Celtis*), and pignut hickory (*Carya*), as well as an understory of dogwood (*Cornus*), American olive (*Osmanthus*), and Carolina laurel cherry (*Prunus*) and shrubs such as beautyberry (*Callicarpa*) and red buckeye (*Aesculus*). On relict dune ridges, a xeric pine woodland develops that contains live, laurel, sand, and turkey oaks (all *Quercus*) and longleaf pine (*Pinus*) (SC DNR, 2005e). Salt-tolerant grasses and herbs may occur in all settings; however, they tend to be at low diversity except in transition zones associated with beach dunes, where maritime grasslands containing sea oats and bitter panic grass (both Poaceae), seabeach evening primrose (*Oenothera*), and dune water pennywort (*Hydrocotyle*) occur, as do maritime scrub thickets composed mainly of stunted wax myrtles (*Morella*), red bay (*Persea*), groundsel (*Baccharis*), saw greenbriar (*Smilax*), and poison ivy (*Toxicodendron*) (SC DNR, 2005e). In ponded areas, a variety of emergent plants occur, including water lilies (*Nymphaea* and *Nuphar*). Invasive tallowtree (*Triadica*) is common in the region.

No information on nectar source plants exists specifically for the Southern Coastal Plain, as Hood (2006) groups it with the Middle Atlantic Coastal Plain. Ayers and Harmon (1992) group it in with Region 12, and Smith (2021) considers it to be part of the Low Country. The present study, while limited, represents the only datapoint specifically for this region.

Hive weight data for the Southern Coastal Plain were limited to a single site (Apiary 3) and was recorded from late April through early November (Figure 4.6). Here, a pattern of steady weight gain punctuated by precipitous drops (possibly harvests) was present from the initiation of readings through mid-August. An increase in weight was also noted in mid-October.

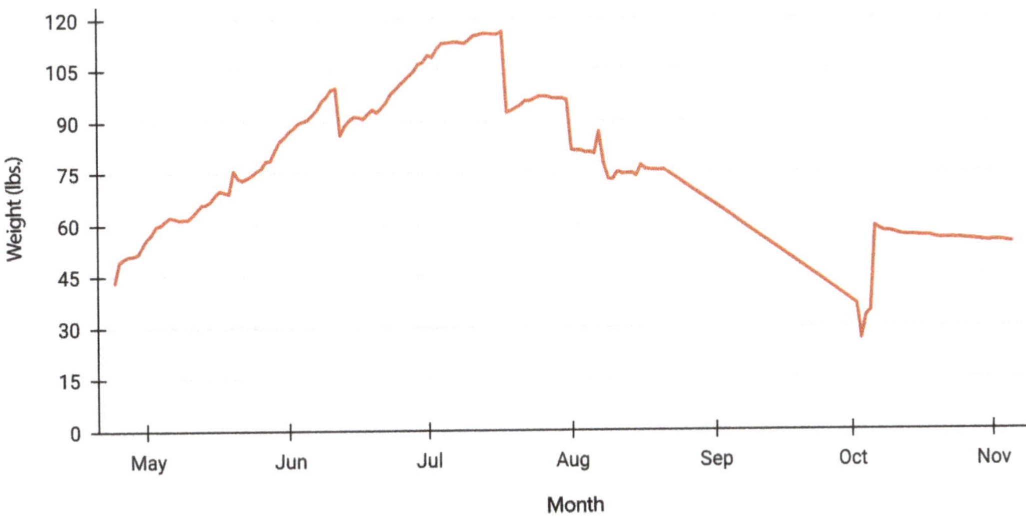

Figure 4.6. An example of BroodMinder hive weight data collected from Apiary 3 in the Southern Coastal Plain.
Note: Labels indicate the first day of each listed month.

The 75 pollen taxa observed in the Southern Coastal Plain ecoregion in this study did not match any of the 21 important taxa noted in Hood (2006) for the Coastal Plain section of South Carolina (Table 4.22). Likewise, little correlation was found with the 56 taxa reported by Ayers and Harmon (1992) or the taxa reported by Smith (2021). This is remarkable considering the similarities in taxa between the reported taxa and those observed in this study for the other four ecoregions. The most obvious hypothesis to explain the discrepancy is that the sampling locations between this study and previous ones did not overlap. This is a hypothesis that could be tested in the future.

Unlike the participating apiaries in the other four ecoregions, neither apiary in the Southern Coastal Plain produced monofloral *Acer* (maple) honey (Tables 4.23 and 4.24). Monofloral honeys are indicated with red color-coding in the apiary pollen occurrence tables. This is not to say that *Acer* (maple) pollen is not present nor important in this ecoregion. *Acer* (maple) pollen was present in the honey from both apiaries participating in this project from mid-January through late April, and the pollen levels were as high as 37.3% in Apiary 3 (Table 4.24)(to be considered "monofloral" the taxon must be present at >45%). But both apiaries did produce several monofloral honeys, with no overlap in taxa. Apiary 1 (Table 4.23) produced one sample of monofloral *Camellia* honey in mid-January,

Table 4.22. Plant Taxa whose Pollen was Observed in Honey Collected from the Southern Coastal Plain Ecoregion during 2022

Southern Coastal Plain	Jan		J/F	Feb			F/M	Mar			M/A	Apr			
Week ▶	4	5	6	7	8	9	10	11	12	13	14	15	16	17	18
Acer (Maple)															
Alyssum															
Amaranthaceae (Goosefoot family)															
Arecaceae (Palm family)															
Asteraceae (Coneflower/Golden Rod/Sunflower family)															
Asteraceae (Dandelion/Cat's Ear/Endive family)															
Asteraceae (Ragweed/Calendula/Solidago family)															
Berberis (Barberry)															
Berchemia (Rattan Vine)															
Brassicaceae (Mustard family)															
Camellia															
Carya (Hickory/Pecan)															
Cassia															
Castanea (Chinkapin/Chestnut)															
Celtis (Hackberry)															
Cephalanthus (Buttonbush)															
Cercis (Redbud)															
Cirsium (Thistle)															
Cocculus															
Cornus (Dogwood)															
Corylus/Carpinus (Hazel/Hornbeam)															
Diospyros (Persimmon)															
Elaeagnus (Autumn/Russian Olive)															
Fabaceae (Pea family)															
Fraxinus (Ash)															
Gelsemium (Yellow Jessamine)															
Hypercicaceae (St. John's Wort family)															
Ilex (Holly)															

while Apiary 3 (Table 4.24) produced two samples of monofloral *Ulmus* (elm) honey in late February and early March respectively, one sample of monofloral Rosaceae (rose/cherry/plum/peach/blackberry family) honey in mid-March, and seven samples of *Lagerstroemia* (crepe myrtle) honey in early May through early July.

If we consider taxa that contributed to honey for at least three straight weeks, there were 29 such taxa in the Southern Coastal Plain (Table 4.22), almost the same number as observed in the Middle Atlantic Coastal Plain (30). Twenty-nine taxa is about one-third (38.7%) of the 75 total taxa observed in this ecoregion (see Chapter 5). These 29 taxa were: *Acer* (maple), Asteraceae (coneflower/goldenrod/sunflower family), Brassicaceae (mustard family), *Camellia, Carya* (hickory/pecan), *Cassia, Corylus/Carpinus* (hazel/hornbeam), Fabaceae (pea family), Hypercicaceae (St. John's wort family), *Ilex* (holly), *Lagerstroemia* (crepe myrtle), Lamiaceae (mint family), *Ligustrum* (privet), Liliaceae (lily family), *Lonicera* (honeysuckle), *Magnolia, Parthenocissus* (Virginia creeper), *Nyssa* (tupelo/black gum), *Pinus* (pine), *Plantago* (plantain), Poaceae (grass family), *Ribes* (currant), Rosaceae (rose/cherry/plum/peach/blackberry family), *Quercus* (oak), *Salix* (willow), *Saururus* (lizard's tail), *Triadica* (Chinese tallow), *Trifolium/Melilotus* (red/white clover), *Ulmus* (elm), and *Vitis* (grape).

May				M/J	Jun			J/J	Jul				J/A	Aug			A/S	Sep			S/O	Oct				O/N	Nov
19	20	21	22	23	24	25	26	27	28	29	30	31	32	33	34	35	36	37	38	39	40	41	42	43	44	45	46

(continued)

Table 4.22. ***(continued)***

Southern Coastal Plain	Jan		J/F	Feb			F/M	Mar			M/A	Apr			
Week ▸	**4**	**5**	**6**	**7**	**8**	**9**	**10**	**11**	**12**	**13**	**14**	**15**	**16**	**17**	**18**
Impatiens (Touch-me-not)			□	□	□	□	■	□	□	□	□	□	□	□	□
Lagerstroemia (Crepe Myrtle)	■		■	■	■	□	■	■	□	■	□	□	■	□	□
Lamiaceae (Mint family)			□	□	□	■	□	■	□	□	□	□	■	□	□
Ligustrum (Privet)	□		■	□	■	□	■	■	□	■	□	□	■	■	■
Liliaceae (Lily family)	□		□	□	□	□	□	□	□	□	□	□	□	□	□
Liquidambar (Sweetgum)	□				■			□		■			□		□
Liriodendron (Tulip Poplar)	□		□	□	□	□	□	□	□	■	□	□	■	□	□
Lonicera (Honeysuckle)			□	□	□	□	□	□	□	□	□	■	■	■	■
Magnolia	□		□	□	□	□	□	□	□	□	□	□	■	■	■
Melaleuca			□	□	□	□	□	□	□	□	□	■	□	□	□
Menispermaceae (Moonseed family)			□	□	□	□	□	□	□	□	□	□	□	□	□
Mitchella			□	□	■	□	□	□	□	□	□	□	□	□	□
Morus (Mulberry)	□		□	□	□	□	■	□	■	□	□	□	■	□	□
Myrtaceae (Myrtle family)	□				□			□		□			■		□
Nuphar (Water Lily)			□	□	■	□	□	□	□	□	□	□	□	□	□
Nyssa (Tupelo/Black Gum)	■				■			□		■			■		■
Ostrya (Hophornbeam)			□	□	□	□	□	□	□	□	□	□	□	□	□
Parthenocissus (Virginia Creeper)	□		□	□	□	□	□	□	□	□	□	□	□	□	□
Phlox			□	□	□	□	■	■	□	□	□	□	□	□	□
Physalis			□	□	□	□	□	□	□	□	□	□	■	□	□
Pinus (Pine)	□				□			■		□			■		□
Plantago (Plantain)			□	□	□	■	■	■	□	□	□	■	□	□	□
Poaceae (Grass family)	□		□	□	■	■	■	■	□	■	□	□	■	□	■
Polygonum/Persicaria (Buckwheat/Knotweed)	□				□			□		■			□		□
Populus (Aspen/Cottonwood/Poplar)			□	□	□	■	■	■	□	□	□	□	□	□	□
Quercus (Oak)	□		□	□	□	□	□	□	■	■	□	■	■	■	■
Ranunculaceae (Buttercup family)			□	□	□	□	□	□	□	□	□	□	□	□	□
Rhamnaceae (Buckthorn family)			□	□	□	□	□	□	□	□	■	■	□	□	□
Rhus/Toxicodendron (Sumac/Poison Ivy)	□		□	□	■	□	□	□	□	■	□	□	■	□	■
Ribes (Currant)			□	□	□	□	□	□	□	□	□	□	■	■	■
Robinia (Locust)			□	□	□	□	□	□	□	□	■	■	□	□	□
Rosaceae (Rose/Cherry/Plum/Peach/Blackberry family)	□		□	■	■	□	■	■	■	■	■	■	■	■	■
Rubus (Blackberry/Dewberry)			□	□	□	□	□	□	□	□	□	□	■	□	□
Salix (Willow)	■		■	■	■	■	■	■	■	■	■	■	■	□	□
Saururus (Lizard's Tail)			■	■	■	■	■	■	□	□	□	■	□	□	■
Triadica (Chinese Tallow)			□	□	□	□	□	□	□	□	□	□	□	□	□
Trifolium/Melilotus (Red/White Clover)			■	□	■	■	■	■	■	■	■	■	■	□	□
Ulmus (Elm)	■		■	■	■	■	■	■	□	■	□	□	■	□	□
Urtica (Nettles)			□	□	□	□	□	□	□	□	□	□	■	□	□
Verbena			□	□	□	□	□	□	□	□	□	□	□	□	□
Vicia (Vetch)			□	□	□	□	□	□	□	□	□	□	□	□	□
Vitis (Grape)	□				■			■		□			□		□
Zamia			□	□	□	□	□	□	□	□	□	□	□	■	■
Unknowns #3 and #9	□				□			■		□			□		□
Unknown #10			■	■	□	■	□	□	□	□	□	□	□	□	□
Unknown #11			■	□	□	□	□	□	□	□	□	□	□	□	□
Unknown #12			□	□	□	□	□	□	□	□	□	□	□	■	■
Various Unknown Palynomorphs	□		□	■	■	■	■	■	□	□	■	■	■	■	■
Broken Palynomorphs			■	■	■	■	■	■	■	■	■	■	■	■	■

Key: *no data collected* (blank) *taxon absent on this week* □ *taxon present on this week* ■

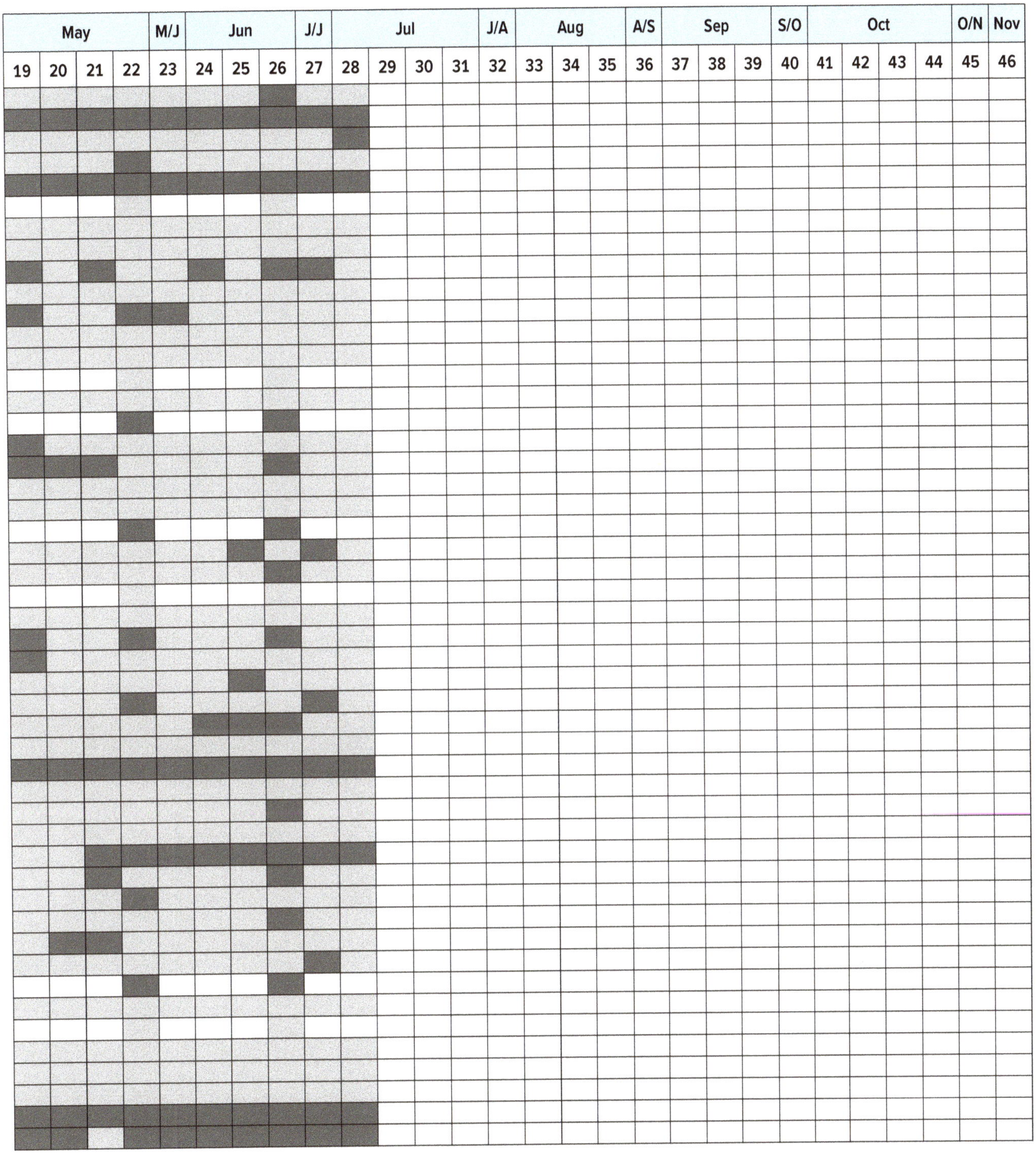

May				M/J	Jun			J/J	Jul				J/A	Aug			A/S	Sep			S/O	Oct				O/N	Nov
19	20	21	22	23	24	25	26	27	28	29	30	31	32	33	34	35	36	37	38	39	40	41	42	43	44	45	46

Table 4.23. Pollen Occurrence in Honey Collected in Apiary 1 in the Southern Coastal Plain Ecoregion during 2022

Apiary 1	Jan			J/F	Feb				F/M	Mar					M/A	Apr					
Week ▶	4		5	6	7	8		9	10	11		12	13		14	15	16		17	18	
Type of Data ▶	%	PC				%	PC			%	PC		%	PC			%	PC		%	PC
Acer (Maple)	18.8	I				5.1	M			16.7	I		10.4	M			0.0			1.8	L
Arecaceae (Palm family)	0.0					0.0				0.0			0.0				0.0			0.0	
Asteraceae (Coneflower/Golden Rod/Sunflower family)	2.9	L				2.5	L			6.9	M		3.3	M			0.0			0.0	
Asteraceae (Ragweed/Calendula/Solidago family)	0.0					0.8	L			16.7	I		2.8	L			1.4	L		0.0	
Brassicaceae (Mustard family)	0.0					1.3	L			1.0	L		1.4	L			2.4	L		0.0	
Camellia	49.0	D				1.3	L			0.0			0.0				0.0			0.0	
Carya (Hickory/Pecan)	0.0					0.0				0.0			0.0				0.0			0.9	L
Cercis (Redbud)	0.0					2.5	L			0.0			0.0				0.0			0.0	
Cornus (Dogwood)	0.0					0.0				0.0			0.0				0.0			0.5	L
Corylus (Hazel)	0.0					0.0				0.0			0.5	L			0.9	L		0.0	
Diospyros (Persimmon)	0.0					0.0				0.0			0.0				0.0			0.0	
Ilex (Holly)	0.0					0.0				0.0			0.0				0.0			0.5	L
Lagerstroemia (Crepe Myrtle)	1.9	L				3.4	M			1.5	L		0.9	L			0.9	L		0.0	
Ligustrum (Privet)	0.0					0.0				3.0	M		1.9	L			0.0			2.3	L
Liliaceae (Lily family)	0.0					0.0				0.0			0.0				0.0			0.0	
Liquidambar (Sweetgum)	0.0					5.1	M			0.0			0.5	L			0.0			0.0	
Liriodendron (Tulip Poplar)	0.0					0.0				0.0			1.4	L			0.0			0.0	
Magnolia	0.0					0.0				0.0			0.0				0.0			0.0	
Morus (Mulberry)	0.0					0.0				0.0			0.0				0.0			0.0	
Myrtaceae (Myrtle family)	0.0					0.0				0.0			0.0				13.3	M		0.0	
Nyssa (Tupelo/Black Gum)	1.0	L				0.8	L			0.0			0.9	L			44.1	I		39.4	I
Parthenocissus (Virginia Creeper)	0.0					0.0				0.0			0.0				0.0			0.0	
Pinus (Pine)	0.0					0.0				0.5	L		0.0				0.9	L		0.0	
Poaceae (Grass family)	0.0					14.3	M			2.0	L		1.9	L			0.9	L		2.8	L
Polygonum/Persicaria (Buckwheat/Knotweed)	0.0					0.0				0.0			1.9	L			0.0			0.0	
Quercus (Oak)	0.0					0.0				0.0			1.4	L			22.3	I		19.3	I
Rhus/Toxicodendron (Sumac/Poison Ivy)	0.0					41.8	I			0.0			25.6	I			3.8	M		0.5	L
Rosaceae (Rose/Cherry/Plum/Peach/Blackberry family)	0.0					10.5	M			3.0	M		2.4	L			7.1	M		30.7	I
Salix (Willow)	24.5	I				3.0	M			14.8	M		25.1	I			0.5	L		0.0	
Ulmus (Elm)	1.9	L				0.4	L			30.0	I		17.5	I			0.9	L		0.0	
Vitis (Grape)	0.0					3.0	M			0.5	L		0.0				0.0			0.0	
Unknowns #3 and #9	0.0					0.0				1.5	L		0.0				0.0			0.0	
Various Unknown Palynomorphs	0.0					4.2	M			2.0	L		0.0				0.5	L		1.4	L
Totals ▶	100					100				100			100				100			100	

Key: *D* (predominant) >45% I (secondary) 16-45% M (important) 3-15% L (minor) <3%

May					M/J	Jun				J/J	Jul				J/A	Aug			A/S	Sep			S/O	Oct				O/N	Nov
19	20	21	22		23	24	25	26		27	28	29	30	31	32	33	34	35	36	37	38	39	40	41	42	43	44	45	46
			%	PC				%	PC																				
			0.0					0.0																					
			0.0					10.8	M																				
			0.0					0.0																					
			0.0					0.0																					
			0.0					0.0																					
			0.0					0.0																					
			1.3	L				0.0																					
			0.0					0.0																					
			0.0					0.0																					
			0.0					0.0																					
			2.6	L				0.0																					
			3.9	M				0.0																					
			0.0					28.6	I																				
			18.6	I				0.0																					
			0.0					0.9	L																				
			0.0					0.0																					
			0.0					0.0																					
			0.0					0.9	L																				
			0.0					0.0																					
			0.0					0.0																					
			40.7	I				1.4	L																				
			0.0					0.5	L																				
			0.4	L				0.9	L																				
			0.0					0.9	L																				
			0.0					0.0																					
			27.7	I				0.5	L																				
			0.9	L				0.0																					
			0.9	L				0.0																					
			0.0					0.5	L																				
			0.4	L				0.0																					
			1.3	L				18.3	I																				
			0.0					0.0																					
			1.3	L				35.7	I																				
			100					100																					

Table 4.24. Pollen Occurrence in Honey Collected in Apiary 3 in the Southern Coastal Plain Ecoregion during 2022

Apiary 3	Jan		J/F		Feb						F/M		Mar						M/A		Apr							
Week ▶	4	5	6		7		8		9		10		11		12		13		14		15		16		17		18	
Type of Data ▶			%	PC	%	PC	%	PC	%	PC	%	PC	%	PC	%	PC	%	PC	%	PC	%	PC	%	PC	%	PC	%	PC
Acer (Maple)			19.9	I	35.0	I	37.3	I	3.1	M	2.7	L	1.3	L	13.6	M	4.8	M	5.4	M	18.2	I	12.8	M	0.6	L	8.0	M
Alyssum			0.0		0.0		0.0		0.0		0.0		0.0		0.0		0.0		0.0		0.0		0.0		0.0		0.0	
Amaranthaceae (Goosefoot family)			7.9	M	0.0		0.0		0.0		0.0		0.0		0.0		0.0		0.0		0.0		0.0		0.0		0.0	
Asteraceae (Coneflower/Golden Rod/Sunflower family)			1.2	L	0.0		5.6	M	12.1	M	5.5	M	8.3	M	0.8	L	0.0		0.0		1.4	L	0.0		0.0		0.0	
Asteraceae (Dandelion/Cat's Ear/Endive family)			0.0		0.0		0.4	L	0.0		0.0		0.4	L	0.3	L	0.0		0.0		0.0		0.0		0.0		0.0	
Asteraceae (Ragweed/Calendula/Solidago family)			0.0		0.0		0.0		0.0		0.0		0.4	L	0.0		0.0		0.0		0.0		0.0		0.0		0.0	
Berberis (Barberry)			0.0		0.0		0.0		0.0		0.0		0.0		0.0		0.4	L	0.0		0.0		0.0		0.0		0.0	
Berchemia (Rattan Vine)			0.0		0.0		0.0		0.0		0.0		0.0		0.0		0.0		0.0		0.0		0.3	L	0.0		0.0	
Brassicaceae (Mustard family)			14.9	M	0.7	L	1.6	L	0.9	L	0.0		0.0		0.3	L	0.0		1.2	L	1.0	L	0.3	L	0.0		1.1	L
Camellia			3.3	M	9.6	M	0.0		0.0		0.0		0.0		0.0		0.0		5.7	M	0.0		0.0		0.0		0.0	
Carya (Hickory/Pecan)			0.0		0.0		0.0		0.0		0.0		0.0		0.0		0.0		0.0		4.9	M	5.6	M	9.4	M	4.3	M
Cassia			0.0		0.0		0.0		0.9	L	0.9	L	0.4	L	0.0		0.0		0.0		0.0		0.0		0.0		0.0	
Castanea (Chinkapin/Chestnut)			0.0		0.0		0.0		0.0		0.0		0.0		0.0		0.0		0.0		0.0		0.3	L	0.0		0.0	
Celtis (Hackberry)			0.0		0.0		0.0		0.4	L	0.0		0.0		0.0		0.0		0.0		0.0		0.0		0.0		0.0	
Cephalanthus (Buttonbush)			0.0		0.0		0.0		0.0		0.0		0.0		0.0		0.0		0.0		0.0		0.0		0.0		0.0	
Cirsium (Thistle)			0.0		0.0		0.0		0.0		0.0		0.0		0.0		0.0		0.3	L	0.0		0.0		0.0		0.0	
Cocculus			0.0		0.0		0.0		0.0		0.0		0.0		0.0		2.9	L	13.1	M	0.0		0.0		0.0		0.0	
Cornus (Dogwood)			0.0		0.0		0.0		0.0		0.0		0.0		0.0		0.0		0.0		0.0		0.6	L	0.0		0.0	
Elaeagnus (Autumn/Russian Olive)			0.0		0.0		0.0		0.4	L	0.0		0.0		0.0		0.0		0.0		0.0		0.0		0.0		0.0	
Fabaceae (Pea family)			0.0		0.0		0.0		0.0		0.0		0.0		0.0		0.0		0.0		0.0		0.0		0.0		0.0	
Fraxinus (Ash)			0.0		0.0		0.0		0.0		0.0		0.0		0.3	L	0.0		0.0		0.0		0.0		0.0		0.0	
Gelsemium (Yellow Jessamine)			0.0		0.3	L	0.0		0.0		0.0		0.4	L	0.0		0.0		0.0		0.0		0.0		0.0		0.0	
Hypercicaceae (St. John's Wort family)			0.0		0.0		7.5	M	1.3	L	0.0		0.0		0.0		6.8	M	19.4	I	0.7	L	1.4	L	0.0		0.6	
Ilex (Holly)			0.0		0.0		0.0		0.4	L	0.0		0.0		0.8	L	0.8	L	0.6	L	10.5	M	10.8	M	14.2	M	12.3	M
Impatiens (Touch-me-not)			0.0		0.0		0.0		0.0		0.9	L	0.0		0.0		0.0		0.0		0.0		0.0		0.0		0.0	
Lagerstroemia (Crepe Myrtle)			0.4	L	0.3	L	0.4	L	0.0		0.5	L	0.4	L	0.0		0.0		0.0		0.0		0.0		0.0		0.0	
Lamiaceae (Mint family)			0.0		0.0		0.0		0.4	L	0.0		0.9	L	0.0		0.0		0.0		0.0		0.3	L	0.0		0.0	
Ligustrum (Privet)			0.8	L	0.0		0.4	L	0.0		0.5	L	0.0		0.0		0.0		0.0		0.0		1.1	L	0.9	L	0.0	
Liliaceae (Lily family)			0.0		0.0		0.0		0.0		0.0		0.0		0.0		0.0		0.0		0.0		0.0		0.0		0.0	
Liriodendron (Tulip Poplar)			0.0		0.0		0.0		0.0		0.0		0.0		0.0		0.0		0.0		0.0		1.7	L	0.0		0.0	
Lonicera (Honeysuckle)			0.0		0.0		0.0		0.0		0.0		0.0		0.0		0.0		0.0		2.1	L	3.6	M	2.2	L	4.0	M
Magnolia			0.0		0.0		0.0		0.0		0.0		0.0		0.0		0.0		0.0		0.0		1.4	L	1.9	L	4.0	M
Melaleuca			0.0		0.0		0.0		0.0		0.0		0.0		0.0		0.0		0.0		13.6	M	0.0		0.0		0.0	
Menispermaceae (Moonseed family)			0.0		0.0		0.0		0.0		0.0		0.0		0.0		0.0		0.0		0.0		0.0		0.0		0.0	
Mitchelia			0.0		0.0		0.4	L	0.0		0.0		0.0		0.0		0.0		0.0		0.0		0.0		0.0		0.0	
Morus (Mulberry)			0.0		0.0		0.0		0.0		0.5	L	0.0		0.3	L	0.0		0.0		0.0		0.3	L	0.0		0.0	
Nuphar (Water Lily)			0.0		0.0		0.4	L	0.0		0.0		0.0		0.0		0.0		0.0		0.0		0.0		0.0		0.0	
Ostrya (Hophornbeam)			0.0		0.0		0.0		0.0		0.0		0.0		0.0		0.0		0.0		0.0		0.0		0.0		0.0	
Parthenocissus (Virginia creeper)			0.0		0.0		0.0		0.0		0.0		0.0		0.0		0.0		0.0		0.0		0.0		0.0		0.0	
Phlox			0.0		0.0		0.0		0.0		0.5	L	0.4	L	0.0		0.0		0.0		0.0		0.0		0.0		0.0	
Physalis			0.0		0.0		0.0		0.0		0.0		0.0		0.0		0.0		0.0		0.0		0.3	L	0.0		0.0	
Plantago (Plantain)			0.0		0.0		0.0		4.0	M	1.4	L	0.4	L	0.0		0.0		0.0		0.7	L	0.0		0.0		0.0	
Poaceae (Grass family)			0.0		0.0		1.2	L	4.0	M	21.8	I	1.8	L	0.0		0.0		0.0		0.0		0.0		0.0		0.0	
Populus (Aspen/Cottonwood/Poplar)			0.0		0.0		0.0		4.0	M	8.6	M	0.9	L	0.0		0.0		0.0		0.0		0.0		0.0		0.0	
Quercus (Oak)			0.0		0.0		0.0		0.0		0.0		0.0		4.2	M	4.1	M	0.0		2.1	L	30.6	I	23.6	I	29.3	I
Ranunculaceae (Buttercup family)			0.0		0.0		0.0		0.0		0.0		0.0		0.0		0.0		0.0		0.0		0.0		0.0		0.0	
Rhamnaceae (Buckthorn family)			0.0		0.0		0.0		0.0		0.0		0.0		0.0		0.0		0.3	L	0.3	L	0.0		0.0		0.0	
Rhus/Toxicodendron (Sumac/Poison Ivy)			0.0		0.0		0.0		0.0		0.0		0.0		0.0		0.0		0.0		0.0		0.0		0.0		0.0	
Ribes (Currant)			0.0		0.0		0.0		0.0		0.0		0.0		0.0		0.0		0.0		0.0		0.6	L	2.8	L	3.4	M
Robinia (Locust)			0.0		0.0		0.0		0.0		0.0		0.0		0.0		0.0		1.2	L	10.5	M	0.0		0.0		0.0	
Rosaceae (Rose/Cherry/Plum/Peach/Blackberry family)			0.0		0.3	L	0.0		0.0		0.5	L	3.1	M	51.0	D	41.7	I	37.6	I	0.7	L	11.7	M	22.3	I	6.8	M
Rubus (Blackberry/Dewberry)			0.0		0.0		0.0		0.0		0.0		0.0		0.0		0.0		0.0		0.0		1.4	L	0.0		0.0	
Salix (Willow)			32.8	I	6.9	M	3.6	M	3.6	M	1.8	L	0.4	L	11.4	M	5.8	M	6.3	M	0.3	L	0.0		0.0		0.0	
Saururus (Lizard's Tail)			0.4	L	0.7	L	1.2	L	1.8	L	1.8	L	0.4	L	0.0		0.0		0.0		0.7	L	0.0		0.0		0.9	L
Triadica (Chinese Tallow)			0.0		0.0		0.0		0.0		0.0		0.0		0.0		0.0		0.0		0.0		0.0		0.0		0.0	

May								M/J		Jun						J/J		Jul				J/A	Aug			A/S	Sep			S/O	Oct				O/N	Nov
19		20		21		22		23		24		25		26		27		28		29	30	31	32	33	34	35	36	37	38	39	40	41	42	43	44	45	46
%	PC	%	PC	%	PC	%	PC	%	PC	%	PC	%	PC	%	PC	%	PC	%	PC																		
0.0		0.0		0.0		0.0		0.0		0.0		0.0		0.0		0.0		0.0																			
0.0		0.0		0.0		0.0		0.0		0.0		4.3	M	1.0	L	0.0		0.0																			
0.0		0.0		0.0		0.0		0.0		0.0		0.0		0.0		0.0		0.0																			
0.0		0.0		0.0		0.0		0.0		0.0		0.0		0.0		0.0		0.0																			
0.0		0.0		0.0		0.0		0.0		0.0		0.0		0.0		0.0		0.0																			
0.0		0.0		0.0		0.0		0.0		0.0		0.0		0.0		0.0		0.0																			
0.0		0.0		0.0		0.0		0.0		0.0		0.0		0.0		0.0		0.0																			
0.0		0.0		0.0		0.0		0.0		0.0		0.0		0.0		0.0		0.0																			
3.8	M	0.0		0.0		0.0		0.0		3.5	M	0.0		0.0		0.0		2.0	L																		
0.0		0.0		0.0		0.0		0.0		0.0		0.0		0.0		0.0		0.0																			
0.0		0.0		0.0		0.0		0.0		0.0		0.0		0.0		0.0		0.0																			
0.0		0.0		0.0		0.0		0.0		0.0		0.0		0.0		0.0		0.0																			
0.0		0.0		0.0		0.0		0.0		0.0		0.0		0.0		0.0		0.0																			
0.0		0.0		0.0		0.0		0.0		0.0		0.0		0.0		0.0		0.0																			
0.0		3.3	M	11.9	M	0.0		0.0		0.0		0.0		0.0		0.0		0.0																			
0.0		0.0		0.0		0.0		0.0		0.0		0.0		0.0		0.0		0.0																			
0.3	L	0.0		0.0		0.0		0.0		0.0		0.0		0.0		0.0		0.0																			
0.0		0.0		0.0		0.0		0.0		0.0		0.0		0.0		0.0		0.0																			
0.0		0.0		0.0		0.0		0.0		0.0		0.0		0.0		0.0		0.0																			
0.0		0.0		8.6	M	7.8	M	1.9	L	15.4	M	1.0	L	1.3	L	2.5	L	0.0																			
0.0		0.0		0.0		0.0		0.0		0.0		0.0		0.0		0.0		0.0																			
0.0		0.0		0.0		0.0		0.0		0.0		0.0		0.0		0.0		0.0																			
1.3	L	0.7	L	0.0		0.0		0.0		0.0		0.0		0.0		0.0		0.0																			
0.0		0.0		0.0		0.0		0.0		0.0		0.0		0.0		0.0		0.0																			
0.0		0.0		0.0		0.0		0.0		0.0		0.0		0.3	L	0.0		0.0																			
64.1	D	77.2	D	33.3	I	48.4	D	47.2	D	34.4	I	63.2	D	58.9	D	34.1	I	47.7	D																		
0.0		0.0		0.0		0.0		0.0		0.0		0.0		0.0		0.0		0.3	L																		
0.0		0.0		0.0		0.0		0.0		0.0		0.0		0.0		0.0		0.0																			
10.8	M	8.8	M	6.2	M	10.1	M	10.0	M	11.9	M	1.4	L	19.4	I	6.6	M	12.3	M																		
0.0		0.0		0.0		0.0		0.0		0.0		0.0		0.0		0.0		0.0																			
0.0		0.0		0.0		0.0		0.0		0.0		0.0		0.0		0.0		0.0																			
0.3	I	0.0		1.4	L	0.0		0.0		3.2	M	0.0		0.3	L	0.3	L	0.0																			
0.0		0.0		0.0		0.0		0.0		0.0		0.0		0.0		0.0		0.0																			
0.3	L	0.0		0.0		4.9	M	3.6	M	0.0		0.0		0.0		0.0		0.0																			
0.0		0.0		0.0		0.0		0.0		0.0		0.0		0.0		0.0		0.0																			
0.0		0.0		0.0		0.0		0.0		0.0		0.0		0.0		0.0		0.0																			
0.0		0.0		0.0		0.0		0.0		0.0		0.0		0.0		0.0		0.0																			
0.3	L	0.0		0.0		0.0		0.0		0.0		0.0		0.0		0.0		0.0																			
2.9	L	1.3	L	0.5	L	0.0		0.0		0.0		0.0		0.0		0.0		0.0																			
0.0		0.0		0.0		0.0		0.0		0.0		0.0		0.0		0.0		0.0																			
0.0		0.0		0.0		0.0		0.0		0.0		0.0		0.0		0.0		0.0																			
0.0		0.0		0.0		0.0		0.0		0.0		1.0	L	0.0		6.3	M	0.0																			
0.0		0.0		0.0		0.0		0.0		0.0		0.0		0.0		0.0		0.0																			
0.0		0.0		0.0		0.0		0.0		0.0		0.0		0.0		0.0		0.0																			
0.6	L	0.0		0.0		0.0		0.0		0.0		0.0		0.0		0.0		0.0																			
0.3	L	0.0		0.0		0.0		0.0		0.0		0.0		0.0		0.0		0.0																			
0.0		0.0		0.0		0.0		0.0		0.0		4.8	M	0.0		0.0		0.0																			
0.0		0.0		0.0		0.0		0.0		0.0		0.0		0.0		2.2	L	0.0																			
0.0		0.0		0.0		0.0		0.0		2.6	L	1.0	L	0.3	L	0.0		0.0																			
0.0		0.0		0.0		0.0		0.0		0.0		0.0		0.0		0.0		0.0																			
10.2	M	0.7	L	1.0	L	0.0		1.6	L	2.6	L	6.7	M	4.2	M	0.9	L	8.6	M																		
0.0		0.0		0.0		0.0		0.0		0.0		0.0		0.0		0.0		0.0																			
0.0		0.0		0.0		0.0		0.0		0.0		0.0		0.0		0.0		0.0																			
0.0		0.0		0.0		0.0		0.0		0.0		0.0		0.0		0.0		0.0																			
0.0		0.0		32.9	I	27.6	I	33.0	I	25.1	I	3.8	M	9.1	M	33.1	I	23.2	I																		

(continued)

Table 4.24. (*continued*)

Apiary 3	Jan		J/F		Feb						F/M		Mar						M/A		Apr							
Week ▸	4	5	6		7		8		9		10		11		12		13		14		15		16		17		18	
Type of Data ▸			%	PC	%	PC	%	PC	%	PC	%	PC	%	PC	%	PC	%	PC	%	PC	%	PC	%	PC	%	PC	%	PC
Trifolium/Melilotus (Red/White Clover)			0.4	L	0.0		12.3	M	1.8	L	1.4	L	0.4	L	0.6	L	1.0	L	0.3	L	0.3	L	0.3	L	0.0		0.0	
Ulmus (Elm)			0.4	L	8.3	M	4.4	M	49.1	D	39.1	I	62.3	D	0.0		0.0		0.0		0.0		0.0		0.0		0.0	
Urtica (Nettles)			0.0		0.0		0.0		0.0		0.0		0.0		0.0		0.0		0.0		0.0		0.3	L	0.0		0.0	
Verbena			0.0		0.0		0.0		0.0		0.0		0.0		0.0		0.0		0.0		0.0		0.0		0.0		0.0	
Vicia			0.0		0.0		0.0		0.0		0.0		0.0		0.0		0.0		0.0		0.0		0.0		0.0		0.0	
Zamia			0.0		0.0		0.0		0.0		0.0		0.0		0.0		0.0		0.0		0.0		0.0		1.9	L	2.8	L
Unknown #10			2.1	L	3.0	M	0.0		0.4	L	0.0		0.0		0.0		0.0		0.0		0.0		0.0		0.0		0.0	
Unknown #11			2.1	L	0.0		0.0		0.0		0.0		0.0		0.0		0.0		0.0		0.0		0.0		0.0		0.0	
Unnkown #12			0.0		0.0		0.0		0.0		0.0		0.0		0.0		0.0		0.0		0.0		0.0		12.6	M	2.0	L
Various Unknown Palynomorphs			0.0		2.3	L	5.6	M	2.2	L	4.5	M	5.3	M	0.0		0.0		0.3	L	2.4	L	1.7	L	3.5	M	6.3	M
Broken Palynomorphs			13.3	M	32.7	I	17.9	I	8.9	M	7.3	M	11.8	M	16.4	I	31.6	I	8.4	M	29.4	I	13.1	M	4.1	M	14.2	M
Totals ▸			100		100		100		100		100		100		100		100		100		100		100		100		100	

Key: *D* (predominant) >45% I (secondary) 16-45% M (important) 3-15% L (minor) <3%

May								M/J		Jun						J/J		Jul					J/A	Aug			A/S	Sep			S/O	Oct				O/N	Nov
19		20		21		22		23		24		25		26		27		28		29	30	31	32	33	34	35	36	37	38	39	40	41	42	43	44	45	46
%	PC	%	PC	%	PC	%	PC	%	PC	%	PC	%	PC	%	PC	%	PC	%	PC																		
0.0		0.0		1.4	L	0.0		0.0		0.0		0.0		0.3	L	0.0		0.0																			
0.0		0.0		0.0		0.0		0.0		0.0		0.0		0.0		0.0		0.0																			
0.0		0.0		0.0		0.0		0.0		0.0		0.0		0.3	L	0.0		0.0																			
0.0		2.9	L	2.4	L	0.0		0.0		0.0		0.0		0.0		0.0		0.0																			
0.0		0.0		0.0		0.0		0.0		0.0		0.0		0.0		10.3	M	0.0																			
0.0		0.0		0.0		0.0		0.0		0.0		0.0		0.0		0.0		0.0																			
0.0		0.0		0.0		0.0		0.0		0.0		0.0		0.0		0.0		0.0																			
0.0		0.0		0.0		0.0		0.0		0.0		0.0		0.0		0.0		0.0																			
0.0		0.0		0.0		0.0		0.0		0.0		0.0		0.0		0.0		0.0																			
3.5	M	3.9	M	0.5	L	0.6	L	1.3	L	0.6	L	2.9	L	1.6	L	1.6	L	2.6	L																		
1.3	L	1.3	L	0.0		0.6	L	1.3	L	0.6	L	10.0	M	2.9	L	2.2	L	3.3	M																		
100		100		100		100		100		100		100		100		100		100																			

CHAPTER

5

Comparison of Ecoregions

We compared the plant taxa observed in the honey from each ecoregion to see if plant taxa tended to be unique to a certain ecoregion or were shared between multiple ecoregions. Only 38% of the plant taxa (53 of 141) were unique to a single ecoregion (Table 5.1).

The plant taxa that were unique to a single ecoregion included native herbaceous plants (like *Baptisia* [wild indigo] and *Tradescantia* [spiderwort]), cultivated plants (like *Pisum sativa* [pea]), ornamental plants (like *Gardenia*), and invasive plants (like *Casuarina* [Australian pine]) (Table 5.2). No trees were unique to a single ecoregion.

Eighteen percent of the plant taxa (26 of 141) were shared by all five of the ecoregions (Table 5.1). Unlike the plant taxa that were unique to a single ecoregion, the plant taxa shared by all the ecoregions included many trees, like *Acer* (maple), *Cornus* (dogwood), and *Salix* (willow). Many of these trees are common throughout the entire Level I ecoregion Eastern Temperate Forest. This group also includes plants that are wind-pollinated, like the Poaceae (grasses) and *Pinus* (pine)(Table 5.2).

The Southeastern Plains had the most unique taxa (34%, or 18 of 53), while the Blue Ridge had no unique taxa (Table 5.1). The results observed for the Blue Ridge may not truly represent the taxa used by the bees in this region, because there was only one apiary in this region (whereas all the other regions included two or more apiaries collecting honey samples).

The two ecoregions that shared the most plant taxa were the Southeastern Plains and the Piedmont (Table 5.1). Those two regions shared 54% of the plant taxa present in two regions. Furthermore, these two regions shared 96% of the plant taxa present in three regions (24%+64%+8%) and 100% of plant taxa present in four regions. It's probably not surprising that these ecoregions shared so many plant taxa in their honey, given that they are adjacent to each other and have significant overlap among characteristic plant communities (Chapter 4).

If you align the pollen occurrence data for the plant taxa that were present in all five ecoregions (Table 5.3), some interesting patterns are revealed. One striking observation is that some taxa, like *Acer* (maple), *Ilex* (holly), *Quercus* (oak), and the Rosaceae (rose/cherry/plum/peach/blackberry family), had pollen present in the honey essentially simultaneously across the state. For example, in Weeks 8, 13, and 15 *Acer* (maple) pollen was observed in honey in all 5 ecoregions. Similarly, *Ilex* (holly) pollen was observed across the state in Weeks 15, 16, 18, and 22. Other plant taxa—like *Lagerstroemia* (crepe myrtle), *Liriodendron* (tulip poplar), *Magnolia*, *Nyssa* (tupelo/black gum), and *Trifolium/Melilotus* (red/white clover)—also occured at similar times, but in a more regional fashion. *Lagerstroemia* (crepe myrtle) pollen was observed in all ecoregions except the Blue Ridge in Weeks 20, 22, 23, 24, and 26. *Trifolium/Melilotus* (red/white clover) pollen was observed at the same time in the Piedmont, Southeastern Plains, and Middle Atlantic Coastal Plain for most of the year. This occurrence also coincided with *Trifolium/Melilotus* (red/white clover) pollen presence in the Southern Coastal Plain during the spring and the Blue Ridge in the fall. It is also striking that even though the pollen from some plant taxa were present in all five ecoregions, they don't always show up at the same times of year in all regions. Pollen from *Pinus* (pine), *Plantago* (plantain), and *Ulmus* (elm), for example, were present in several honey samples in each ecoregion through the year, but there was little overlap in the timing of their occurrences between the ecoregions.

Aligning the pollen occurrence data for the plant taxa that were present in all five ecoregions also allows the overlap in plant taxa between the Southeastern Plains and the Piedmont (as described above) to stand out (Table 5.3). This is also true for seeing the overlaps in plant taxa between the Southeastern Plains, Piedmont, and Middle Atlantic Coastal Plain.

The analyses already described focus on comparing plant taxa whose pollen was present in the honey between the ecoregions and the timing of their presence. Another useful way to analyze the data is to ask, "In what ecoregion was a particular plant pollen found?" The answer to this question can be found in Table 5.4. This table contains a complete list of all plant taxa observed in the 302 honey samples we examined. For each plant taxon, its presence in all five ecoregions is indicated. It should be noted that no distinction is made of whether only one pollen grain was observed in only one honey sample for the ecoregion or if there were multiple honey samples with dozens of pollen grains in numerous hives for the ecoregion. You should refer to the apiary pollen occurrence tables for more detailed information.

Table 5.1. Analysis of the Overlap of Plant Taxa Between Ecoregions

Plant taxa distribution

Ecoregion	Number of Taxa	Taxa Percentage
found in only 1 ecoregion	53	37.6%
found in 2 ecoregions	24	17.0%
found in 3 ecoregions	25	17.7%
found in 4 ecoregions	13	9.2%
found in all 5 ecoregions	26	18.4%
	141	100.0%

Plant taxa with no overlap between ecoregions

Ecoregion	Number of Taxa	Taxa Percentage
BR	0	0.0%
P	13	24.5%
* **SP**	**18**	**34.0%**
MASP	7	13.2%
SCP	15	28.3%
	53	100.0%

Ecoregions

Blue Ridge
Piedmont
Southeastern Plains
Middle Atlantic Coastal Plain
Southern Coastal Plain

Plant taxa with overlap between 2 ecoregions

Ecoregion	Ecoregion	Number of Taxa	Taxa Percentage
* **SP**	**P**	**13**	**54.2%**
SP	MACP	1	4.2%
SP	BR	1	4.2%
SP	SCP	5	20.8%
P	SCP	2	8.3%
P	MACP	1	4.2%
P	BR	1	4.2%
		24	100.0%

Plant taxa with overlap between 3 ecoregions

Ecoregion	Ecoregion	Ecoregion	Number of Taxa	Taxa Percentage
* **SP**	**P**	**SCP**	**16**	64.0%
SP	P	MACP	6	**24.0%**
SP	P	BR	2	8.0%
SCP	SP	MACP	1	4.0%
			25	100.0%

Plant taxa with overlap between 4 ecoregions

Ecoregion	Ecoregion	Ecoregion	Ecoregion	Number of Taxa	Taxa Percentage
* **SCP**	**MACP**	**P**	**SP**	9	69.2%
SCP	P	SP	BR	1	**7.7%**
MACP	P	SP	BR	3	23.1%
				13	100.0%

Table 5.2. Lists of Plant Taxa Organized by Their Overlap Between Ecoregions

Plant taxa found in only 1 ecoregion
Abelia
Acanthus (Bear's breeches)
Alyssum
Artemisia (Wormwood/Absinthe)
Baptisia (Wild Indigo)
Boraginaceae (Forget-me-not family)
Canna
Carex
Cassia
Casuarina (invasive Australian Pine)
Centaurea (Knapweed)
Cirsium (Thistle)
Cnidoscolus/Croton (Bull Nettle/Croton)
Cocculus
Cucumis (Cucumber/Melon)
Erodium
Gardenia
Gossypium (Cotton)
Hyssopus (Hyssop)
Linaria
Malus (Apple/Crabapple)
Malvaviscus (Wax Mallow)
Melaleuca
Menispermaceae (Moonseed family)
Mitchelia
Narcissus (Daffodil)
Nerium (Oleander)
Ostrya (Hophornbeam)
Onagraceae (Primrose family)
Phoardendron
Physalis
Pisum sativum (Pea)
Portulaca (Purslanes)
Primulaceae (Primrose family)
Ranunculaceae (Buttercup family)
Rhamnaceae (Buckthorn family)
Richardia
Rumex
Sambucus (Elderberry)
Sarcocapnos (Poppy)
Spathiphyllum
Symphoricarpos (Snowberry)
Tillandsia
Tradescantia (Spiderwort)
Typha
Veronica
Wisteria
Zamia
Zanthoxylum (Prickly Ash)
Unknown #10
Unknown #11
Unknown #12
Unknown #13
53

Plant taxa found in 2 ecoregions
Aesculus (Buckeye)
Apiaceae (Carrot/Parsely/ Queen Anne's Lace family)
Berchemia (Rattan Vine)
Bidens (Tickseed)
Buddleja (Butterfly Bush)
Caryophyllaceae (Carnation family)
Ceanothus
Cephalanthus (Buttonbush)
Cyperaceae (Sedge family)
Dalea (Prairie Clover)
Fagopyrum (Buckwheat)
Fraxinus (Ash)
Hamamelis (Witch Hazel)
Impatiens (Touch-me-not)
Juglans (Walnut)
Ludwigia
Mimosa (Sensitive Plant)
Nuphar (Water Lily)
Solanum (Tomato)
Triadica (Chinese Tallow)
Urtica (Nettles)
Vernonia
Viola (Violet)
Unknown #1
24

Plant taxa found in 3 ecoregions
Alnus (Alder)
Alternanthera (Joyweed)
Amaryllidaceae (Amaryllis/Onion family)
Arecaceae (Palm family)
Berberis (Barberry)
Betula (Birch)
Castanea (Chinkapin/Chestnut)
Celtis (Hackberry)
Cercis (Redbud)
Corylus/Carpinus (Hazel/Hornbeam)
Fabaceae (Pea family)
Fagus (Beech)
Lamiaceae (Mint family)
Morus (Mulberry)
Myrtaceae (Myrtle family)
Oxydendrum (Sourwood)
Phlox
Ribes (Currant)
Robinia (Locust)
Rubus (Blackberry/Dewberry)
Saururus (Lizard's Tail)
Tilia (Basswood)
Verbena
Vicia (Vetch)
Zea mays (Corn)
25

Ecoregions
Blue Ridge
Piedmont
Southeastern Plains
Middle Atlantic Coastal Plain
Southern Coastal Plain

Plant taxa found in 4 ecoregions

Plant taxa found in 4 ecoregions
Camellia
Carya (Hickory/Pecan)
Elaeagnus (Autumn/Russian Olive)
Ericaceae (Heath family)
Gelsemium (Yellow Jessamine)
Glycine max (Soybean)
Hexasepalum teres (Poorjoe/Rough Buttonweed)
Hypercicaceae (St. John's Wort family)
Liliaceae (Lily family)
Parthenocissus (Virginia Creeper)
Polygonum/Persicaria (Buckwheat/Knotweed)
Populus (Aspen/Cottonwood/Poplar)
Unknowns #3 and #9

13

Plant taxa found in all 5 ecoregions

Plant taxa found in all 5 ecoregions
Acer (Maple)
Amaranthaceae (Goosefoot family)
Asteraceae (Coneflower/Golden Rod/Sunflower family)
Asteraceae (Dandelion/Cat's Ear/Endive family)
Asteraceae (Ragweed/Calendula/Solidago family)
Brassicaceae (Mustard family)
Cornus (Dogwood)
Diospyros (Persimmon)
Ilex (Holly)
Lagerstroemia (Crepe Myrtle)
Ligustrum (Privet)
Liquidambar (Sweetgum)
Liriodendron (Tulip Poplar)
Lonicera (Honeysuckle)
Magnolia
Nyssa (Tupelo/Black Gum)
Pinus (Pine)
Plantago (Plantain)
Poaceae (Grass family)
Quercus (Oak)
Rhus/Toxicodendron (Sumac/Poison Ivy)
Rosaceae (Rose/Cherry/Plum/Peach/Blackberry family)
Salix (Willow)
Trifolium/Melilotus (Red/White Clover)
Ulmus (Elm)
Vitis (Grape)

26

Table 5.3. Comparison of the Timing of Pollen Occurrence in the Honey for Plant Taxa Present in all 5 Ecoregions

	Jan		J/F	Feb			F/M	Mar			M/A	Apr				May			
Week ▶	4	5	6	7	8	9	10	11	12	13	14	15	16	17	18	19	20	21	22
Acer (Maple)																			
Acer (Maple)																			
Acer (Maple)																			
Acer (Maple)																			
Acer (Maple)																			
Amaranthaceae (Goosefoot family)																			
Amaranthaceae (Goosefoot family)																			
Amaranthaceae (Goosefoot family)																			
Amaranthaceae (Goosefoot family)																			
Amaranthaceae (Goosefoot family)																			
Asteraceae (Coneflower/Golden Rod/Sunflower family)																			
Asteraceae (Coneflower/Golden Rod/Sunflower family)																			
Asteraceae (Coneflower/Golden Rod/Sunflower family)																			
Asteraceae (Coneflower/Golden Rod/Sunflower family)																			
Asteraceae (Coneflower/Golden Rod/Sunflower family)																			
Asteraceae (Dandelion/Cat's Ear/Endive family)																			
Asteraceae (Dandelion/Cat's Ear/Endive family)																			
Asteraceae (Dandelion/Cat's Ear/Endive family)																			
Asteraceae (Dandelion/Cat's Ear/Endive family)																			
Asteraceae (Dandelion/Cat's Ear/Endive family)																			
Asteraceae (Ragweed/Calendula/Solidago family)																			
Asteraceae (Ragweed/Calendula/Solidago family)																			
Asteraceae (Ragweed/Calendula/Solidago family)																			
Asteraceae (Ragweed/Calendula/Solidago family)																			
Asteraceae (Ragweed/Calendula/Solidago family)																			
Brassicaceae (Mustard family)																			
Brassicaceae (Mustard family)																			
Brassicaceae (Mustard family)																			
Brassicaceae (Mustard family)																			
Brassicaceae (Mustard family)																			
Cornus (Dogwood)																			
Cornus (Dogwood)																			
Cornus (Dogwood)																			
Cornus (Dogwood)																			
Cornus (Dogwood)																			
Diospyros (Persimmon)																			
Diospyros (Persimmon)																			
Diospyros (Persimmon)																			
Diospyros (Persimmon)																			
Diospyros (Persimmon)																			
Ilex (Holly)																			
Ilex (Holly)																			
Ilex (Holly)																			
Ilex (Holly)																			
Ilex (Holly)																			
Lagerstroemia (Crepe Myrtle)																			
Lagerstroemia (Crepe Myrtle)																			
Lagerstroemia (Crepe Myrtle)																			
Lagerstroemia (Crepe Myrtle)																			
Lagerstroemia (Crepe Myrtle)																			
Ligustrum (Privet)																			
Ligustrum (Privet)																			
Ligustrum (Privet)																			
Ligustrum (Privet)																			
Ligustrum (Privet)																			

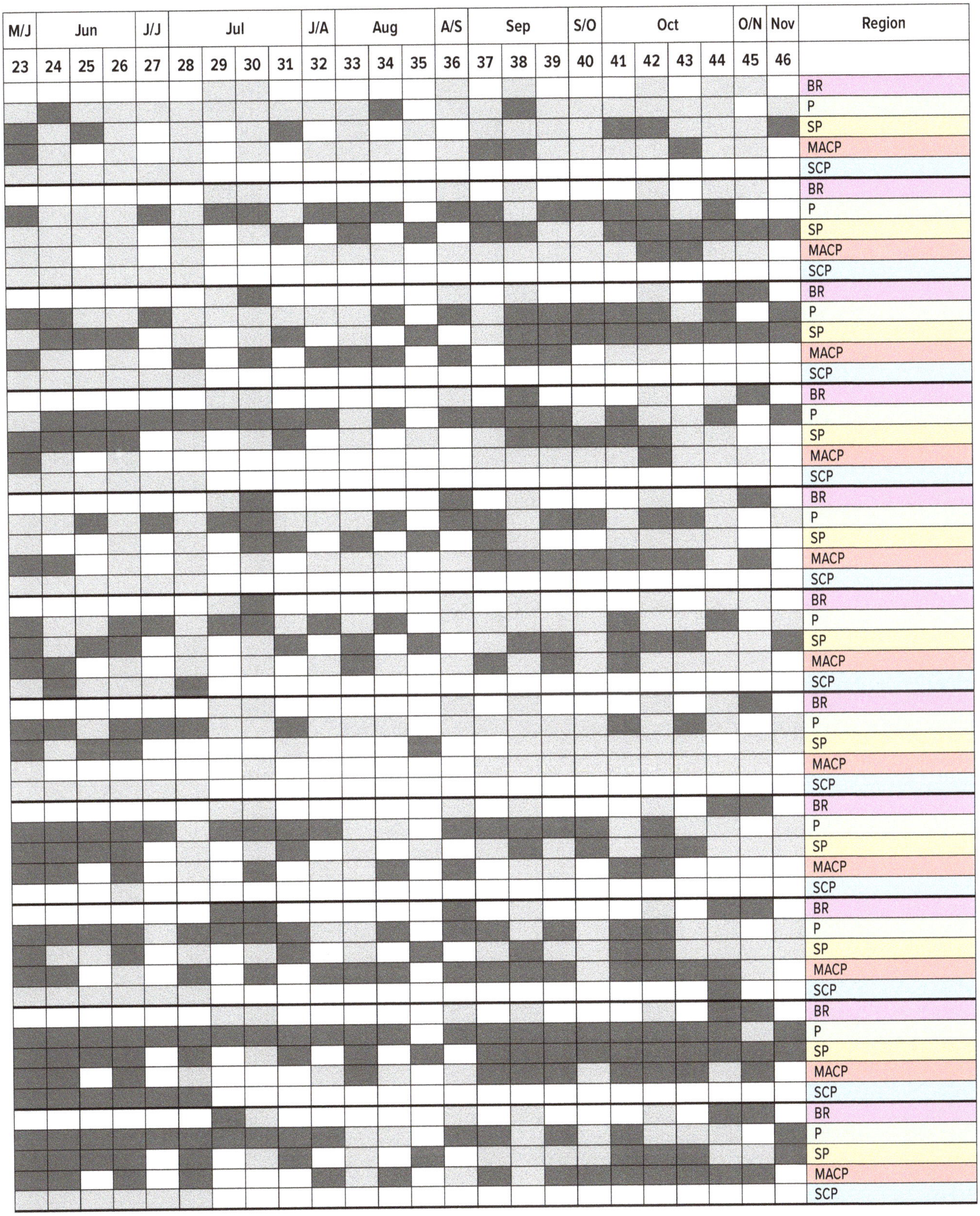

M/J	Jun			J/J	Jul				J/A	Aug			A/S	Sep			S/O	Oct				O/N	Nov	Region
23	24	25	26	27	28	29	30	31	32	33	34	35	36	37	38	39	40	41	42	43	44	45	46	
																								BR
																								P
																								SP
																								MACP
																								SCP
																								BR
																								P
																								SP
																								MACP
																								SCP
																								BR
																								P
																								SP
																								MACP
																								SCP
																								BR
																								P
																								SP
																								MACP
																								SCP
																								BR
																								P
																								SP
																								MACP
																								SCP
																								BR
																								P
																								SP
																								MACP
																								SCP
																								BR
																								P
																								SP
																								MACP
																								SCP
																								BR
																								P
																								SP
																								MACP
																								SCP
																								BR
																								P
																								SP
																								MACP
																								SCP
																								BR
																								P
																								SP
																								MACP
																								SCP
																								BR
																								P
																								SP
																								MACP
																								SCP

(continued)

Table 5.3. (*continued*)

	Jan		J/F	Feb			F/M	Mar			M/A	Apr				May			
Week ▶	4	5	6	7	8	9	10	11	12	13	14	15	16	17	18	19	20	21	22
Liquidambar (Sweetgum)																			
Liquidambar (Sweetgum)																			
Liquidambar (Sweetgum)																			
Liquidambar (Sweetgum)																			
Liquidambar (Sweetgum)																			
Liriodendron (Tulip Poplar)																			
Liriodendron (Tulip Poplar)																			
Liriodendron (Tulip Poplar)																			
Liriodendron (Tulip Poplar)																			
Liriodendron (Tulip Poplar)																			
Lonicera (Honeysuckle)																			
Lonicera (Honeysuckle)																			
Lonicera (Honeysuckle)																			
Lonicera (Honeysuckle)																			
Lonicera (Honeysuckle)																			
Magnolia																			
Magnolia																			
Magnolia																			
Magnolia																			
Magnolia																			
Nyssa (Tupelo/Black Gum)																			
Nyssa (Tupelo/Black Gum)																			
Nyssa (Tupelo/Black Gum)																			
Nyssa (Tupelo/Black Gum)																			
Nyssa (Tupelo/Black Gum)																			
Pinus (Pine)																			
Pinus (Pine)																			
Pinus (Pine)																			
Pinus (Pine)																			
Pinus (Pine)																			
Plantago (Plantain)																			
Plantago (Plantain)																			
Plantago (Plantain)																			
Plantago (Plantain)																			
Plantago (Plantain)																			
Poaceae (Grass family)																			
Poaceae (Grass family)																			
Poaceae (Grass family)																			
Poaceae (Grass family)																			
Poaceae (Grass family)																			
Quercus (Oak)																			
Quercus (Oak)																			
Quercus (Oak)																			
Quercus (Oak)																			
Quercus (Oak)																			
Rhus/Toxicodendron (Sumac/Poison Ivy)																			
Rhus/Toxicodendron (Sumac/Poison Ivy)																			
Rhus/Toxicodendron (Sumac/Poison Ivy)																			
Rhus/Toxicodendron (Sumac/Poison Ivy)																			
Rhus/Toxicodendron (Sumac/Poison Ivy)																			
Rosaceae (Rose/Cherry/Plum/Peach/Blackberry family)																			
Rosaceae (Rose/Cherry/Plum/Peach/Blackberry family)																			
Rosaceae (Rose/Cherry/Plum/Peach/Blackberry family)																			
Rosaceae (Rose/Cherry/Plum/Peach/Blackberry family)																			
Rosaceae (Rose/Cherry/Plum/Peach/Blackberry family)																			

M/J	Jun			J/J	Jul				J/A	Aug			A/S	Sep			S/O	Oct				O/N	Nov	Region
23	24	25	26	27	28	29	30	31	32	33	34	35	36	37	38	39	40	41	42	43	44	45	46	
																								BR
																								P
																								SP
																								MACP
																								SCP
																								BR
																								P
																								SP
																								MACP
																								SCP
																								BR
																								P
																								SP
																								MACP
																								SCP
																								BR
																								P
																								SP
																								MACP
																								SCP
																								BR
																								P
																								SP
																								MACP
																								SCP
																								BR
																								P
																								SP
																								MACP
																								SCP
																								BR
																								P
																								SP
																								MACP
																								SCP
																								BR
																								P
																								SP
																								MACP
																								SCP
																								BR
																								P
																								SP
																								MACP
																								SCP
																								BR
																								P
																								SP
																								MACP
																								SCP
																								BR
																								P
																								SP
																								MACP
																								SCP

(*continued*)

Table 5.3. *(continued)*

	Jan		J/F	Feb			F/M	Mar			M/A	Apr				May			
Week ▶	4	5	6	7	8	9	10	11	12	13	14	15	16	17	18	19	20	21	22
Salix (Willow)																			
Salix (Willow)																			
Salix (Willow)																			
Salix (Willow)																			
Salix (Willow)																			
Trifolium/Melilotus (Red/White Clover)																			
Trifolium/Melilotus (Red/White Clover)																			
Trifolium/Melilotus (Red/White Clover)																			
Trifolium/Melilotus (Red/White Clover)																			
Trifolium/Melilotus (Red/White Clover)																			
Ulmus (Elm)																			
Ulmus (Elm)																			
Ulmus (Elm)																			
Ulmus (Elm)																			
Ulmus (Elm)																			
Vitis (Grape)																			
Vitis (Grape)																			
Vitis (Grape)																			
Vitis (Grape)																			
Vitis (Grape)																			

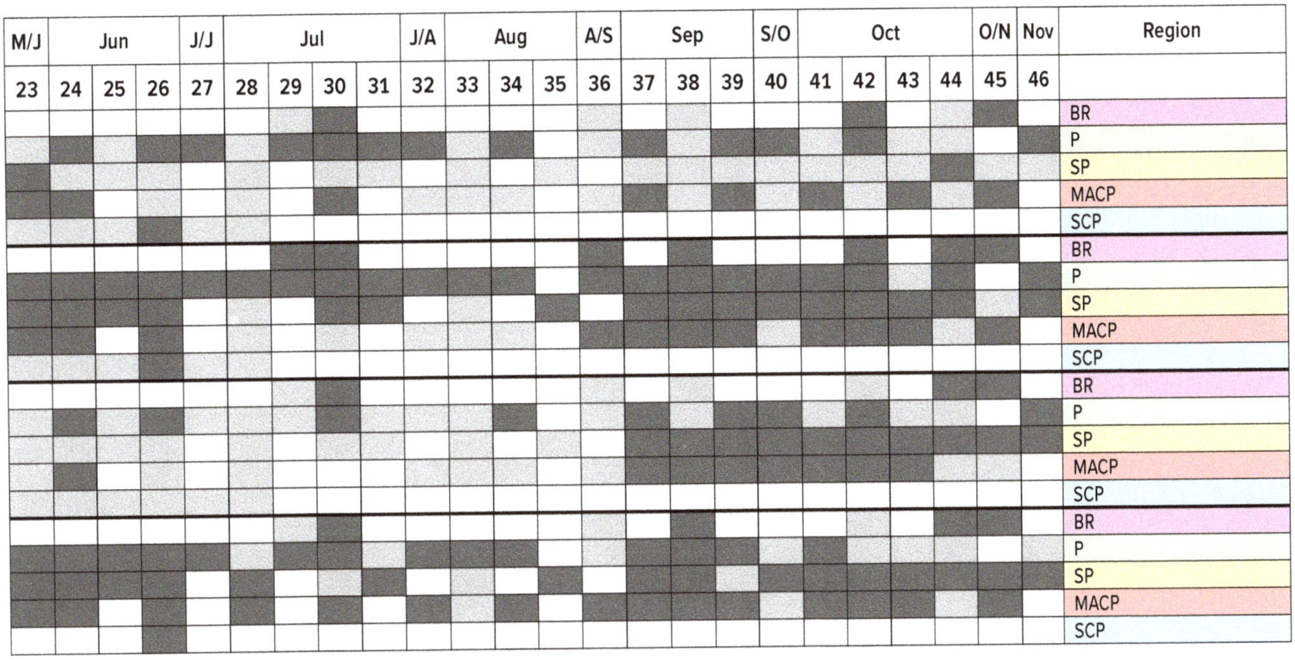

M/J	Jun			J/J	Jul				J/A	Aug			A/S	Sep			S/O	Oct				O/N	Nov	Region
23	24	25	26	27	28	29	30	31	32	33	34	35	36	37	38	39	40	41	42	43	44	45	46	
																								BR
																								P
																								SP
																								MACP
																								SCP
																								BR
																								P
																								SP
																								MACP
																								SCP
																								BR
																								P
																								SP
																								MACP
																								SCP
																								BR
																								P
																								SP
																								MACP
																								SCP

Table 5.4. Complete Plant Taxa List Indicating Occurrence in Ecoregions

Taxa	BR	P	SP	MACP	SCP
Abelia					
Acanthus (Bear's breeches)					
Acer (Maple)					
Aesculus (Buckeye)					
Alnus (Alder)					
Alternanthera (Joyweed)					
Alyssum					
Amaranthaceae (Goosefoot family)					
Amaryllidaceae (Amaryllis/Onion family)					
Apiaceae (Carrot/Parsely/Queen Anne's Lace family)					
Arecaceae (Palm family)					
Artemisia (Wormwood/Absinthe)					
Asteraceae (Coneflower/Golden Rod/Sunflower family)					
Asteraceae (Dandelion/Cat's Ear/Endive family)					
Asteraceae (Ragweed/Calendula/Solidago family)					
Baptisia (Wild Indigo)					
Berberis (Barberry)					
Berchemia (Rattan Vine)					
Betula (Birch)					
Bidens (Tickseed)					
Boraginaceae (Forget-me-not family)					
Brassicaceae (Mustard family)					
Buddleja (Butterfly Bush)					
Camellia					
Canna					
Carex					
Carya (Hickory/Pecan)					
Caryophyllaceae (Carnation family)					
Cassia					
Castanea (Chinkapin/Chestnut)					
Casuarina (invasive Australian Pine)					
Ceanothus					
Celtis (Hackberry)					
Centaurea (Knapweed)					
Cephalanthus (Buttonbush)					
Cercis (Redbud)					
Cirsium (Thistle)					
Cnidoscolus/Croton (Bull Nettle/Croton)					
Cocculus					
Cornus (Dogwood)					
Corylus/Carpinus (Hazel/Hornbeam)					
Cucumis (Cucumber/Melon)					
Cyperaceae (Sedge family)					
Dalea (Prairie Clover)					
Diospyros (Persimmon)					
Elaeagnus (Autumn/Russian Olive)					
Ericaceae (Heath family)					
Erodium					
Fabaceae (Pea family)					
Fagopyrum (Buckwheat)					
Fagus (Beech)					
Fraxinus (Ash)					
Gardenia					
Gelsemium (Yellow Jessamine)					
Glycine max (Soybean)					
Gossypium (Cotton)					
Hamamelis (Witch Hazel)					
Hexasepalum teres (Poorjoe/Rough Buttonweed)					
Hypercicaceae (St. John's Wort family)					

(continued)

Table 5.4. (*continued*)

Taxa	BR	P	SP	MACP	SCP
Hyssopus (Hyssop)					
Ilex (Holly)					
Impatiens (Touch-me-not)					
Juglans (Walnut)					
Lagerstroemia (Crepe Myrtle)					
Lamiaceae (Mint family)					
Ligustrum (Privet)					
Liliaceae (Lily family)					
Linaria					
Liquidambar (Sweetgum)					
Liriodendron (Tulip Poplar)					
Lonicera (Honeysuckle)					
Ludwigia					
Magnolia					
Malus (Apple/Crabapple)					
Malvaviscus (Wax Mallow)					
Melaleuca					
Menispermaceae (Moonseed family)					
Mimosa (Sensitive Plant)					
Mitchelia					
Morus (Mulberry)					
Myrtaceae (Myrtle family)					
Narcissus (Daffodil)					
Nerium (Oleander)					
Nuphar (Water Lily)					
Nyssa (Tupelo/Black Gum)					
Onagraceae (Primrose family)					
Ostrya (Hophornbeam)					
Oxydendrum (Sourwood)					
Parthenocissus (Virginia Creeper)					
Phlox					
Phoardendron					
Physalis					
Pinus (Pine)					
Pisum sativum (Pea)					
Plantago (Plantain)					
Poaceae (Grass family)					
Polygonum/Persicaria (Buckwheat/Knotweed)					
Populus (Aspen/Cottonwood/Poplar)					
Portulaca (Purslanes)					
Primulaceae (Primrose family)					
Quercus (Oak)					
Ranunculaceae (Buttercup family)					
Rhamnaceae (Buckthorn family)					
Rhus/Toxicodendron (Sumac/Poison Ivy)					
Ribes (Currant)					
Richardia					
Robinia (Locust)					
Rosaceae (Rose/Cherry/Plum/Peach/Blackberry family)					
Rubus (Blackberry/Dewberry)					
Rumex					
Salix (Willow)					
Sambucus (Elderberry)					
Sarcocapnos (Poppy)					
Saururus (Lizard's Tail)					
Solanum (Tomato)					
Spathiphyllum					
Symphoricarpos (Snowberry)					
Tilia (Basswood)					

(*continued*)

Table 5.4. (*continued*)

Taxa	BR	P	SP	MACP	SCP
Tillandsia					
Tradescantia (Spiderwort)					
Triadica (Chinese Tallow)					
Trifolium/Melilotus (Red/White Clover)					
Typha					
Ulmus (Elm)					
Urtica (Nettles)					
Verbena					
Vernonia					
Veronica					
Vicia (Vetch)					
Viola (Violet)					
Vitis (Grape)					
Wisteria					
Zamia					
Zanthoxylum (Prickly Ash)					
Zea mays (Corn)					
Unknown #1					
Unknowns #3 and #9					
Unknown #10					
Unknown #11					
Unknown #12					
Unknown #13					
Total Taxa in each Ecoregion	34	93	102	54	75

CHAPTER 6

Conclusions

The purpose of this book is to provide information to beekeepers as well as homeowners or bee-enthusiasts who want to grow bee-friendly plants.

Our objectives were three-fold:

1. Determine **what** plants are nectar sources for honeybees by region across the state of South Carolina via analysis of pollen from collected honey.
2. Determine **when** these nectar plants bloom across the state by regular, routine collection of honey samples for one calendar year.
3. Determine whether the maple (*Acer*) bloom (at the end of January/early February) is **both** a pollen source and a honey/nectar source.

To meet these objectives, we set out to conduct a pollen study of unprecedented scale, with a goal of collecting weekly fresh honey samples from 19 individual hive sites throughout the year 2022, during honey flow. The beekeepers participating in this year-long project collected a total of 302 samples, submitting honey samples while honeybees were actively collecting nectar distributed among South Carolina's five Level III ecoregions: the Blue Ridge, the Piedmont, the Southeastern Plains, the Middle Atlantic Coastal Plain, and the Southern Coastal Plain. Three palynologists (O'Keefe, Warny, and Wymer) worked for two years—through 2022 and 2023—to conduct all necessary detailed palynological analyses in their various laboratories.

The results allow us to say with confidence that *Acer* (maple) pollen, already known to be a protein source for bees in late winter/early spring, also appears in the honey. This is true in all five ecoregions. Indeed, in many cases, the early season honey produced by bees in South Carolina is monofloral *Acer* (maple) honey, suggesting that this plant is a very important nectar source.

When the study began in 2022, published lists of nectar-producing plants utilized by honeybees for South Carolina were very old and relatively short. We demonstrate that honeybees in South Carolina collect nectar from a wide variety of trees, shrubs, and herbaceous plants in the region, including both native and non-native plants. The honey samples collected included 141 distinct plant taxa. Some of these plants are very important across the state, including *Acer* (maple), *Cornus* (dogwood), *Ilex* (holly), *Lagerstroemia* (crepe myrtle), *Liriodendron* (tulip poplar), *Magnolia*, *Nyssa* (tupelo/black gum),

Rosaceae (rose/cherry/plum/peach/blackberry family), and *Salix* (willow). Others are very restricted in their range, such as *Baptisia* (wild indigo), *Tradescantia* (spiderwort), *Pisum sativa* (pea), and *Gardenia*.

Many pollen types appear in honey near-simultaneously across the state (with the same timing and in all ecoregions), while others appear in more nuanced patterns, likely corresponding to both ecosystem dynamics and overall local weather conditions that influence bloom times of individual plants.

This study, the first of its kind for South Carolina and among very few of its kind performed in the United States, serves as a baseline for determining what bees are likely using for nectar forage. In the process, it has highlighted probable timing and types of monofloral honey varietals produced in South Carolina.

In the Blue Ridge, monofloral honeys were produced intermittently from March through October and included monofloral *Acer* (maple), Rosaceae (rose/cherry/plum/peach/blackberry family), *Salix* (willow), and *Trifolium/Melilotus* (red/white clover) honeys (Table 6.1).

In the Piedmont, 11 monofloral honeys were encountered from the January/February transition through late September, with the majority produced prior to mid-June. These monofloral honeys included *Acer* (maple), Amaranthaceae (goosefoot family), *Cornus* (dogwood), *Diospyros* (persimmon), *Lagerstroemia* (crepe myrtle), Lamiaceae (mint family), *Ligustrum* (privet), *Robinia* (locust), Rosaceae (rose/cherry/plum/peach/blackberry family), *Salix* (willow), and *Trifolium/Melilotus* (red/white clover) honeys (Table 6.2).

Table 6.1. Monofloral (Varietal) Honeys Observed in the Blue Ridge Ecoregion During 2022

Blue Ridge			Jan		J/F	Feb			F/M	Mar			M/A	Apr			
	Apiary	Week ▸	4	5	6	7	8	9	10	11	12	13	14	15	16	17	18
Acer (Maple)	17																
Rosaceae (Rose/Cherry/Plum/Peach/Blackberry family)	17																
Salix (Willow)	17																
Trifolium/Melilotus (Red/White Clover)	17																

Monofloral honeys were produced in the Southeastern Plains ecoregion from the January/February transition through mid-October (Table 6.3). In keeping with the Southeastern Plains' high plant diversity, this ecoregion produced many types of monofloral honey with monofloral honeys from 13 plant taxa: *Acer* (maple), Amaranthaceae (goosefoot family), Asteraceae (coneflower/golden rod/sunflower family), Brassicaceae (mustard family), *Cercis* (redbud), *Gelsemium* (yellow jessamine), *Glycine max* (soybean), *Liriodendron* (tulip poplar), *Nyssa* (tupelo/black gum), *Robinia* (locust), Rosaceae (rose/cherry/plum/peach/blackberry family), *Ulmus* (elm), and *Vitis* (grape).

Monofloral honeys were produced in the Middle Atlantic Coastal Plain in mid-February through late October (Table 6.4). Despite having half the plant taxa as the Southeastern Plains (54 compared to 102 [Table 5.2]), the same number of monofloral honeys were produced here: 13. The monofloral honeys produced were from: *Acer* (maple), Asteraceae (ragweed/calendula/solidago family), *Cornus* (dogwood), *Diospyros* (persimmon), *Glycine max* (soybean), *Ilex* (holly), *Magnolia*, *Nyssa* (tupelo/black gum), Poaceae (grass family), Rosaceae (rose/cherry/plum/peach/blackberry family), *Trifolium/Melilotus* (red/white clover), *Ulmus* (elm), and Unknowns #3 and #9.

Finally, the apiaries in the Southern Coastal Plain also produced several monofloral honeys, including a monofloral *Camellia* honey in mid-January, some *Ulmus* (elm) honey in late February and early March, a monofloral Rosaceae (rose/cherry/plum/peach/blackberry family) honey in mid-March, and *Lagerstroemia* (crepe myrtle) honey from early May through early July (Table 6.5).

Our team hopes that this information will assist all honey and bee enthusiasts in carefully selecting the plants surrounding their apiaries in the future.

May				M/J	Jun			J/J	Jul				J/A	Aug			A/S	Sep			S/O	Oct				O/N	Nov
19	20	21	22	23	24	25	26	27	28	29	30	31	32	33	34	35	36	37	38	39	40	41	42	43	44	45	46

Table 6.2. Monofloral (Varietal) Honeys Observed in the Piedmont Ecoregion During 2022

Piedmont			Jan		J/F	Feb			F/M	Mar			M/A	Apr			
	Apiary	Week ▶	4	5	6	7	8	9	10	11	12	13	14	15	16	17	18
Acer (*Maple*)	9																
	12																
	14																
	15																
	16																
	18																
	19																
Amaranthaceae (Goosefoot family)	12																
Cornus (Dogwood)	16																
Diospyros (Persimmon)	12																
Lagerstroemia (Crepe Myrtle)	12																
	18																
Lamiaceae (Mint family)	18																
Ligustrum (Privet)	12																
Robinia (Locust)	18																
Rosaceae (Rose/Cherry/Plum/Peach/Blackberry family)	12																
	14																
	15																
	16																
	18																
	19																
Salix (Willow)	12																
	19																
Trifolium/Melilotus (Red/White Clover)	9																
	12																
	14																

Table 6.3. Monofloral (Varietal) Honeys Observed in the Southeastern Plains Ecoregion During 2022

Southeastern Plains			Jan		J/F	Feb			F/M	Mar			M/A	Apr			
	Apiary	Week ▶	4	5	6	7	8	9	10	11	12	13	14	15	16	17	18
Acer (Maple)	6																
	10																
	11																
	13																
Amaranthaceae (Goosefoot family)	10																
Asteraceae (Coneflower/Golden Rod/Sunflower family)	7																
Brassicaceae (Mustard family)	6																
Cercis (Redbud)	13																
Gelsemium (Yellow Jessamine)	10																
Glycine max (Soybean)	10																
Liriodendron (Tulip Poplar)	7																
Nyssa (Tupelo/Black Gum)	6																
	7																
	10																
Robinia (Locust)	7																
Rosaceae (Rose/Cherry/Plum/Peach/Blackberry family)	7																
	10																
Ulmus (Elm)	7																
Vitis (Grape)	10																

May				M/J	Jun			J/J	Jul				J/A	Aug			A/S	Sep			S/O	Oct				O/N	Nov
19	20	21	22	23	24	25	26	27	28	29	30	31	32	33	34	35	36	37	38	39	40	41	42	43	44	45	46

May				M/J	Jun			J/J	Jul				J/A	Aug			A/S	Sep			S/O	Oct				O/N	Nov
19	20	21	22	23	24	25	26	27	28	29	30	31	32	33	34	35	36	37	38	39	40	41	42	43	44	45	46

Table 6.4. Monofloral (Varietal) Honeys Observed in the Middle Atlantic Coastal Plain Ecoregion During 2022

Middle Atlantic Coastal Plain			Jan		J/F	Feb			F/M	Mar			M/A	Apr			
	Apiary	Week ▶	4	5	6	7	8	9	10	11	12	13	14	15	16	17	18
Acer (Maple)	2											■					
	4						■					■	■				
	5						■										
	8						■		■								
Asteraceae (Ragweed/Calendula/Solidago family)	5																
Cornus (Dogwood)	5													■			
Diospyros (Persimmon)	4																
Glycine max (Soybean)	5																
Ilex (Holly)	2																
	8																
Magnolia	8																■
Nyssa (Tupelo/Black Gum)	5																
Poaceae (Grass family)	8																
Rosaceae (Rose/Cherry/Plum/Peach/Blackberry family)	2																
	5								■		■	■					
	8												■				
Trifolium/Melilotus (Red/White Clover)	4																■
Ulmus (Elm)	5																
Unknowns 3 and 9	4														■		
	5																

Table 6.5. Monofloral (Varietal) Honeys Observed in the Southern Coastal Plain Ecoregion During 2022

Southern Coastal Plain			Jan		J/F	Feb			F/M	Mar			M/A	Apr			
	Apiary	Week ▶	4	5	6	7	8	9	10	11	12	13	14	15	16	17	18
Camellia	1		■														
Lagerstroemia (Crepe Myrtle)	3																
Rosaceae (Rose/Cherry/Plum/Peach/Blackberry family)	3										■						
Ulmus (Elm)	3							■		■							

May				M/J	Jun			J/J	Jul				J/A	Aug			A/S	Sep			S/O	Oct				O/N	Nov
19	20	21	22	23	24	25	26	27	28	29	30	31	32	33	34	35	36	37	38	39	40	41	42	43	44	45	46

May				M/J	Jun			J/J	Jul				J/A	Aug			A/S	Sep			S/O	Oct				O/N	Nov
19	20	21	22	23	24	25	26	27	28	29	30	31	32	33	34	35	36	37	38	39	40	41	42	43	44	45	46

References

Abella, S. R. (2003). Quantifying ecosystem geomorphology of the Southern Appalachian Mountains. *Physical Geography 24*(6), 488–501. https://digitalscholarship.unlv.edu/sea_fac_articles/55

Abella, S. R., Sherburne, V. B., & MacDonald, N. W. (2003). Multifactor classification of forest landscape ecosystems of Jocassee Gorges, Southern Appalachian Mountains, South Carolina. *Canadian Journal of Forest Research 33*(10), 1933–1946. www.doi.org/0.1139/x03-116

Aldrich, A. (2015). *A history of honey in Georgia and the Carolinas*. History Press. 127p.

Arias-Calluari, K., Colin, T., Latty, T., Myerscough, M., & Altmann, E. G. (2023). Modelling daily weight variation in honey bee hives. *PLoS Computational Biology, 19*(3), e1010880. https://doi.org/10.1371/journal.pcbi.1010880

Ayers, G. S., & Harmon, J. (1992). Chapter 11: Bee forage of North America and the Potential for Planting for Bees. In J. M. Graham (Ed.), *The hive and the honey bee* (pp. 437–535). Dadant & Sons, Inc.

Bailey, L. (1952). The action of the proventriculus of the worker honeybee, *APIS MELLIFERA* L. *Journal of Experimental Biology, 29*(2), 310–327. https://doi.org/10.1242/jeb.29.2.310.

Benninghoff, W. S. (1962). *Report: Calculation of pollen and spore density in sediments by addition of exotic pollen in known quantities*. Pollen et Spores.

Birks, H. J. B. & Birks, H. H. (1980). *Quaternary palaeoecology*. Academic Press.

Bottero, I., Dominik, C., Schweiger, O., Albrecht, M., Attridge, E., Brown, M. J. F., Cini, E., Costa, C., De la Rúa, P., de Miranda, J. R., Di Prisco, G., Dzul Uuh, D., Hodge, S., Ivarsson, K., Knauer, A. C., Klein, A.-M., Mänd, M., Martínez-López, V., Medrzycki, P., ... Stout, J.C. (2023). Impact of landscape configuration and composition on pollinator communities across different European biogeographic regions. *Frontiers in Ecology and Evolution, 11*. https://doi.org/10.3389/fevo.2023.1128228

Bryant, V. (2015, July). Filtering honey: What is best?. *Bee Culture*. https://www.beeculture.com/filtering-honey-what-is-best/

Bryant, V. (2017, March). Searching for pollen in honey. *Bee Culture*. https://www.beeculture.com/searching-pollen-honey/

Bryant, V. (2017, September). Filtering honey almost every filter removes some pollen. *Bee Culture*. https://www.beeculture.com/filtering-honey-almost-every-filter-removes-pollen/

Caffrey, M. A. & Horn, S. P. (2012). Technical note—Buying and maintaining nail lacquer for laboratory use: A practical guide for palynologists. *AASP – The Palynological Society Newsletter, 45*(1), 24–26. https://palynology.org/wp-content/uploads/2015/05/2012NL45-1.pdf

de Almeida-Muradian, L. B., Barth, O. M., Dietemann, V., Eyer, M., da Silva de Freitas, A., Martel, A.-C., Marcazzan, G. L., Marchese, C. M., Mucignat-Caretta, C., Pascual-Maté, A., Reybroek, W., Sancho, M. T., & Gasparotto Sattler, J. A. (2020). Standard methods for *Apis mellifera* honey research. *Journal of Apicultural Research, 59*(3), 1–62. https://doi.org/10.1080/00218839.2020.1738135

Eckert, J. E. (1933). The flight range of the honeybee. *Journal of Agricultural Research, 47*, 257–285.

Food and Drug Administration (2018). *Proper labeling of honey and honey products: Guidance for industry.* FDA Center for Food Safety and Applied Nutrition. https://www.fda.gov/files/food/published/PDF---Guidance-for-Industry--Proper-Labeling-of-Honey-and-Honey-Products.pdf

Goddard Space Flight Center (2007). *Honey bee forage map.* Honey Bee Net. Retrieved January 20, 2024, from https://honeybeenet.gsfc.nasa.gov/Honeybees/Forage.htm.

Griffith, G. E., Omernik, J. M., Comstock, J. A., Schfale, M. P., McNab, W. H., Lenat, D. R., Glover, J. B., & Sherlburne, V. B. (2002). *Ecoregions of North Carolina and South Carolina.* U.S. Geological Survey. https://www.researchgate.net/publication/267633398_Ecoregions_of_North_Carolina_and_South_Carolina

Halbritter, H., Ulrich, S., Grímsson, F., Weber, M., Zetter, R., Hesse, M., Buchner, R., Svojtka, M., & Frosch-Radivo, A. (2018). *Illustrated pollen terminology.* Springer. https://doi.org/10.1007/978-3-319-71365-6.

Hood, W. M. (ed.). (2006). *Beekeeping in South Carolina.* Clemson Cooperative Extension, Clemson, South Carolina, Bulletin EB 122, revised February 2006. 34p.

Hood, W. M., Purser, W. H., & Howard, F. J. (1990). Beekeeping in South Carolina. Clemson Cooperative Extension, Bulletin 122, Rev., 28p.

Jones, G. D. & Bryant, V. M., Jr. (1992). Melissopalynology in the United States: A review and critique. *Palynology, 16,* 63–71. https://www.jstor.org/stable/3687653

Jones, G. D. & Bryant, V. M., Jr. (1998). Are all counts created equal?. In V. M. Bryant, Jr. & J. H. Wrenn (Eds.), *New developments in palynomorph sampling, extraction, and analysis: AASP contributions series #33* (pp. 115–120). American Association of Stratigraphic Palynologists Foundation.

Jones, G. D., & Bryant, V. M., Jr. (2004). The use of ETOH for the dilution of honey. *Grana, 43*(3), 174–182. www.doi.org/10.1080/00173130410019497

Jones, G. D., & Bryant, V. M., Jr. (2007). A comparison of pollen counts: Light versus scanning electron microscopy. *Grana, 46*(1), 20–33. https://doi.org/10.1080/00173130601173897

Jones, G. D., & Bryant, V. M., Jr. (2014). Pollen studies of East Texas honey. *Palynology, 38*(2), 242–58. http://www.bioone.org/doi/full/10.1080/01916122.2014.899276

Jones, G. D., Bryant, V. M., Jr., & Wrenn, J. H. (1998). Pollen recovery from honey. In V. M. Bryant, Jr., & J. H. Wrenn (Eds.), *New developments in palynomorph sampling, extraction, and analysis: AASP contributions series #33* (pp. 107–114). American Association of Stratigraphic Palynologists Foundation.

Kapp, R. O., Davis, O. K., & King, J. E. (2000). *Ronald O. Kapp's pollen and spores* (2nd ed.). American Association of Stratigraphic Palynologists Foundation.

Lau, P., Bryant, V., & Rangel, J. (2018). Determining the minimum number of pollen grains needed for accurate honey bee (*Apis mellifera*) colony pollen pellet analysis. *Palynology 42(*1), 36–42. www.doi.org/10.1080/01916122.2017.1306810

Lau, P., Lense, P., Grebenok, R. J., Rangel, J., & Behmer, S. T. (2022). Assessing pollen nutrient content: A unifying approach to the study of bee nutritional ecology. *Philosophical Transactions of the Royal Society B, 377*(1853). https://doi.org/10.1098/rstb.2021.0510

Lau, P., Sgolastra, F., Williams, G. R., & Straub, L. (2023). Editorial: Insect pollinators in the Anthropocene: How multiple environmental stressors are shaping pollinator health. *Frontiers in Ecology and Evolution, 11.* https://doi.org/10.3389/fevo.2023.1279774

Lecocq, A., Kryger, P., Vejsnæs, F., & Brunn Jensen, A. (2015). Weight watching and the effect of landscape on honeybee colony productivity: Investigating the value of colony weight monitoring for the beekeeping industry. *PLoS One 10*(7). https://doi.org/10.1371/journal.pone.0132473.

Lieux, M. E. H. (1970). *A palynological investigation of Louisiana honeys* [Unpublished doctoral dissertation]. Louisiana State University.

Lieux, M. H. (1972). A melissopalynological study of 54 Louisiana (U.S.A.) honeys. *Review of Palaeobotany and Palynology, 13*(2), 95–124. www.doi.org/10.1016/0034-6667(72)90039-5

Lieux, M. H. (1975). Dominant pollen types recovered from commercial Louisiana honeys. *Economic Botany, 29,* 87–96. https://doi.org/10.1007/BF02861258

Lieux, M. H. (1977). Secondary pollen types characteristic of Louisiana honeys. *Economic Botany, 31,* 111–119. https://doi.org/10.1007/BF02866580

Lieux, M. H. (1978). Minor honeybee plants of Louisiana indicated by pollen analysis. *Economic Botany, 32,* 418–432. https://doi.org/10.1007/BF02907941

Lieux, M. H. (1980). Acetolysis applied to microscopical honey analysis. *Grana, 19*(1), 57–61. https://doi.org/10.1080/00173138009424988

Lieux, M. H. (1981). An analysis of Mississippi (U.S.A.) honey: Pollen, color, and moisture. *Apidologie, 12*(2), 137–158. https://hal.science/hal-00890542/document

Louveaux, J., Maurizio, A., & Vorwohl, G. (1970). Internationale kommission für bienenbotanik der I.U.B.S. methodik der melissopalynologie. *Apidologie, 1*(2), 193–209. https://doi.org/10.1051/apido:19700205.

Louveaux, J., Maurizio, A., & Vorwohl, G. (1978). Methods of melissopalynology. *Bee World, 59*(4), 139–157. https://doi.org/10.1080/0005772X.1978.11097714

Maher, L. J., Jr. (1981). Statistics for microfossil concentration measurements employing samples spiked with marker grains. *Review of Palaeobotany and Palynology, 32*(2–3), 153–91. www.doi.org/10.1016/0034-6667(81)90002-6

Maurizio, A. (1975). Microscopy of honey. In E. Crane (Ed.), *Honey. A comprehensive survey* (pp. 240–257). Heinemann.

McLellan, A. R. (1977). Honeybee colony weight as an index of honey production and nectar flow: A critical evaluation. *Journal of Applied Ecology, 14*(2), 401–408. www.doi.org/10.2307/2402553

McNeil, M. E. A. (2012, October). Vaughn Bryant, honey sleuth. *Bee Culture Magazine.* https://meamcneil.com/a914/images/Mea_PDF/Bryant.pdf

Meikle, W. G., Holst, N., Colin, T., Weiss, M., Carroll, M. J., McFrederick, Q. S., & Barron, A. B. (2018). Using within-day hive weight changes to measure environmental effects on honey bee colonies. *PLoS One, 13*(5). https://doi.org/10.1371/journal.pone.0197589

Merschat, A. J., Bream, B. R., Huebner, M. T., Hatcher, R. D., Jr., & Miller, C. F. (2017). Temporal and spatial distribution of Paleozoic metamorphism in the southern Appalachian Blue Ridge and Inner Piedmont delimited by ion microprobe U-Pb ages of metamorphic zircon. In R.D. Law, J. R. Thigpen, A. J. Merschat, & H. H. Stowell (Eds.). *Memoir 213 series: Linkages and feedbacks in orogenic systems* (pp. 155–254). Geological Society of America. https://doi.org/10.1130/2017.1213(10)

Nelson, E. V. (1971). History of beekeeping in the United States. In *Agriculture handbook 335: Beekeeping in the United States* (pp. 2–4).

Oertel, E. (1980). History of beekeeping in the United States. In *Agriculture handbook 335: Beekeeping in the United States* (pp. 2–9). United States Department of Agriculture. https://www.ars.usda.gov/ARSUserFiles/60500500/PDFFiles/1-100/015-USDA-%20Beekeeping%20in%20the.pdf

O'Keefe, J. M. K., & Wymer, C. L. (2015). An alternative to acetolysis: application of an enzyme-based method for the palynological preparation of fresh pollen, honey samples and bee capsules. *Palynology,* 41(1), 117–120. https://doi.org/10.1080/01916122.2015.1103321

Overstreet, W. C. & Bell, H., III (1965). *Geological survey bulletin 1183: The crystalline rocks of South Carolina.* United States Government Printing Office. https://pubs.usgs.gov/bul/1183/report.pdf

Pammel, L. H. & King, C. M. (1930). *Iowa geological survey bulletin no. 7: Honey plants of Iowa.* Iowa Geological Society. https://publications.iowa.gov/49940/1/iowa_geological_survey_honey_plants_1930_OCR_RED_.pdf

Paredes, R. & Bryant, V. M. (2020). Pollen analysis of honey samples from the Peruvian Amazon. *Palynology, 44*(2), 344–354. https://doi.org/10.1080/01916122.2019.1604447

Pellett, F. C. (1930). *American honey plants; together with those which are of special value to the beekeeper as sources of pollen.* American Bee Journal. https://doi.org/10.5962/bhl.title.13578

Peng, Y.-S. & Marston, J. M. (1986). Filtering mechanism of the honey bee proventriculus. *Physiological Entomology, 11*(4), 433–439. https://doi.org/10.1111/j.1365-3032.1986.tb00434.x.

Powell, B. (2022, April 4). CAPPings – Mar/Apr 2022. *Clemson Apiculure and Pollinator Program.* Retrieved February 11, 2024, from https://blogs.clemson.edu/clemsonpollinators/2022/04/04/cappings-mar-apr-2022/

Purser, W. H. & Sparks, L. M. (1962). *Bulletin no. 122: South Carolina beekeeping.* Clemson Cooperative Extension.

Rull, V. (1987). A note on pollen counting in palaeoecology. *Pollen et Spores, 29,* 471–480.

Runkle, J., Kunkel, K. E., Stevens, L. E., Frankson, R., Stewart, B. C., Sweet, W., & Rayne, S. (2022). *State climate summaries 2022: South Carolina, NOAA technical report NESDIS 150-SC.* NOAA/NESDIS. https://statesummaries.ncics.org/chapter/sc/

Schneider, A. (2011, November 7). *Tests show most store honey isn't honey.* Food Safety News. https://www.foodsafetynews.com/2011/11/tests-show-most-store-honey-isnt-honey/

Smith, W. B. (2021). *Honey bee pollen timing chart (factsheet KGIC 1740).* Clemson Cooperative Extension Home & Garden Information Center. https://hgic.clemson.edu/factsheet/honey-bee-pollen-timing-chart/

South Carolina Department of Natural Resources (2005a). *Blue Ridge ecoregion terrestrial habitats.* https://dc.statelibrary.sc.gov/server/api/core/bitstreams/12d7cdae-f772-432e-b064-43dfa4427250/content

South Carolina Department of Natural Resources (2005b). *Piedmont ecoregion terrestrial habitats.* https://www.nrc.gov/docs/ML1002/ML100211054.pdf

South Carolina Department of Natural Resources (2005c). *Sandhills ecoregion terrestrial habitats.* https://dc.statelibrary.sc.gov/items/7b23f7f3-5f01-4117-9538-73d17665fc9f

South Carolina Department of Natural Resources (2005d). *Coastal Plain ecoregion terrestrial habitats.* https://dc.statelibrary.sc.gov/bitstream/handle/10827/26420/DNR_Coastal_Plain_Ecoregion_Terres trial_Habitats_2005.pdf

South Carolina Department of Natural Resources (2005e). *Coastal Zone and marine ecoregion terrestrial and aquatic habitats.* https://dc.statelibrary.sc.gov/items/06a88043-59b8-4649-915d-d56495f7f3c2

Sponsler, D. B., Shump, D., Richardson, R. T., & Grozinger, C. M. (2020). Characterizing the floral resources of a North American metropolis using honey bee foraging assay. *Ecosphere, 11*(4). https://doi.org/10.1002/ecs2.3102.

Thigpen, J. R., Moecher, D. P., Stowell, H. H., Merschat, A., Hatcher, R. D., Jr., Powell, N. E., Spencer, B. M., Mako, C. A., Bollen, E. M., & Kylander-Clark, A. (2022). Defining the timing, extent, and conditions of Paleozoic metamorphism in the southern Appalachian Blue Ridge terranes of Tennessee, North Carolina, and northern Georgia. *Tectonics, 41*(10). https://doi.org/10.1029/2022TC007406

Todd, F. E. & Vansell, G. H. (1942). Pollen grains in nectar and honey. *Journal of Economic Entomology, 35,* 728–731.

Trimboli, S. R. (2018). *Plants honey bees use in the Ohio and Tennessee valleys.* Solidago Press.

Young, W. J. (1908). A microscopical study of honey pollen. In C. A. Browne (Ed.) *Bulletin 110: Analysis and composition of American honeys* (pp. 70–88). U.S. Department of Agriculture, Bureau of Chemistry.

Von Der Ohe, W., Persano Oddo, L., Piana, M. L., Morlot, M., & Martin, P. (2004). Harmonized methods of melissopalynology. *Apidologie, 35*(Supplement 1), S18-S25. https://doi.org/10.1051/apido:2004050.

Wang, R. & Dobritsa, A. A. (2018). Exine and aperture patterns on the pollen surface: Their formation and roles in plant reproduction. *Annual Plant Reviews, 1*(2). https://doi.org/10.1002/9781119312994.apr0625.

Warny, S., Ferguson, S., Hafner, M. S., & Escarguel, G. (2020). Using museum pelt collections to generate pollen prints from high-risk regions: A new palynological forensic strategy for geolocation. *Forensic Science International, 306.* https://doi.org/10.1016/j.forsciint.2019.110061

www.ingramcontent.com/pod-product-compliance
Lightning Source LLC
Chambersburg PA
CBHW050040220326
41599CB00044B/7229
9781638041672